Introduction to
THERMODYNAMICS

Introduction to
THERMODYNAMICS

Kurt C. Rolle

University of Dayton

Charles E. Merrill Publishing Company
A Bell & Howell Company
Columbus, Ohio

Published by
Charles E. Merrill Publishing Co.
A Bell & Howell Company
Columbus, Ohio 43216

ISBN: 0–675–08994–8

Library of Congress Catalog Number: 72–95278

4 5 6 7 8 9 — 79 78 77

PRINTED IN THE UNITED STATES OF AMERICA

To my wife

Joy

and to our children

Kurt, Loreli, Timothy, and Heidi

Their encouragement, understanding,
patience, and love made this book possible

Contents

12 The Closed Steam Power Cycle and the Rankine Cycle 345

13 The Steam Engine 389

14 Refrigeration Cycles 413

Introduction to
THERMODYNAMICS

1

Introduction

**1.1
Energy and
Thermo-
dynamics**

The most significant contribution to the developing and maintaining of a modern technological society has been man's ability to extract large amounts of energy from nature, thereby gaining capacities for work or power. The science which explains and predicts this extraction is called *thermodynamics*. We will be concerned with energy (*therme*, heat) as it changes or changes something else (dynamic). From studies in physics, we know that there are two forms of energy: *kinetic* and *potential*. However, there is at least one other form, *internal energy*, which helps to explain the hotness and coldness of a body. We will investigate these three forms of energy more fully in later chapters, but for now let's define energy.

Energy: *The capacity of a given body to produce physical effects external to that body.*

An explanation is in order regarding the meaning of "effects." In this text physical effects will mean movement or changes in size, color, temperature, or numerous other changes in the physical character of objects.

1

We define our subject as follows:

Thermodynamics: *That branch of science which treats (1) the conversion of energy from one form to another, and (2) the conveyance of energy from one place to another.*

We will, in fact, be more concerned with converting energy than with conveying it. For example, we will want to know how much of the energy in one pound of coal can be converted into electrical power at the district power plant, how much electrical power is needed to keep a refrigerator at 38°F, how much energy is needed to boil one pound of water, or how much energy of one gallon of gasoline we can extract to move an automobile. We can cite other equally valid examples of questions needing answers in both technical and social problems. These questions will be answered with the tools of thermodynamics exposed in the following chapters.

We will be treating ideas and areas such as those just mentioned on a macroscopic scale, and for our purposes, we have arbitrarily divided the physical world into the following three regions:

(1) A microscopic world of atoms, protons, electrons, bacteria, viruses, and generally those bodies invisible to the human senses.

(2) A macroscopic world, the size of which man can appreciate with his senses. All our directly usable quantities here on earth are in this region.

(3) A universal or infinite world: the earth's solar system, the universe, galaxy, and cosmos. This world, due to the efforts of astronomers and astro-physicists, is becoming more comprehensible, but it is infinite in size and numbers.

An understanding of the tremendous effects that these three worlds have on each other is necessary. In particular, we will need some intuitive appreciation of the microscopic (or atomic) structure of matter as it affects the macroscopic world. Thermodynamics is equally at home in all three worlds, but here we will concern ourselves with macroscopic quantities.

Sometimes the precise subject we are studying is called *engineering thermodynamics, general thermodynamics,* or *thermostatics.* There are two other fields which need to be mentioned: *statistical thermodynamics* and *irreversible thermodynamics.* Statistical thermodynamics is the application of mathematical statistics in conjunction with thermodynamics to the microscopic atomic world. The results from this endeavor are enlightening and useful in thermodynamics as well, but we do not have the space or background to present it here. Irreversible thermodynamics concerns itself with systems in states far removed from a macroscopic balance. Examples such as capillary action of atoms creeping up a narrow tube (an apparent contradiction of natural law), diffusion of atoms through a fine plug, severe shock waves induced by supersonic aircraft, and electron flow through a thin wire are a few phenomena which are treated through the concepts of irreversible thermodynamics.

A firm understanding of the tools and concepts of thermodynamics, however, is a prerequisite to proceed intelligently to the seemingly more exotic realms of statistical and irreversible thermodynamics. More importantly, an awareness of the concepts of thermodynamics is important and mandatory for the solution of everyday problems encountered in many technical fields.

1.2 Historical Background of Thermodynamics

The development of thermodynamics can be traced back to the earliest recorded dates in man's history. Central to the theme of this development is man's desire to ease or replace his manual efforts with additional animate or inanimate sources of power. What follows in this section is a brief sketch of the history of the present science of thermodynamics. It is obviously not a complete exposition; it is meant only to give you some historical insight into how the ideas of thermodynamics originated and expanded.

Man's use of animate power, such as horses and oxen, began around 4000 B.C. and represented the major source of energy through the nineteenth century A.D. By 3500 B.C., wheeled vehicles were used in Mesopotamia to ease the burden of man and beast. Watermills, steam jets, and various mechanical devices were in use during Christ's lifetime, and around 150 A.D. Hero's turbine was invented. This turbine was a globe containing water from which hot steam could escape through two nozzles, as shown in figure 1–1. A fire placed under the device boiled water in the flask, and the steam traveled up the vertical tubes and into the globe. Once in the globe, the steam was expelled through nozzles, thus causing the globe to rotate. It was really nothing but a novelty toy at the time, but it represents a thermodynamic concept of converting inanimate energy from a fuel into an effect (motion).

The science of thermodynamics probably began around 1592 when Galileo used a thermometer to make the first measurement of temper-

Figure 1–1   Hero's turbine or aeolipile. Revised from A. Sinclair, *Development of the Locomotive Engine* (Cambridge, 1970) p. 2; with permission of The MIT Press.

ature. The inaccurate and fickle human sense of touch was thus circumvented and replaced by quantitatively describing the hotness or coldness of objects. Meanwhile, the first use of steam for furnishing significant amounts of power for social needs occurred in the late seventeenth century. In 1698, Thomas Savery devised an arrangement of tanks and hand operated valves to utilize steam and its energy to pump water from a well. (See figure 1–2.) In the pump steam was produced in a boiler (a) and conducted to the two reservoirs (b) through hand operated valves (c). The steam was furnished to the reservoirs alternately, in turn pushing water in the reservoir out through pipe (d) and to the top. Valve (c) would then be closed and a trickle of cold water would condense the steam in the reservoir, thus causing a vacuum to be created. This vacuum would allow water to flow up pipe (e) from a low water supply (f) into the reservoir. Valve (c) would be then opened again to repeat the cycle. By alternating between the right and left reservoirs a continuous flow of water was created at the top. While representing a historical first this device was little improvement over animate power. Thomas Newcomen, in 1712, developed a steam-piston engine which was a logical replace-

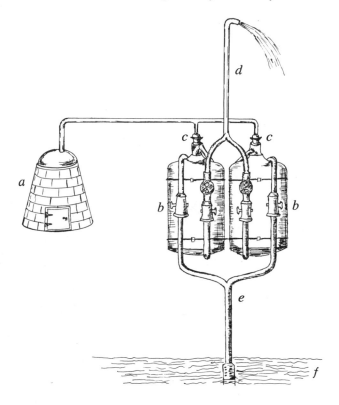

Figure 1–2   Savery's pistonless steam-vacuum pump (c. 1698). Revised from figure of Photo. Science Museum, London, with permission of The Science Museum, London, England.

ment of animate power for pumping water. This arrangement, shown in figure 1–3, provided a cycling motion which we call a *heat engine*. Known as the "atmospheric engine" because a vacuum and atmospheric pressure combined to provide the power stroke, this engine was used for pumping water. Steam produced in the boiler (a) was conducted through a hand valve (b) to the piston-cylinder (c). The steam would push the piston up to the position shown, allowing the pump rod (d) to descend into a water supply. The valve (e) was then opened to allow a spray of water to condense the steam in the cylinder, causing a vacuum to be created. The piston was then pushed down by atmospheric pressure, the pump rod raised, and water pumped up out of the water supply (f). Valve (e) was closed, valve (b) opened, and the process repeated. Line (g) was intermittently opened to allow the condensed steam to flow out of the cylinder.

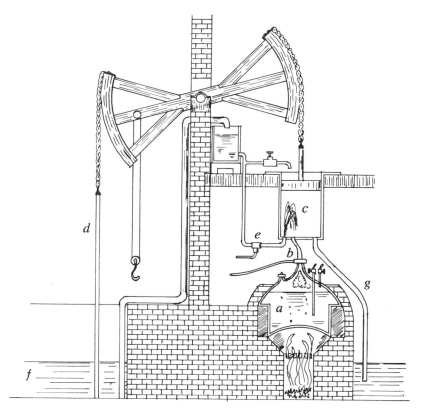

Figure 1–3   Newcomen's steam engine (1712). Revised, with permission, from Sinclair, *Development of the Locomotive Engine*, p. 7.

With improved machining techniques available for manufacturing parts, James Watt developed a steam-piston engine which represented a significant improvement over Newcomen's. Watt's engine, first operated in 1775 to pump water, represents the forerunner to the steam engines used for railroads, ships, and numerous other applications (figure 1–4).

Unlike Savery's or Newcomen's devices which condensed steam inside the working cylinder chamber, Watt's steam engine condensed the spent steam external to the cylinder (a). Steam was furnished from a boiler through a pipe (b). Valve (c), controlled from a tappet rod (d), allowed steam to enter the top side of the piston (e). This pushed the piston down and, through the walking beam (f), raised the pump rods (g) and (h). This motion drew water out of the reservoir (i), through the pipe (j) and from a reservoir (k) to reservoir (i). Valve (l) was then shifted to allow steam to enter the bottom of the piston; thus equilibrated, the piston moved to the top to begin a new cycle.

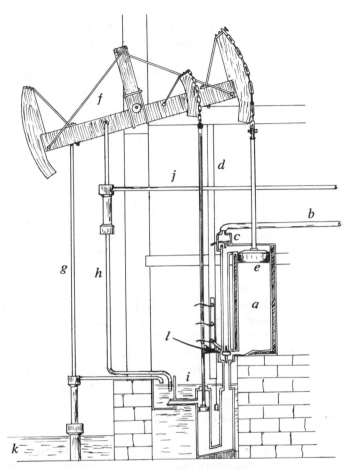

Figure 1–4   Watt's steam engine. Revised, with permission, from Sinclair, *Development of the Locomotive Engine*, p. 10.

Thermodynamic theory began to take form in 1693, when G. W. Liebnitz pronounced the conservation of mechanical energy (kinetic and potential). Sadi Carnot published a treatise in 1824 which described cycling devices or heat engines and which alluded to the first and second laws of thermodynamics. Twenty-six years later, in 1850, Rudolph Clausius formally stated these two laws of thermodynamics, and in 1854 he identified and defined the property now called *entropy*.

From 1840 to 1848, James Joule experimentally proved the equivalence of heat and work, thus making thermodynamics a quantitative science in the best tradition of Galileo. The internal combustion gasoline engine, later used to provide power for automobiles, trucks, and numerous other devices, was developed around 1860 by Lenoir. This type of engine was first used in vehicles around 1876 by Otto and Benz.

Around 1884, Parson introduced a steam turbine capable of developing significant amounts of power. This type of device, utilizing the popular medium steam, has been a most durable power generator and seems more popular today than ever. In the early 1900s, Nernst and Planck separately enunciated the earliest definition of the third law of thermodynamics. These statements have since been refined and revised by various theoreticians.

These advancements in both theory and technology reflect the applicability of thermodynamics to practical efforts; this usefulness has stimulated much of the interest in further expanding that body of knowledge known as the *science of thermodynamics*.

We see that thermodynamics developed through theory, experiments, and practice. Theoretical advancements came from the giants of thought; many men such as Joseph Black, Lord Kelvin, J. W. Gibbs, James Maxwell, L. Boltzmann, H. L. F. Helmholtz, and Albert Einstein contributed to thermodynamics at least as significantly as those mentioned previously. However, without the experimentation, design, creativity, and artisan ability to machine and fabricate parts precisely, useful engines providing significant amounts of power and devices to use this power would be nonexistent.

Our concern in this book is to understand and use the concepts of thermodynamics, clarified by theoreticians both past and present, for the solution of engineering and technological problems.

1.3 Summary

Since you now are introduced to what thermodynamics is and how it evolved, let's see how the material that is to follow is presented, and let's give you some idea of what to expect.

Chapter 2 introduces the concept of taking a specific part of nature (called the *thermodynamic system*) and describing it with numbers or properties. Chapter 3 describes the interaction of this system with its surroundings. The first law of thermodynamics and the conservation principles are introduced in chapter 4, followed by some classic uses of the first law in chapters 5 and 6. The important thermodynamic property, entropy, is introduced and explained in chapter 7. The idea of a cycling device, or a motor which runs as long as we wish, is presented in chapter 8, and the second law of thermodynamics is presented in chapter 9. In chapter 9 are also shown some significant implications of the second law which must be considered in any complete analysis of an energy-producing or energy-expending device.

Chapters 10 through 15 are devoted to some of the more traditional applications of thermodynamics. Each chapter could be read and studied separately, and each contains topics of direct application, such as the internal combustion engine, steam and gas turbines, the jet engine, refrigerators, and mixtures of two or more different liquids or gases. Chapter 16 introduces some other applications of thermodynamics to stimulate your imagination.

Questions for Discussion

1.1. Describe as many energy conversion devices as you can.

1.2. From your previous knowledge of physics, can you define *work?* Refer to chapter 3 and see if you agree.

1.3. From your introduction to physics, define *heat*. Compare with the definition in chapter 3.

1.4. Define *energy* without referring to section 1.1.

2

The Thermodynamic System

In any scientific or technical analysis the first steps should be to focus attention on an object and identify it quantitatively. The science of thermodynamics is no exception to this rule. In this chapter, we will focus mainly on the concept of a *system* and will discuss how to establish the parameters necessary to measure a system. We will be defining terms extensively, including the necessary terminology for treating systems as they change in time or space, that is, as systems interact with their surroundings.

2.1 System

The thermodynamic system: *Any region which occupies a volume and has a boundary.*

In solving a technical problem through thermodynamics, you must identify the system and its boundaries. For instance, suppose you wish to know the power required to operate a refrigerator. In this circumstance the system boundary would be the outside surface of the refrigerator;

11

everything included inside this surface would be the system. On the other hand, if you are concerned only with the operation of a compressor within the refrigerator, then the compressor itself is the system.

As another example, let's consider the internal combustion reciprocating engine of the present day automobile. If you are interested in the total operation of the automobile, your system might contain the whole vehicle, including the engine, fuel tank, battery, controls, and maybe even the passengers. However, if you wish to study the detailed manner in which power is extracted from the fuel and converted into mechanical energy, the system might be only one cylinder of the engine itself and not even the actual surfaces of the cylinder. This one-cylinder system or *piston-cylinder* for our purposes, is shown in figure 2–1, with the boundary indicated by a dashed line. This figure illustrates some important characteristics that a boundary has, and as you look at it, you should recognize that it represents a dynamic system where the piston is in motion at all times. In addition, the valves open and close at opportune instances, either allowing fuel and air to enter the system or exhausting the burned gases. Now, obviously the boundary can move (since the piston and valves move), and we can even shuffle fuel, air, and exhaust gases across the boundary; but not all system boundaries have this capability, and we will see in section 3.8 that the difference between two types of systems — open and closed — will be determined by whether matter crosses the boundary or not. Open and closed systems will then be an important part of the thermodynamic analysis, each handled in a slightly different manner.

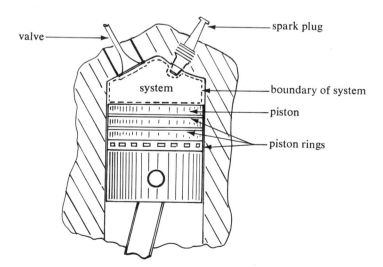

Figure 2–1   Piston-cylinder system.

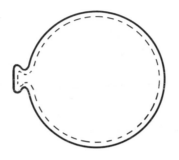 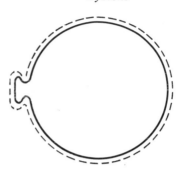

Figure 2-2   Inflated balloon with system defined by boundary inside or outside balloon material.

Another system is identified in figure 2–2. This is a balloon made of thin
rubber which can stretch and contract. The system boundary of the right
side configuration is drawn outside the balloon surface and envelopes two
separate materials: air in the balloon and the rubber balloon itself.
In analyzing this system, you must be careful. In figure 2–1 the system
was homogeneous at any given time, but the balloon in figure 2–2 is
not. By *homogeneous*, we mean that there is one and only one distinct
uniform material throughout the system at any instant in time. In figure
2–2 we have two different materials, the balloon and the air, each of which
has distinct ways of handling energy, thereby affecting the analysis
differently.

In defining or establishing a system, whenever possible, select a boundary
which will enclose homogeneous materials. In the balloon problem,
we were concerned with rubber as well as the air, so that ignoring all but
the air would be undesirable. If we wished, however, to specify a homogeneous system in the balloon we could do so by denoting a boundary
inside the balloon, thus defining a system composed of air alone, as
indicated in the left side configuration of figure 2–2.

As we have seen, the boundary and the enclosed volume of a system can
be arbitrary, and the possible variations are infinite. Let's allude to the
physical world. The volume outside the boundary we will call the
surroundings of the system, and the sum of the system and its surroundings we will call the *universe*. The universe, if we are completely rigorous
and correct, is infinite, but we need not in this book concern ourselves
with this aspect. We will consider the surroundings as only that physical
part of the universe which is capable of affecting our system during the
time period in which we are concerned. The point, though, is that we will
directly concern ourselves only with the system and its boundary, thereby
circumventing the problem of infinite universes.

In addition, once we have assured ourselves of either a homogeneous system or a simple composition of homogeneous systems (like the balloon problem), then the internal details of the dynamics or design of the system are not needed. That is, for the piston-cylinder system of figure 2–1, the materials, dimensional sizes, or velocities and accelerations of the valves, pistons, spark plugs, or cylinder wall are not important in the thermodynamic analysis, provided we know that the system is some homogeneous gas inside the cylinder.

2.2 Elementary Matter Theory

The thermodynamic system is nothing more than some volume of space and invariably contains matter or some material. If the volume has no matter, it is, of course, a perfect vacuum, and a very attractive thermodynamic system in terms of producing work or power. However, we will reserve any discussion on this aspect for chapter 9 and here consider only a system containing matter. It is matter which contains, collects, sorts, transfers, and gives off energy, and a brief description of its structure should be beneficial.

Matter is composed of *atoms*, each of which in turn is composed of a cluster (one or more) of *protons* and *neutrons* very tightly compressed into a *nucleus*. The nucleus is surrounded by a cloud of one or more very small particles, called *electrons*, found in different concentric levels, or shells, encircling the nucleus. Protons have a positive charge; electrons, a negative charge; and neutrons, as the name implies, have no charge, that is, they are neutral. The electrons, protons, and neutrons of different atoms vary in number and arrangement, and each unique arrangement is called an *element*. There are both naturally occurring and manmade elements. (See table B.18 of appendix B.) A simplistic diagram of an atom is shown in figure 2–3, a sketch of a hydrogen atom.

Each shell of an atom can hold a certain maximum number of electrons, for example, the first shell can hold two electrons, and the second, eight. When the outer shell, that is, the last shell that contains electrons, contains the maximum number it can hold, that atom is in a stable state and does not readily combine with other atoms. For example, neon atoms have two electrons in their first shell and eight in their second and are thus stable. However, if an atom has an incomplete outer shell and does not contain the maximum of electrons it can hold, it is unstable and will readily combine with other atoms to make its outer shell complete. When two atoms combine, one gains the electrons which the other loses; in combining, atoms form different types of chemical bonds, such as ionic or metallic (figure 2–4). When atoms combine, they form what is called *molecules*.

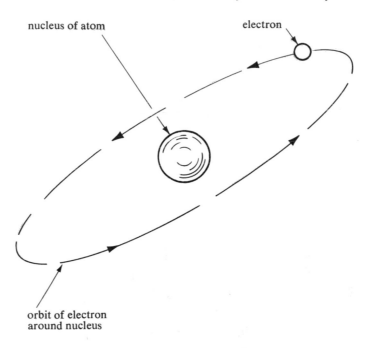

Figure 2–3   Simplified visualization of hydrogen atom.

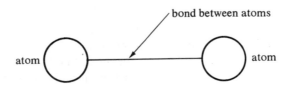

Figure 2–4   Simplified view of metallic or ionic bond in molecule composed of two atoms.

In addition to gaining or losing electrons, atoms can form bonds by "sharing" one or more pairs of electrons; this type of bond is called *covalent*. In this sharing of electron pairs, each electron in a pair comes from a different atom. A hydrogen molecule is a good example of covalent bonding, or electron sharing (figure 2–5). Each hydrogen molecule contains two atoms, each of which when it exists singly without the other, has only one electron in its outer shell and needs one more to make it complete. However, when the atoms combine to form a molecule, each shares the other's electron so that each has two electrons and thus a completed outer shell.

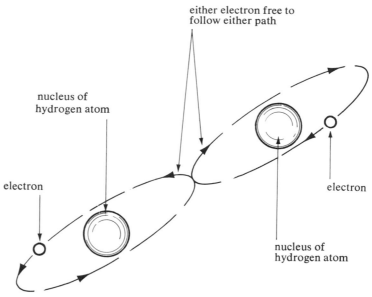

Figure 2–5   Hydrogen molecule.

Atoms and molecules have mass, which is measured in *atomic weight*. Notice in table B.18 that each element has an atomic weight. This number represents the mass of one gram-mole of the element. The gram-mole is a mass which contains 6×10^{23} atoms of a given element under standard atmospheric conditions and based on the mass units of grams. This number is called Avogadro's number and is an important chemical quantity. For instance, 6×10^{23} molecules of hydrogen make up 1 gram-mole of hydrogen and have a mass of 1 gram. Similarly, 6×10^{23} atoms of carbon make up 1 gram-mole of carbon, and they have a mass of 14 grams. We can also speak of lb-moles of an element, which is the mass of an element based on the lb-mass unit of mass. (Refer to section 2.7 for a discussion of the lb-mass unit.) Since there are 454 grams in one lb-mass of any material, we see that there are 6×10^{23} atoms $\times$ 454 = 2.724×10^{24} molecules of hydrogen, or any other element in one lb-mass mole (written 1 lbm-mole).

In our approach to thermodynamics we will not be concerned with counting or otherwise describing individual atoms or molecules, but we will consider very large amounts of atoms or molecules mixed together, called a *substance*. If the substance is composed of only one type of molecule, we will call this a *pure substance*. Water, composed of molecules made up of two hydrogen (H) atoms and one oxygen (O) atom and written H_2O, is an example of a pure substance.

The atoms and molecules of substances behave differently, depending on the conditions of the surroundings in which the substance is immersed. Consequently, we will want to know the arrangement and interaction of these particles, the *phase* of the element or atoms. We will concern ourselves with three phases of a substance: solid, liquid, and gaseous. The solid phase of a substance is characterized by rigid bonds between atoms or molecules, bonds which can generally stretch, bend, and break (figure 2–6). Also, the solid phase will, under many conditions, be characterized by a crystal structure.

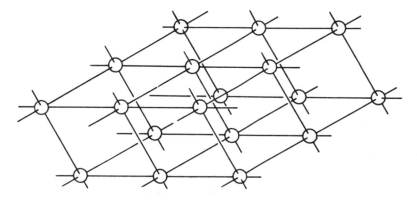

Figure 2–6   Solid phase — Typical "lattice" composed of metallic or ionic bonds and atoms in a rigid configuration.

The gaseous phase has free-floating molecules with some, although rather insignificant force between them (figure 2–7). We will be discussing a substance called a *perfect gas*, which is a gas that has no forces, other than collision forces, between the individual atoms or molecules. The collision for a perfect gas is assumed to be "perfect" in the sense that no distortion occurs among the particles before or after colliding.

The liquid phase is behaviorally like a mixture of the solid and gaseous phases. The molecules are more closely packed than in a gas, but the bonds between the molecules are passive. That is, the molecules will adhere to one another under very light forces but can easily slide apart and shift arrangements without great forces, as shown in figure 2–8. We could characterize liquids as being able to retain some volume (one gallon for instance), but not shape. If left alone in space, gaseous substances cannot even retain their volume, let alone their shape, whereas solids retain volume and shape.

At a low temperature, materials will almost invariably pass from a solid to a liquid phase; as the temperature rises, they will pass to a gaseous phase. We will see that "high" or "low" temperature has no absolute value thermodynamically, so that some materials, such as air, do indeed have liquid and solid phases at their "low" temperatures and that some solids have liquid and gaseous phases at their "high" temperatures.

In this text we will generally treat gas and liquid phases and mixtures of the two. In addition, we will treat pure substances or homogeneous mixtures of pure substances, as mentioned earlier. We will not discuss solids because they go beyond the scope and depth of this text.

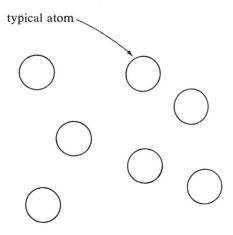

typical atom

Figure 2–7   Gaseous Phase — Little or no evidence of bonds between atoms: atoms free to move independently in any direction.

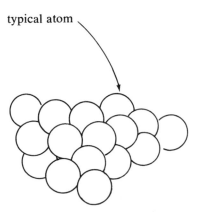

typical atom

Figure 2–8   Liquid Phase — Atoms attracted by bonds which allow sliding between atoms but prevent separation of individual atoms.

**2.3
Property**

In order to describe and analyze a system, we must know some of the quantities that are characteristic of it. These quantities are called *properties*, and include volume, mass, weight, pressure, temperature, density, shape, position in space, velocity, energy, specific heat, color, taste, and odor. The list could go on and on, and the longer the list, the better description we could give our system.

We will separate properties into two general classifications: *intensive properties* and *extensive properties*. An intensive property is independent of the mass or total amount of the system, for example, color, taste, odor, velocity, density, temperature, and pressure. Extensive properties are all those properties which are dependent on the total amount of the system, for example, mass, weight, energy, and volume. It is worth noting that any extensive property can be made intensive by dividing by the mass. In this book, we will denote extensive properties by upper case letters and intensive by lower case letters.

**2.4
State of a
System**

A complete list of the properties of a system describes its state. In order that a list does not get too long and confusing, we will generally assume that the system is composed of pure substances of one phase, or of simple inert mixtures of the three phases, gaseous, liquid, and solid. We will see that only a few properties need to be known under these conditions, and these are generally as follows:

> Type of substance (element)
> Volume
> Weight or mass
> Pressure
> Density or specific volume
> Temperature
> Energy

**2.5
Process**

The primary reason for describing the state of a system is to analyze the system as a power-producing or power-consuming quantity. *Process* is a change of state, which can occur in a number of ways. For example, if we change one or more of the properties such as energy, pressure, temperature, or volume, we will have gone through a process. Our concern will be evaluating the state immediately before and after this change. An important point to remember is that work, power, and heat can occur only during processes and only across the boundary of the system. We will return to this later.

2.6
Cycle

Having a system which changes its state, thereby producing or using work and/or heat is all well and good, but if we want an engine that runs continuously without seemingly changing its state further and further, then we must occasionally return to our stabilizing point or initial state.

Cycle: *A combination of two or more processes which, when completed, return the system to its initial state; a system operating on a cycle is called a* **cyclic device** *or a* **heat engine.**

Notice that there is no change in energy or property over a complete cycle. There can be, however, work and heat added or extracted, which is the essence of the power-producing and power-consuming devices of our society—we return the engine (or cyclic device) to its initial condition periodically (for example, 300 times per minute if we have an engine running at 300 revolutions/minute), and it still continues to produce work.

2.7
Weight and
Mass

In describing a system, one of the first properties which we think of is weight. We will denote weight by the symbol W and express it in the dimensions of pounds-force, lbf.

When we measure the weight of a system, we use a weight scale and determine the deflection, or "stretch," of a spring. This stretch is directly proportional to a force applied to the spring. Figure 2–9(a) shows a typical weight scale which is composed of a spring, a pointer attached to the spring, and a stationary scale. Figure 2–9(b) shows this same device after a test weight (our system) has been attached. In order to determine the system weight, note that the spring has extended in length (a distance proportional to the system weight W), and the pointer has moved from a position indicating no weight (0 on the stationary scale) to a new position where we say the weight is 10 pounds. If the system had weighed 20 pounds, then, of course, the pointer would have indicated 20 pounds on the stationary scale, and any other weight would as well register that particular number on the stationary scale.

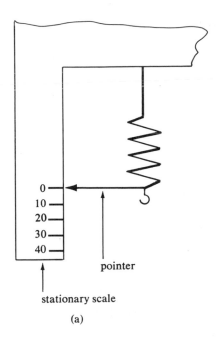

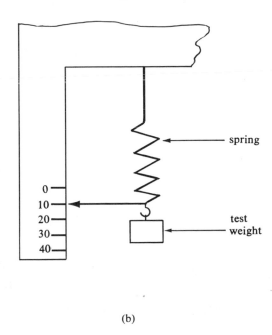

Figure 2-9   Principle of weight measurement.

Note in figure 2–9 that we are not directly measuring the weight of a system. First, we are measuring spring deflection, which is proportional to the force on the spring. This force is equal to the weight of the system since there is no motion after the spring has deflected. Figure 2–10 shows the forces acting of the system for the situation of figure 2–9. From physics mechanics we know that under static (motionless) conditions the sum of all forces equals zero.

$$\Sigma \text{ Forces} = 0$$

or

$$F - W = 0 \qquad\qquad (2\text{--}1)$$

and

$$F = W$$

This result was stated before. Now, in order to understand that weight is a force, we will indicate the units or dimensions of W by pounds-force (lbf) rather than just pounds.

In addition, since any force is a vector (a vector has magnitude and direction), we see that weight is a vector.

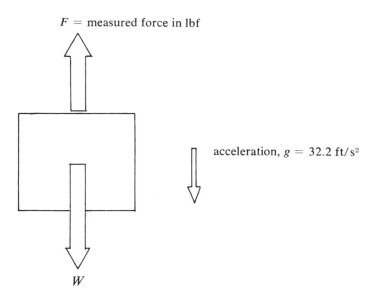

F = measured force in lbf

acceleration, $g = 32.2$ ft/s^2

W

Figure 2-10   Forces acting on system when measuring weight.

Mass, m, is a scalar (magnitude only) quantity and is, along with the volume of a system, the primary measure of the amount of matter in the system. (See section 2.2.) Mass can be related to the weight by Newton's second law of motion which says that an external force applied to a body is proportional to the mass of that body times the acceleration, and that the force will have the same direction as the acceleration. Algebraically this is

$$F \backsim ma \qquad\qquad (2\text{-}2)$$

where

m = mass of the block
a = acceleration of the block

and

F = external force applied to the block

We can easily make equation (2-2) an equality by inserting a constant C so that

$$F = Cma \qquad\qquad (2\text{-}3)$$

In figure 2-10 we have a static condition which means that no acceleration is present. Since the block is immersed in a gravitational field which tends to accelerate the block *down*, the external measured force F

accelerates the block *up* the same amount; we will call this gravitational acceleration *g*. Combining the results of equations (2–1) and (2–3), we get

$$W = Cmg \qquad (2\text{--}4)$$

Note that *C* is a constant for any condition whereas *g* is a variable. On the earth we find that $g \approx 32.2$ ft/s^2, but that *g* decreases as we move away from the earth. For reference, table B.17 in appendix B lists values of *g* for various conditions. If we were in a situation (like space travelers) where $g = 0$ at some points in our excursions, we would have no weight since

$$W = Cm(0) = 0$$

By tradition, mass, along with weight, has been measured in pounds, which has caused all sorts of confusion. We will call the mass unit *pounds-mass* (lbm) to differentiate it from weight units (lbf). The reader should do this in problem solutions, too. At the earth's sea level surface, we define lbf and lbm, on the average, as follows:

$$1 \text{ lbm} = 1 \text{ lbf} \quad \text{where} \quad g = 32.174 \text{ ft/s}^2$$

This convention gives us a value for *C* from equation (2–4):

$$1 \text{ lbf} = C \times 1 \text{ lbm} \times 32.174 \text{ ft/s}^2$$

And then

$$C = \frac{1 \text{ lbf}}{1 \text{ lbm} \times 32.174 \text{ ft/s}^2} \qquad (2\text{--}5)$$

For convenience we will say

$$C = \frac{1}{g_c} \qquad (2\text{--}6)$$

and call g_c the universal gravitational constant equal to 32.174 ft/s^2 lbm/lbf.

Then, if we use lbm and lbf units, equation (2–4) is

$$W = \frac{1}{g_c} mg \qquad (2\text{--}7)$$

and Newton's law, equation (2–3), becomes

$$F = \frac{1}{g_c} ma \qquad (2\text{--}8)$$

or, from equation (2–7),

$$F = \frac{W}{g} a \qquad (2\text{--}9)$$

Note that equation (2–9) is not defined for weightless conditions since

$$W = 0$$

and

$$g = 0$$

Another mass unit defined by 32.174 times the mass in pounds-mass is called the *slug*. That is

$$32.174 \text{ lbm} \equiv 1 \text{ slug} \qquad (2\text{–}10)$$

The constant of proportionality in equations (2–3), (2–4), and (2–5) can then be written in terms of slug units, where

$$\begin{aligned} C &= \frac{1 \text{ lbf/lbm}}{32.174 \text{ ft/s}^2} \times 32.174 \text{ lbm/slug} \\ &= 1 \frac{\text{lbf/slug}}{\text{ft/s}^2} \end{aligned}$$

And we see that in terms of mass units of slugs, equations (2–3) and (2–4) become

$$F = ma \qquad (2\text{–}11)$$

and

$$W = mg \qquad (2\text{–}12)$$

respectively.

Example 2.1

What is the weight W of a system having a mass of 1 slug when the system is:
(a)   at sea level and 40° latitude.
(b)   at 1,000 ft above sea level and 40° latitude.

Solution

(a)   The value of g is 32.158 ft/s² at sea level and 40° latitude. (See table B.17.)
Then from equation (2–12)

$$\begin{aligned} W &= mg \\ &= 1 \text{ slug} \times 32.158 \text{ ft/s}^2 \\ &= 32.158 \text{ lbf} \qquad\qquad \textit{Answer} \end{aligned}$$

We see that

$$1 \text{ slug} = 1 \frac{\text{lbf s}^2}{\text{ft}}$$

(b) At 1,000 ft we see from table B.17 that g is 32.155 ft/s² so

$$W = 1 \text{ slug} \times 32.155 \text{ ft/s}^2$$
$$= 32.155 \text{ lbf} \qquad\qquad Answer$$

Alternatively, the mass of the system in lbm units is

$$m = 1 \text{ slug} \times 32.174 \text{ lbm/slug}$$

from equation (2–10), and from equation (2–7)

$$W = \frac{1}{32.174 \text{ ft/s}^2 \times \text{lbm/lbf}} \times 1 \text{ slug} \times 32.174 \text{ lbm/slug} \times 32.155 \text{ ft/s}^2$$
$$= 32.155 \text{ lbf}$$

While the mass unit of slugs provides somewhat simplified equations by eliminating a factor of $1/g_c$ from the force-mass relationships as shown by equations (2–11) and (2–12) as compared to (2–9) and (2–7), we will generally use pound-mass (lbm) units unless otherwise specified. The slug unit is used in some reference literature, but the pound-mass unit is generally used for reporting thermodynamic data. This is our primary reason for using the pound-force and pound-mass units.

One other item should be mentioned regarding units. In this text we will be using the English engineering system of units and occasionally the metric system of units. The basic mass units in the metric system are *grams* and *kilograms*, while force is expressed in *dynes* or *newtons*. We will reserve further discussion of the conversions and basis for these two-dimensional systems for section 2.12.

2.8 Volume, Density, and Pressure

Volume

We have seen that volume and mass are the primary measures of the quantity of matter in a system.

Volume: *An extensive and geometric property having a value characterized by a length times a height times a width, simply "length cubed."*

The units associated with volume will be cubic feet (ft³), cubic inches (in³), gallons (gal), or possibly the metric unit of cubic centimeters (cm³). Volume will be denoted in this text by the letter V.

Specific Volume   Many times we will want to know the volume occupied by a unit mass of the system constituent. This is called the *specific volume* and is an intensive property denoted by the letter v. If the matter is homogeneous then

$$v = \frac{V}{m} \qquad (2\text{--}13)$$

The units describing specific volume are characteristically volume per unit mass as indicated by equation (2–13). Some of the units which will be used in this text to describe specific volume are ft^3/lbm, in^3/lbm, $ft^3/slug$, gal/lbm, and $cm^3/grams$.

Density   Density is a property used frequently in describing the mass of a system. It is denoted in this book by the Greek letter ρ, and for homogeneous materials we have

$$\rho = \frac{m}{V} \qquad (2\text{--}14)$$

Note that density is an intensive property and is the inverse of specific volume, that is

$$\rho = \frac{1}{v} \qquad (2\text{--}15)$$

and the units of density are lbm/ft^3, lbm/in^3, $slug/ft^3$, lbm/gal, and $grams/cm^3$. Table 2–1 lists the densities of various liquids at common atmospheric conditions.

Table 2–1
Density and specific gravity of some liquids at 68°F and 14.7 Psia.

Liquid	Density ρ, rho, lbm/ft^3	Specific Gravity sg
Benzene	54.9	0.8790
Carbon Tetrachloride	99.1	1.5950
Ethyl Alcohol	49.2	0.7893
Glycerine	78.6	1.2600
Mercury	845.0	13.5460
Methyl Alcohol	49.5	0.7928
Water	62.3	0.9982

Specific Weight   Occasionally a property given by the weight per unit volume is desired. This we call the *specific weight*, denoted by the Greek letter γ. Specific weight for homogeneous materials is calculated from

$$\gamma = \frac{W}{V} \tag{2-16}$$

The units of specific weight are lbf/ft³, lbf/in³, lbf/gal, and dynes/cm³. The numerical value of specific weight is very nearly equal to density on the earth's surface. In general though, using equations (2–5), (2–14), and (2–15), we see that

$$\gamma = \left(\frac{g}{g_c}\right)\left(\frac{m}{V}\right) \tag{2-17}$$

$$= \frac{g}{g_c}\rho$$

Specific Gravity

Liquids are sometimes described by their specific gravity. This quantity, denoted by the symbol sg, is the ratio of the density of the described fluid to the density of water where the water is at 39.2°F. The density of water at this temperature has been found to be 62.43 lbm/ft³ so that we can write, for the specific gravity of a liquid

$$sg = \frac{\rho}{62.43 \text{ lbm/ft}^3} \tag{2-18}$$

Note that, when density of the liquid is in lbm/ft³, the specific gravity is unitless, representing nothing more than a ratio of densities. Table 2–1 lists values of specific gravity for some common liquids.

Pressure

One of the properties which we will use extensively in this text is pressure. For our purposes we will consider pressure to be a force per unit area. Specifically, we will define it as

$$p = \frac{F}{A} \tag{2-19}$$

where p denotes the pressure and F is the perpendicular or normal force to the area A as indicated in figure 2–11.

From an important physical principle known as Pascal's law it is known that in gases or liquids a pressure at one point in one direction (such as pressure p_1 in figure 2–12) induces a pressure in all directions at that same point as indicated in figure 2–12. We can also note that the above definition for pressure corresponds to a definition for stress in solid materials. There are, however, conditions which make Pascal's law for solids much more complicated than for liquids and gases.

To understand how pressure can differ within a system, let's look at a vertical tube of cross-sectional area A, and height (1), shown in figure 2–13. The tube is filled with water at 39.2°F and the pressure at the top of

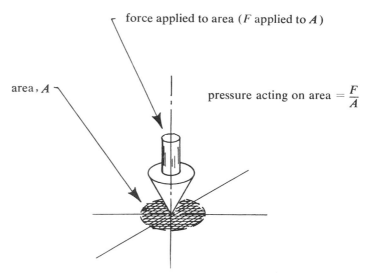

Figure 2–11   Pictorial representation of pressure.

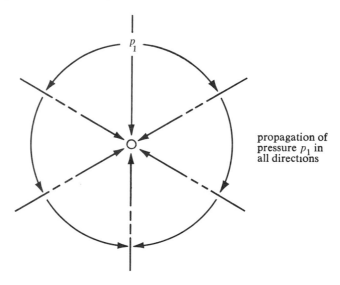

Figure 2–12   Pascal's law for liquids and gases.

the tube is observed to be the weight of a volume of air (the atmosphere) divided by A. This pressure we call atmospheric pressure, p_a and its value is near 14.7 lbf/in² on the surface of the earth. Table B.16 gives the U.S. standard day values of atmospheric pressure for various locations on or near the earth. Standard day is defined as annual averages at 45° latitude within the U.S. (probably never realized exactly).

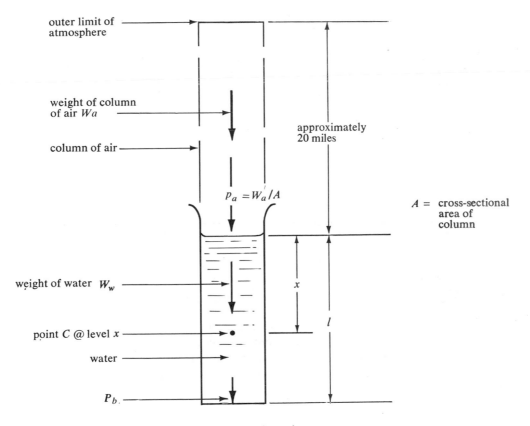

Figure 2-13   Pressures acting on water in tube.

At the bottom of the water column now, we see that the pressure, p_b, is the pressure due to the weight of the water column, W_w, plus p_a.

$$p_b = \frac{W_w}{A} + p_a \qquad (2\text{–}20)$$

We could also substitute for W_w

$$W_w = \gamma_w \times l \times A$$

where γ_w = specific weight of water, yielding

$$p_b = \gamma_w l + p_a \qquad (2\text{–}21)$$

Additionally, if we look at some other point in the water level, say point C at x in figure 2–13, the pressure will be

$$p_c = \gamma_w x + p_a \qquad (2\text{–}22)$$

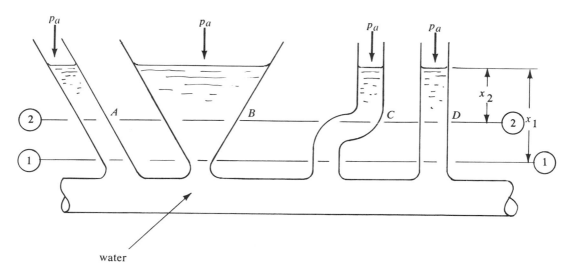

water

Figure 2-14   Illustration of complete dependence of pressure to elevation.

Interestingly, a consequence of Pascal's law is that the pressure in a gas or liquid is dependent only on the vertical depth (x in figure 2–13) called the *pressure head* and the *atmospheric pressure*. In figure 2–14, the pressure at level (1) is the same in all the columns—*A*, *B*, *C*, and *D*.

$$p_1 = \gamma_w x_1 + p_a$$

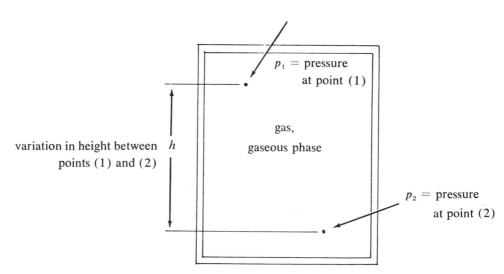

p_1 = pressure at point (1)

gas, gaseous phase

variation in height between h points (1) and (2)

p_2 = pressure at point (2)

Figure 2-15   Illustration of independence of pressure with elevation in gaseous systems.

Similarly, at level (2)

$$p_2 = \gamma_w x_2 + p_a \text{ for all four columns.}$$

For systems which contain gases, we will generally neglect the pressure changes due to elevation differences because the specific weight times the elevation change ($\gamma \times h$ term) will be several orders of magnitude less than the system pressure at the top; that is, we will say

$$p_1 = p_2 = \text{constant as identified in the system of figure 2–15.}$$

There are various schemes for measuring pressure of a system, but we will look only at the U-tube manometer since it is simple and contains the essence of all pressure-measuring devices. It is merely a glass or transparent tube having a constant cross section area A and formed into a "U" as shown in figure 2–16(a). This tube can contain water or any other

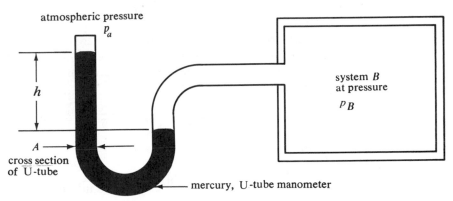

Figure 2–16(a)   Pressure measurement of system.

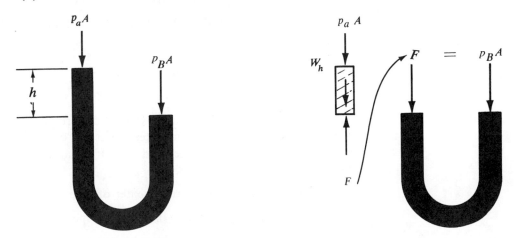

Figure 2–16(b)   Forces acting on mercury in manometer of figure 2–16 (a).

liquid, but more commonly mercury (Hg) is used. Now, let's use this gage to measure the pressure inside a closed container (system B in figure 2–16 a). If the pressure in the container, p_B, is greater than the atmospheric pressure p_a we see from figure 2–16(b) that

$$p_B A = p_a A + \gamma_{Hg} h A \quad \text{or} \quad p_B = p_a + \gamma_{Hg} h$$

Notice that we are physically determining the term h in this measurement, but we will call the term γh the gage pressure, p_g, so that

$$p = \text{pressure of a system}$$

or

$$p = p_g + p_a \tag{2–23}$$

where

$$p_g = \gamma h \tag{2–24}$$

The term h is described in inches of mercury (in Hg), and to convert this to pounds per square inch (psi) we see, using table 2–1, that the density of mercury is 845 lbm/ft³. The specific weight of mercury near the earth is given by equation (2–17)

$$\gamma_{Hg} = \frac{g}{g_c} \rho \cong \frac{32.2}{32.2} \, 845 = 845 \, \text{lbf/ft}^3$$

or

$$\gamma_{Hg} = 845 \, \text{lbf/ft}^3 \times \frac{1}{1728 \, \text{in}^3/\text{ft}^3} = 0.49 \, \text{lbf/in}^3$$

If h is 1 inch of mercury we have, from equation (2–24) that

$$P_g = 0.49 \, \text{lbf/in}^3 \times 1 \, \text{in} = 0.49 \, \text{lbf/in}^2$$

so that

$$1 \, \text{in Hg} = 0.49 \, \text{lbf/in}^2$$

Example 2.2

Two pipes A and B are used to convey water and air respectively as shown in figure 2–17. Pressure gage A indicates a pressure of 1.01 psig (pounds per square inch gage) in the water-filled pipe. If the atmospheric pressure is 14.69 psi, find

(a) Absolute pressure at center of pipe A.

(b) Absolute pressure in air-filled pipe B.

(c) Gage pressure measured by pressure gage B.

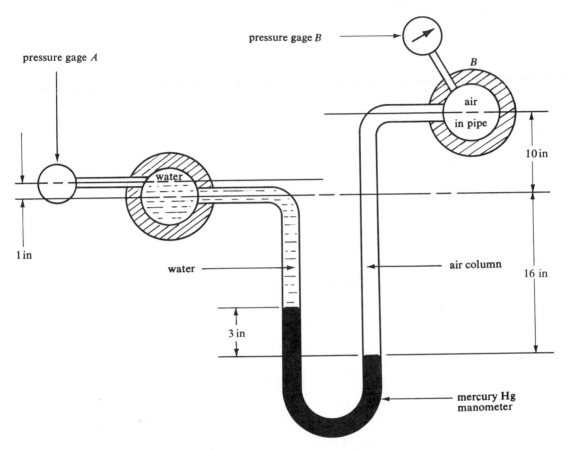

pressure gage B

pressure gage A

B

air
in pipe

10 in

1 in

water

water

3 in

air column

16 in

mercury Hg
manometer

Figure 2–17

Solution

Example 2.2

(a) Pressure gage A (which may be a mercury U-tube manometer) is located 1 inch above the center of the pipe. The absolute pressure at the gage location is, from equation (2–23)

$$p = p_g + 14.69 \text{ psi} = 1.01 \text{ psig} + 14.69 \text{ psi}$$

or

$$p = 15.70 \text{ psia}$$

Corrected to the center of the pipe A, the pressure is, from equation (2–22)

$$p_A = 15.70 \text{ psia} + \gamma_w \times 1 \text{ in}$$

and where γ_w is the specific weight of the water in pipe A. We have

$$\gamma_w = 62.4 \text{ lbf/ft}^3 \cong 0.036 \text{ lbf/in}^3$$

so that

$$p_A = 15.70 + 0.036 \times 1$$

$$= 15.736 \text{ psia} \qquad\qquad Answer$$

(b) To determine the pressure of the air in pipe B, we analyze the mercury manometer. The pressure p_B, sensed by gage B, while not exactly the pressure at the center of the pipe B due to the height variation, is close enough for engineering or most scientific work since air is a gas having very low specific weight:

$$p_B = \text{constant in pipe } B$$

Also, the pressure acting on the right side of the U-tube mercury manometer, p'_B is given by

$$p'_B = p_B + \gamma_{\text{air}} \times 23 \text{ in}$$

For air at 68°F and low pressure

$$\gamma_{\text{air}} \approx 0.00004 \text{ lbf/in}^3$$

so that for our case

$$p'_B \approx p_B$$

From figure 2–18, showing pressures acting on the mercury manometer, we have

$$p'_B = p_A + \gamma_w \times 13 \text{ in water} + \gamma_{\text{Hg}} \times 3 \text{ in Hg}$$

$$= 15.74 \text{ psia} + 0.036 \text{ lbf/in}^2 \times 10 \text{ in} + 0.491 \text{ lbf/in}^3 \times 3 \text{ in}$$

$$= 15.74 + 0.47 + 1.47$$

$$= 17.68 \text{ psia}$$

The pressure in pipe B is, therefore

$$p_B = 17.68 \text{ psia} \qquad\qquad Answer$$

(c) The gage pressure in pipe B must be, from equation (2–23)

$$p = p_B - 14.69 = 2.99 \text{ psig} \qquad\qquad Answer$$

If the system pressure, p, is less than p_a, then we have a vacuum. Referring to figure 2–16, we would then physically have a case of p_B and p_A interchanged so that

$$p_a A = p_B A + \gamma_{\text{Hg}} h A \qquad\qquad (2\text{–}25)$$

or

$$p_a = p + \gamma_{\text{Hg}} h = p + p_{gv}$$

The pressure in a vacuum chamber is then written

$$p = p_a - p_{gv} \tag{2-26}$$

where p_{gv} is the vacuum gage pressure. Notice that this term, p_{gv}, can be no more than p_a; if

$$p = 0$$

then $p_a = p_{gv}$ atmospheric pressure and p cannot be less than zero.

Remember that the gage or vacuum gage pressure is the quantity measured in all pressure gages. The atmospheric pressure must be included as shown by equations (2–23) or (2–26) to determine the absolute or true pressure of the system. In these equations, the system pressure is calculated at the interface with the mercury or gage fluid. If the system contains a liquid then the pressure, of course, varies inside the system as predicted by equation (2–22).

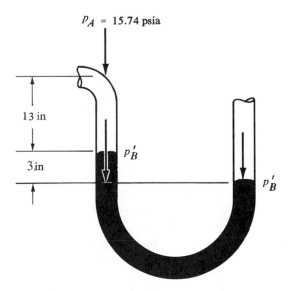

Figure 2–18   Pressures acting on mercury manometer in example 2.2.

2.9
Equilibrium and the Zeroth Law of Thermodynamics

In mechanics we utilize equilibrium to determine forces acting on bodies. This type of equilibrium we call *mechanical equilibrium*, and it is indeed an important part of our scientific efforts. In thermodynamics we will be concerned with *thermal equilibrium*. The property which determines this type of equilibrium we call *temperature* and we postulate the following:

> **Zeroth law of thermodynamics:** *Two separate bodies which are in thermal equilibrium with a third body are also in thermal equilibrium with each other.*

This law tells us that we can measure temperature through thermal equilibrium of bodies and be assured that it is independent of the materials involved.

Remember that if two separate bodies at different temperatures are brought into contact with each other, thermal equilibrium will be reached and retained when the temperature is the same in both bodies. Interestingly, temperature equilizes in this condition, but the energies of the two bodies do not necessarily equalize.

2.10 Temperature and Thermometers

When the temperature is determined for a system (such as the air temperature in a room), the characteristic method involves allowing a temperature meter (thermometer) to come to thermal equilibrium with the observed system and subsequently measuring the change in some property of the thermometer which we will call the thermometric property. Table 2–2 tabulates a few of the more prevalent types of thermometers with their associated thermometric properties.

Table 2–2

Thermometric Properties for Temperature Measurement

Type of Thermometer	Thermometric Property
Mercury-in-glass	Volume
Thermistor	Electrical resistance
Thermocouple	Voltage
Constant volume gas	Pressure of gas
Constant pressure gas	Volume of gas

The mercury-in-glass is the most common, as it is used to measure human body temperatures, normal atmospheric temperatures, and temperatures of objects used in everyday activities. The temperature is recorded in degrees Fahrenheit (°F) or degrees Celsius (°C). These scales are arbitrary, but are both characterized by being defined at the boiling and freezing (or melting) points of pure water at a pressure of 14.696 psia. These are as follows: Boiling point is 212° Fahrenheit which equals 100° Celsius,

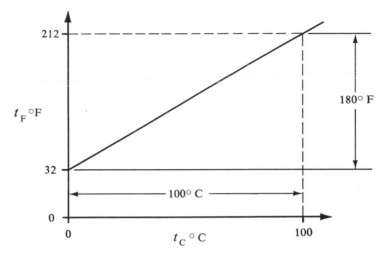

Figure 2-19   Relation of Celsius and Fahrenheit temperature scales.

and the freezing point is 32°F which equals 0° Celsius. The setting t_F is the temperature in °F, and t_C is the temperature in °C.

We see by plotting t_F vs. t_C as shown in figure 2–19 that

$$t_F = \frac{180°F}{100°F} t_C + 32°$$

$$= \frac{9}{5} t_C + 32° \qquad (2\text{--}27)$$

or

$$t_C = \frac{5}{9}(t_F - 32°) \qquad (2\text{--}28)$$

If we were to correlate the two mechanical properties, pressure and volume, of a material with the thermal property of temperature, we would find that for many vapors or gases, the equation

$$pV = mRT \qquad (2\text{--}29)$$

is very descriptive of the material. This equation is called the *perfect gas equation* and is the simplest example of an *equation of state* Other examples of equations of state and a further discussion of the perfect gas will be given in chapter 5. At this point, however, note that if we fill a rigid chamber (constant volume container) with a perfect gas, following equation (2–29), and if we seal the chamber except for a manometer tube as shown in figure 2–20, we have a gas thermometer. For this apparatus we then apply equation (2–29)

$$pV = mRT$$

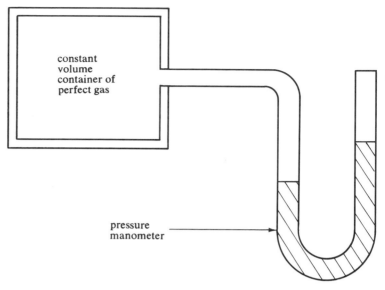

Figure 2–20   Gas thermometer.

Solving for the pressure we obtain

$$p = \frac{m}{V} RT = \rho RT \qquad (2\text{–}30)$$

where ρ is the density defined in equation (2–14). But density and the gas constant R are constant so that we may write

$$p = (\text{constant})T \qquad (2\text{–}31)$$

Suppose now we use the apparatus of figure 2–20 to measure temperature-pressure relationships of three gases, A, B, and C. We could easily generate pressure-temperature data points shown in figure 2–21. Some of the data points fit straight lines, and the point where these lines intersect with the vertical axis (point Z in figure 2–21) we define as the *absolute zero point* of temperature. If the temperature scale in figure 2–21 had been in degrees Fahrenheit, we would have found

$$Z = -459.4°\text{F}$$

Similarly, if we had used degrees Celsius, then

$$Z = -273°\text{C}$$

From these values of absolute zero, we then say the temperature of a body, T, is

$$T = (t_\text{F} + 459.4) \text{ degrees Rankine} \qquad (2\text{–}32)$$

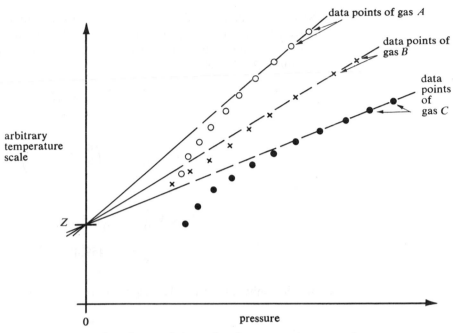

data points of gas *A*

data points of gas *B*

data points of gas *C*

arbitrary temperature scale

Z

0   pressure

Figure 2–21   A method for determining absolute zero temperature.

or

$$T = (t_C + 273) \text{ degrees Kelvin} \qquad \textbf{(2–33)}$$

The reason some of the data points at the lower temperatures do not align with the appropriate straight lines is that all gases depart from a perfect gas at the lower temperatures. Thus equation (2–30) is not a correct description of the physical situation at a low temperature but is correct at higher temperatures.

In this book we will generally use the Rankine scale of temperature because of its prevalence in engineering work.

Example 2.3   A thermodynamic system composed of liquid nitrogen is at 72° Kelvin. Convert this to degrees Celsius, Fahrenheit, and Rankine.

Solution   From equation (2–33), we have

$$t_C = T - 273°K$$
$$= 72 - 273$$

So that

$$t_C = -201°C \qquad \qquad Answer$$

Now t_C can be easily converted to °F by using equation (2–27). Let us, however, approach this problem somewhat differently and, consequently, observe an interesting relationship. Substituting equation (2–28) into equation (2–33) yields

$$T°K = \frac{5}{9}(t_F - 32) + 273$$

or

$$T°K = \frac{5}{9}t_F + \frac{5}{9}\left(\frac{9}{5}273 - 32\right)$$

then,

$$T°K = \frac{5}{9}t_F + \frac{5}{9}(459.4)$$
$$= \frac{5}{9}(t_F + 459.4)$$

But the right side of this last equation is just equal to

$$\frac{5}{9} \times T°R$$

as can be seen from equation (2–32).

So, we have

$$T°K = \frac{5}{9}T°R \qquad\qquad (2\text{–}34)$$

which is a useful relationship between absolute temperature scales. In our problem

$$T°K = 72°K$$

so, from equation (2–34)

$$72°K = \frac{5}{9}T°R$$

which gives us

$$T°R = \frac{9}{5} \times 72°K = 129.6°R \qquad\qquad Answer$$

To find the Fahrenheit temperature, we use equation (2–32) and solve algebraically for t_F

$$t_F = T°R - 459.4$$

Substituting for T we have

$$t_F = 129.6 - 459.4$$

or

$$t_F = -329.8°F \qquad \qquad Answer$$

2.11
Energy

In chapter 1 we defined energy as a certain capacity of a body; that is, energy is a property of the system or body in which it resides. At this point, let us seek a better grasp of energy in quantitive terms. There are many forms of energy, but we will be primarily concerned with *kinetic*, *potential*, and *thermal*, or *internal energy*.

Kinetic Energy

A body or particle has energy by virtue of its movement. As an example, a stone lying on the ground is motionless and incapable of effecting change. Pick the stone up, however, and throw it against a window. This will produce a change directly attributed to the motion of the stone. This energy is called *total kinetic energy* (KE) and is given by

$$KE = \frac{1}{2} m \bar{V}^2 \qquad \qquad \textbf{(2–35)}$$

where $\bar{V}$ is the velocity of the stone. If the result of equation (2–4) is substituted into equation (2–35) for mass, m, *in slugs*, i.e.

$$m = \frac{W}{g} \text{ slugs} \qquad \qquad \textbf{(2–4)}$$

then

$$KE = \frac{W}{2g} \bar{V}^2 \qquad \qquad \textbf{(2–36)}$$

the units of which are ft-lbf.

The kinetic energy we write as

$$ke = \frac{KE}{m} \qquad \qquad \textbf{(2–37)}$$

and substituting equation (2–7)

$$m = \frac{g_c}{g} W$$

and equation (2–36) into this equation we obtain

$$ke = \frac{W}{2g} \bar{V}^2 \times \frac{g}{g_c W}$$

or

$$\text{ke} = \frac{\bar{V}^2}{2g_c} \qquad\qquad (2\text{–}38)$$

which has units of ft-lbf/lbm.

Potential Energy

Objects tend to attract each other because they all have mass. This attracting force due to mass is the essence of Newton's law of gravitation. Here on the earth the attraction between the earth and a system or an object represents a "potential" of motion. That is, the earth and the object are tending to move closer together; we call this capability *total potential energy*, denote it by PE and calculate it from

$$\text{PE} = Wz \qquad\qquad (2\text{–}39)$$

where z is a vertical reference distance between the center of the earth and the center of the object which we say has the potential energy. The units of potential energy are easily seen to be ft-lbf and the potential energy, pe, given by

$$\text{pe} = \frac{\text{PE}}{m} = \frac{Wz}{m} \qquad\qquad (2\text{–}40)$$

has units of ft-lbf/lbm. From equation (2–7), we have

$$\frac{W}{m} = \frac{g}{g_c}$$

and equation (2–40) becomes

$$\text{pe} = \frac{g}{g_c} z \qquad\qquad (2\text{–}41)$$

Internal Energy

The hotness or coldness of a body can physically affect the surroundings of that body. This capacity, sometimes called *heat* or *thermal energy* is quite naturally indicated by the temperature of the body. In this text we will call this property *total internal energy* and designate it by U. Internal energy will thus be the measure of the "hotness" of a body; "coldness" will represent the absence of internal energy. We will not, under any circumstance, use the word *heat* to describe this property but may on occasion use the term *thermal energy*. *Heat* will be assigned a much different meaning in the next chapter.

Being a form of energy, internal energy has units of ft-lbf, but in chapter 3 we will see that the British thermal unit (Btu) and the calorie are the units which are commonly used to describe internal energy, and they are conceptually equivalent to the ft-lbf unit. We will frequently use the internal energy, u, given by

$$u = \frac{U}{m} \tag{2-42}$$

and which has units of ft-lbf/lbm or Btu/lbm.

Internal energy is frequently described as the property which reflects mechanical energy of the molecules and atoms of the material. Generally the contributions to internal energy are as follows:

(a) Translational or kinetic energy of the atoms or molecules as indicated in figure 2–22(a).

(b) Vibrational energy of the individual molecules due to straining of the atomic bonds at increasing temperatures [figure 2–22(b)].

(c) Rotational energy of those molecules which spin about an axis as shown in figure 2–22(c).

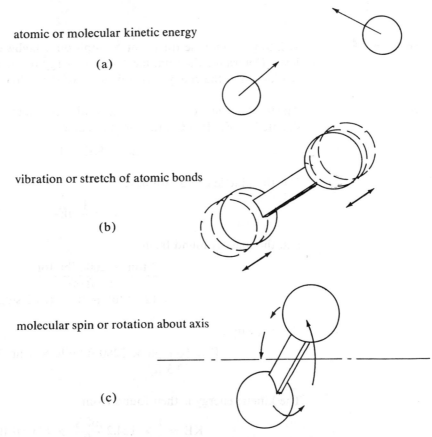

atomic or molecular kinetic energy

(a)

vibration or stretch of atomic bonds

(b)

molecular spin or rotation about axis

(c)

Figure 2–22   Molecular motions contributing to internal energy.

There are other forms of energy, such as electromagnetic energy, chemical energy, and strain energy (caused by the stretching of solid materials), which must be included in any complete analysis of a thermodynamic problem; but in this text we will normally consider the total energy E of a system to be given by

$$E = KE + PE + U \qquad (2\text{–}43)$$

and the energy e,

$$e = ke + pe + u \qquad (2\text{–}44)$$

If the above equations are substituted into the appropriate terms of equation (2–44) we have

$$e = \frac{\bar{V}^2}{2g_c} + \frac{g}{g_c}z + u \qquad (2\text{–}45)$$

Example 2.4

A 2-ton automobile travels at 50 mph on a highway 1000 ft above sea level. Determine the total mechanical energy of the auto with respect to sea level and the energy per unit mass of the auto.

Solution

The total mechanical energy is the total energy minus the internal energy, denoted as E_m. It is found from the equation

$$E_m = KE + PE$$

For the kinetic energy we have

$$KE = \frac{1}{2}m\bar{V}^2$$

and the mass is found from

$$m = \frac{2 \text{ ton} \times 2000 \text{ lbf/ton}}{32.2 \text{ ft/s}^2}$$
$$= 124.2 \text{ lbf s}^2/\text{ft} = 124.2 \text{ slugs}$$

The velocity is

$$\bar{V} = 50 \text{ mph} \times 5280 \text{ ft/mile} \times 1 \text{ hr/3600 s}$$
$$= 73.3 \text{ ft/s}$$

The kinetic energy is then found from

$$KE = \frac{1}{2} \times 124.2 \frac{\text{lbf s}^2}{\text{ft}} \times (73.3)^2 \text{ ft}^2/\text{s}^2$$
$$= 333,700 \text{ ft-lbf}$$

We now observe that the potential energy with respect to the sea level is given by

$$PE = Wz$$

where z has the value of 1000 feet. The potential energy is then equal to

$$PE = 2 \text{ ton} \times 2000 \text{ lbf/ton} \times 1000 \text{ ft}$$
$$= 4,000,000 \text{ ft-lbf}$$

and the total mechanical energy of the auto is

$$E_m = (333,700 + 4,000,000)\text{ft-lbf}$$
$$= 4,333,700 \text{ ft-lbf} \qquad\qquad \textit{Answer}$$

The specific kinetic energy is given by equation (2–38), and substituting values into this equation, we have

$$ke = \frac{(73.3)^2 \text{ ft}^2/\text{s}^2}{2 \times 32.2 \text{ ft-lbm/s}^2\text{-lbf}}$$
$$= 83.4 \text{ ft-lbf/lbm}$$

We could have calculated this result from equation (2–37), which is

$$ke = \frac{KE}{m}$$
$$= \frac{333,700 \text{ ft-lbf}}{124.2 \text{ slugs}}$$
$$= 2,687 \text{ ft-lbf/slug}$$
$$= \frac{2,687 \text{ ft-lbf/slug}}{32.2 \text{ lbm/slug}}$$
$$= 83.4 \text{ ft-lbf/lbm}$$

which agrees with our previous result. Similarly, for potential energy we have

$$pe = \frac{g}{g_0} z$$
$$= \frac{32.2 \text{ ft/s}^2}{32.2 \text{ ft-lbm/s}^2\text{-lbf}} \times 1000 \text{ ft}$$
$$= 1000 \text{ ft-lbf/lbm}$$

and, again, an alternate method of solution is

$$pe = \frac{PE}{m}$$
$$= \frac{4,000,000 \text{ ft-lbf}}{124.2 \text{ slugs} \times 32.2 \text{ lbm/slug}}$$
$$= 1000 \text{ ft-lbf/lbm}$$

Finally, the mechanical energy for the automobile is

$$\frac{E_m}{m} = 83.4 \text{ ft-lbf} + 1000 \text{ ft-lbf/lbm}$$

$$= 1083.4 \text{ ft-lbf/lbm} \qquad\qquad \textit{Answer}$$

In this example problem let us observe an important principle. Suppose the highway on which the auto was traveling sloped down to sea level and the brakes were not applied in the auto as it moved down the highway. What would happen in this condition? The auto would lose its potential energy as it rolled down the highway, but it would increase its velocity and thereby increase its kinetic energy. The point is that the total mechanical energy of the auto remains constant, but the potential energy is converted into kinetic energy. We will see in Chapter 4 that energy is conserved in any process; this principle is the first law of thermodynamics.

2.12 Units

In sections 2.7, 2.8, 2.10, and 2.11, important thermodynamic properties were introduced, and invariably the unit used to describe the given property was stated. The system of units which was used is called the English engineering system. This is a system composed of five fundamental dimensions: Length (L) in foot units; Force (F) in pound-force units; mass (m) in pound-mass units; time (τ) tau in seconds; and temperature (t and T) in degrees Fahrenheit and Rankine. All other units are then combinations of these five basic units. There are, of course, many other equally valid systems of units but only one other will be mentioned here. This is the absolute metric system of units, and it is presented in table 2–3. The equivalent energy units in the two systems of units are also given in table 2–3.

Table 2–3

Quantity	Absolute Metric	English Engineering
length	centimeter (cm)	foot
force	dyne	pound-force
mass	grams	pound-mass
time	second	second
temperature	°C/°K	°F/°R
energy	dyne-cm	ft-lbf

Many times a conversion from one system of units to another is required. To facilitate this, the following equivalents are useful:

Length: 1 inch = 2.54 cm
            1 foot = 30.4801 cm

Force: 1 lbf = 444,805 dynes

Mass: 1 lbm = 453.59 grams
            1 lbm = 0.454 kilograms (kg)

A more complete listing of conversions is given in appendix B.

Example 2.5   Convert 6 ft-lbf of kinetic energy to the absolute metric units.

Solution   Note that here we must utilize more than one conversion factor. We see that

$$1 \text{ ft} = 30.4801 \text{ cm}$$

or

$$30.4801 \text{ cm/ft} = 1$$

similarly

$$1 \text{ lbf} = 444,805 \text{ dynes}$$

or

$$444,805 \text{ dynes/lbf} = 1$$

We can always multiply by a value of one so that;

6 ft-lbf × 30.4801 cm/ft × 444,805 dynes/lbf
$$= 8.135 \times 10^7 \text{ dyne-cm} \quad \textit{Answer}$$

Note also that we have another equivalent which is useful

30.4801 × 444,805 dyne-cm/ft-lbf
$$= 1.35577 \times 10^7 \text{ dyne-cm/ft-lbf} \quad \textit{Answer}$$

In the absolute metric system the dyne-cm is called the *erg* so we see that

$$1.35577 \times 10^7 \text{ ergs} = 1 \text{ ft-lbf}$$

As was mentioned previously, there are many systems of units equally clear and useful as the English engineering, but all serve the primary function of providing consistency. Try to convert to the standard units whenever odd units arise, and, by all means, indicate units on all intermediate steps leading to the solution itself. This practice not only provides for consistent units, but allows for a check of equation correctness, that is, the units must be the same on both sides of an equation.

2.13 Summary

In this chapter, the concept of the system has been introduced and identifying the boundaries of this system has been stressed. Once the system has been identified (not too easy a task in many cases), its quantitative description is determined through the system properties, the total list of which describes the state of the system. The important properties for a thermodynamic analysis are as follows: mass, weight, density, volume, specific volume, pressure, temperature, and energy. Energy has been typified as potential, kinetic, or internal.

The following list of formulas describes the important points of this chapter. The symbols correspond to those on table 2–4.

mass/weight:

$$F = \frac{1}{g_c} ma \qquad\qquad m \text{ in lbm units}$$

$$W = \frac{1}{g_c} mg \qquad\qquad m \text{ in lbm units}$$

$$w = mg \qquad\qquad m \text{ in slug units}$$

density, volume/pressure:

$$v = \frac{1}{\rho} = \frac{V}{m}$$

$$p = p_g + p_a$$

$$p = p_a - p_{gv}$$

temperature/energy:

$$T°R = t°F + 460°$$

$$T°K = t°C + 273°$$

$$T°R = \frac{9}{5} T°K$$

$$\text{ke} = \frac{\text{KE}}{m} = \frac{1}{2} \frac{\bar{V}^2}{g_c}$$

$$\text{pe} = \frac{\text{PE}}{m} = Wz$$

$$u = \frac{U}{m}$$

Table 2–4 should be very helpful in clarifying notations, units, and concepts of this book. On the far right of the table, notice that the chapter in which the given symbol or concept is first introduced is listed.

Table 2–4
Thermodynamic Notation

Symbol	Variable	Units	Chapter Introduced
a	acceleration	ft/s²	2
A	area	ft²	2
b	spring modulus	lbf/in	3
C, c	constants	varied	
C_v	total specific heat @ constant volume	Btu/°R	5
c_v	specific heat @ constant volume	Btu/lbm°R	5
C_p	total specific heat @ constant pressure	Btu/°R	5
c_p	specific heat @ constant pressure	Btu/lbm°R	5
D	diameter	ft	
E	total energy	Btu or ft-lbf	2
e	energy	Btu/lbm or ft-lbf/lbm	2
F	force	lbf	2
g	gram		
g	local gravitational acceleration	ft/s²	2
g_c	gravitational constant	lbm-ft/lbf-s²	2
H	total enthalpy	Btu or ft-lbf	2
h	enthalpy	Btu/lbm or ft-lbf/lbm	2
J	mechanical-thermal conversion factor	778 ft-lbf/Btu	3
k	c_p/c_v	unitless	5
K_h	coefficient of heat transfer	Btu/hr-ft²-°F	
KE	total kinetic energy	Btu or ft-lbf	2
ke	kinetic energy	Btu/lbm or ft-lbf/lbm	2
I, L	length	ft or in	2
m	mass	lbm	2
$\dot{m}$	mass flow rate	lbm/s	4
n	polytropic exponent	unitless	6
p_a	atmospheric pressure	psia	2

Symbol	Variable	Units	Chapter Introduced
p_g	gage pressure	psig	2
p_{gv}	vacuum gage pressure	psiv	2
p	pressure	psia	2
PE	total potential energy	Btu or ft-lbf	2
pe	potential energy	Btu/lbm or ft-lbf/lbm	2
Q	total heat, total heat transfer	Btu or ft-lbf	3
q	heat or heat transfer	Btu/lbm or ft-lbf/lbm	3
$\dot{Q}$	total heat transfer rate	Btu/s or Btu/hr	3
R	gas constant	ft-lbf/lbm°R	2
R_u	universal gas constant	ft-lbf/lbm°R	2
r	radius	ft or in	3
r_p	pressure ratio	unitless	10
r_v	compression ratio	unitless	10
S	entropy	Btu/°R	7
s	specific entropy	Btu/lbm°R	7
SE	strain energy	ft-lbf	6
sg	specific gravity	unitless	2
T	absolute temperature	°R	2
t	thermometric temperature	°F	2
U	total internal energy	Btu or ft-lbf	2
u	internal energy	Btu/lbm or ft-lbf/lbm	2
V	volume	ft³	2
v	specific volume	ft³/lbm	2
$\dot{V}$	volume flow rate	ft³/s	4
$\overline{V}$	velocity	ft/s	2
W	weight	lbf	2
Wk	total work	ft-lbf or Btu	3
Wk_{cs}	total closed system work	ft-lbf or Btu	3
Wk_{os}	total open system work	ft-lbf or Btu	3
wk	work	ft-lbf/lbm or Btu/lbm	3
$\dot{Wk}$	power	ft-lbf/s	4
x, y	length	ft	
Y	Young's modulus	psi	6
z	reference elevation above plane of zero potential energy	ft	2

Symbol	Variable	Units	Chapter Introduced
alpha, α	Seebeck coefficient	Joules/coulomb	16
beta, β	relative humidity	percent	15
gamma, γ	specific weight	lbf/ft³	2
epsilon, ϵ	emissivity	unitless	3
eta, η	efficiency	percent	8
delta, Δ	change in variable		2
kappa, κ	thermal conductivity	Btu/hr-ft-°F	3
lambda, Λ	function of availability		9
mu, μ	chemical potential	Btu/lbm	15
rho, ρ	density	lbm/ft³	2
sigma, σ	Stefan-Boltzmann constant	Btu/hr-ft²-°R⁴	3
tau, τ	time	seconds	3
Phi, Φ	total availability	Btu or ft-lbf	9
phi, ϕ	availability	Btu/lbm or ft-lbf/lbm	9
chi, χ	quality	percent	8
omega, ω	specific humidity	lbm vapor/lbm dry air	15
xi, ξ	strain		6
$\mathcal{D}$	diffusivity	ft²/s	15
$\mathcal{E}$	electric potential	volts	16
$\mathcal{F}$	Faraday's constant	coulombs/g-mole	16
G′	total Gibbs free energy	Btu or ft-lbf	9
g′	Gibbs free energy	Btu/lbm or ft-lbf/lbm	9
H′	total Helmholtz free energy	Btu or ft-lbf	9
h′	Helmholtz free energy	Btu/lbm or ft-lbf/lbm	9
$\mathcal{I}$	electric current	amperes	16
$\mathcal{Q}$	electric charge	coulombs	16
$\mathcal{R}$	resistance	ohms	16
$\mathcal{T}$	Thomson coefficient	Joules/coulomb°K	16
$\mathcal{V}$	voltage	volts	16

Practice Problems

Section 2.7

2.1. A 2-lbm cube of gold is weighed in two locations. The first location has a local gravitational acceleration of 31.98 ft/s² and the second has a gravitational acceleration of 31.0 ft/s². Determine at which location the gold is heaviest.

2.2. Three slugs of water are vaporized in a boiler. What is the weight of this water in lbf, if the local acceleration of gravity is 32.1 ft/s²?

2.3. A weight scale located on the earth's sea level indicates a gallon of water weighs 8.333 lbf. Determine the mass of 1 gallon in units of slugs and lbm.

2.4. A battery weighing 32 lbf on earth's sea level is transported to the moon. Determine the weight of the battery on the moon where $g = 5.47$ ft/s².

2.5. Referring to necessary tables, convert the following masses:
  (a) 1 lbm to grams.
  (b) 2 lbm to kilograms.
  (c) 20 slugs to lbf on earth's sea level.
  (d) 100 grams to dynes on earth's sea level.
  (e) 200 kilograms to lbf on earth's sea level.

Section 2.8
2.6. A cylindrical tank filled with a fluid has a diameter of 3 ft and a length of 4 ft. It weighs 1600 lbf where the gravitational acceleration is 32.2 ft/s². Determine
  (a) Volume occupied by the fluid.
  (b) Specific weight of the fluid.
  (c) Density of the fluid.
  (d) Specific gravity of the fluid.

2.7. A balloon is filled with a gas having a specific volume of 15 ft³/lbm. If the inside volume of the balloon is 300 in³, what is the weight of the gas-filled balloon? Ignore the weight of the balloon itself and assume $g = 31.8$ ft/s².

2.8. Steam has a specific volume of 10.07 ft³/lbm at a pressure of 80 psia and 900°F. At this state, assuming $g = 32.1$ ft/s², determine the following:
  (a) Density.
  (b) Specific weight.
  (c) Specific gravity.

2.9. A tank is filled with hydrogen gas and a gage measures the pressure in the tank at 20 lb/ft².
  (a) If the atmospheric pressure is 14.7 psi, what is the absolute pressure in the tank?
  (b) What is the pressure in the tank if the atmospheric pressure is 31.2 inches Hg?

2.10. Two tanks, A and B, contain air (figure 2–23). Tank A is at a pressure of 20 psig and tank B is at a pressure of 18 psig. If the two tanks are connected by a U-tube manometer, as shown, what is the difference in height of the mercury column, h? Note: Specific weight of Hg = 845 lbf/ft³ and specific weight of air = 0.076 lbf/ft³.

2.11. A hydraulic pump produces a pressure of 250 psig in an oil line as shown (figure 2–24). What force in lbf will be applied to the 1-ft diameter piston?

2.12. A tank 15 ft high is half full of water, and air at 2 psig is occupying the remaining volume. The tank is 5 ft in diameter and the contents are at 70°F.
  (a) What is the gage pressure at the top of the water?
  (b) What is the gage pressure at the bottom of the tank?
  (c) Assume the atmospheric pressure to be 14.8 psia, and then find the absolute pressure of (a) and (b).

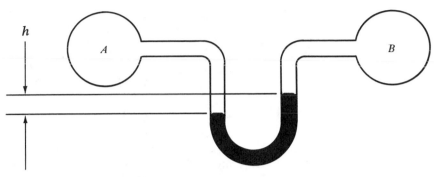

Figure 2–23   Problem 2.10.

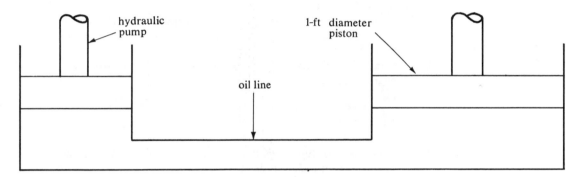

Figure 2–24   Problem 2.11.

2.13. Determine the maximum vacuum pressure possible inside a tank with the following surrounding conditions:
   (a) An atmospheric pressure of 14.8 psia.
   (b) An atmosphere at a pressure of 14 inches Hg.

2.14. Convert the following pressures:
   (a) 14.7 psia to inches Hg.
   (b) 300 inches Hg to lbf/ft².
   (c) 2 inches Hg vacuum to psia if the atmospheric pressure is 29.8 inches Hg.

Sections 2.9 and 2.10

2.15. A thermometer indicates 70°F when in thermal equilibrium with Block A, and 140°F when in thermal equilibrium with Block B. Are Blocks A and B in thermal equilibrium? Give a condition when Blocks A and B are in thermal equilibrium (figure 2–25).

2.16. Convert the following temperatures:
   (a) 140 degrees Fahrenheit (°F), to degrees Rankine (°R).
   (b) 88°F to °R.
   (c) 230 degrees Fahrenheit to degrees Celsius (°C).
   (d) 87 degrees Kelvin (°K) to degrees Rankine.

Figure 2–25   Problem 2.15.

2.17. A temperature scale is proposed which has values of inverse Rankine temperature. If we call this scale the "down scale" and identify it as T_D, then $T_D = 1/T°R$ as proposed. Plot T_D vs. T on regular graph paper between values of $1/10°R$ and $10°R$. What is the slope of the curve at a point where $T = 1°R = 1°T_D$?

2.18. A temperature scale called an L scale is suggested, which has the values of $T_L = \log T°R$. Determine the value of T_L in terms of the Kelvin scale °K.

2.19. Convert the following temperature values to both degrees Rankine and degrees Kelvin.
(a)   412°F
(b)   32°F
(c)   117°C
(d)   72°C

2.20. A new temperature scale, the N scale, based on the melting and freezing of the element cesium is defined as:
$$0°N = \text{melting point of cesium} = 28.5°C$$
$$100°N = \text{boiling point of cesium} = 690°C$$
What temperature is absolute zero in this new scale?

Section 2.11

Note: For problems 2.21 through 2.34, assume that $g_c = 32.2$ ft-lbm/s²-lbf.

2.21. A 100,000-lbm aircraft travels at 800 mph at a 10,000-ft altitude. The local gravitational acceleration at this altitude is 30.10 ft/s². Determine the following:

(a)   Total kinetic energy of the aircraft.
(b)   Total potential energy of the aircraft if the sea level is considered as the plane of zero potential energy.

2.22. Assume that you are taking a trip on a train. You carry a filled suitcase which weighs 45 lbf, and during the course of your trip you are told that the train is moving at 90 mph.
(a)   What do you observe the kinetic energy of the suitcase to be, before you are told the velocity of the train?
(b)   What do you interpret to be the total kinetic energy of the suitcases?

2.23. A 1-lbm piece of wood and a 1-lbm piece of steel are dropped simultaneously from a bridge which is 120 ft high into 80-ft deep water. What is the change in potential energy, or the available potential energy of (a) the wood and (b) the steel?

2.24. A pump is used to remove water from a well into a tank. If the well is 250 ft deep, what energy per lbm of water must be supplied by the pump in this process? Assume $g = 32.2$ ft/s^2.

2.25. A balloon weighing 10 ounces at a location where $g = 31.7$ ft/s^2 is released at 20° latitude and floats to an altitude of 6,000 ft above sea level. If we released the balloon from 1000 ft above sea level, and defined this elevation as where potential energy is zero, determine the following:

(a) Total potential energy of balloon when floating.
(b) Total potential energy of balloon if it had been at sea level.
(c) Total potential energy of balloon at release.

2.26. Steam flows through a 5-inch diameter pipe with a velocity of 80 ft/s. What is the kinetic energy of the steam, ke?

2.27. One lbm of mercury at 426°F has 150 Btu of internal energy under a certain condition, while it also has 28 Btu of KE and 2 Btu of PE. Determine the total energy of the mercury.

2.28. During a windstorm, the velocity of the wind is measured at 70 mph. What is the kinetic energy of the air under this condition?

2.29. A 2-lbm piston in an internal combustion gasoline engine travels at 60 ft/s at an instant. What total and specific kinetic energy does the piston possess?

2.30. Angular kinetic energy (AKE) is a mechanical form of energy, which is not accounted for in the normal kinetic energy. A system has the following amounts of energy:

AKE = 150 ft-lbf
KE = 100 ft-lbf
PE = 20 ft-lbf
U = 35 ft-lbf

(a) Determine the total energy of the system.
(b) Determine the total mechanical energy of the system.

2.31. Ten lbm of steam flowing through a pipe is found to have 15,000 Btu of internal energy U. If the kinetic energy ke is 500 Btu/lbm, and the potential energy pe is 100 Btu/lbm, determine the total energy of the 10-lbm steam, E and the energy e.

2.32. Freon flows through a circuit of pipes and components shown in the figure 2–26. If the freon has a velocity of 2 ft/s at both points (1) and (2) in the circuit, what is the difference in energy per lbm of the freon between (1) and (2), if internal or thermal energy is neglected? Assume $g = g_c$.

2.33. For problem 2.32, determine the kinetic energy at points (1) and (2) (figure 2–26).

2.34. For problem 2.32, determine the specific potential energy at points (1) and (2), if the plane of zero potential energy is assumed to be 100 ft below point (2).

Section 2.12

2.35. The Reynold's number, R_e, is a dimensionless (unitless) number frequently used in fluid mechanics, thermodynamics, and physics. This number is found by the defining equation

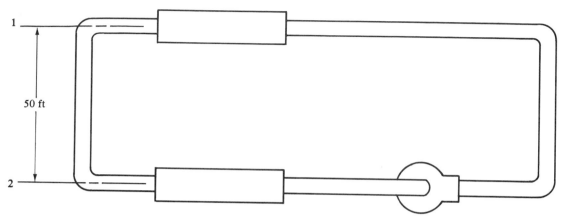

50 ft

Figure 2–26   Problems 2.32 and 2.33.

$$R_e = \frac{\rho \bar{V} D}{\mu}$$

where D is a characteristic length having units of feet, and μ is the viscosity. In English engineering units, what dimensions does viscosity have?

2.36. In terms of English engineering units, determine the dimension of x in the following equations, using table 2–4 when necessary:

(a) $x = T\Delta s/\Delta \tau$
(b) $x = \Delta h/\Delta T$
(c) $x = T\Delta S$
(d) $x = RT/(v - c)^2$
(e) $x = pv/T$

2.37. Show that both sides of the following equation agree dimensionally:

$$\frac{\Delta p}{L} = \frac{\rho \bar{V}^2}{Dg_c} \left[C \frac{\mu}{D\bar{V}p} \right]$$

where C = dimensionless constant and L has units of length. See problem 2.35 for definitions of μ and D.

2.38. What are the dimensions of C in the following equations?

(a) $C = pv^{1.7}$
(b) $C = pv^{1.3}$
(c) $C = pv/v^{2.3}$
(d) $C = p$
(e) $C = T$

3

Work, Heat, and Reversibility

In this chapter we will define *work* and develop some useful relationships for calculating its magnitude. We will make some calculations that require the use of calculus and some that do not; those which do require calculus will be clearly indicated in the text. The mechanisms of heat transfer — *conduction*, *convection*, and *radiation* — will then be presented. The concept of *reversibility* will be introduced, along with the items or causes for irreversibilities in processes. Following this, the concept of the equivalence of work and heat will be briefly discussed to provide some rationale for the conversion factor J = 778 ft-lbf/Btu. The thermodynamic system, which was introduced in chapter 2, will here be classified into three general types: *open*, *closed*, and *isolated*. This classification is introduced now to allow for better understanding of the motivation behind typifying systems. The chapter will end with a tabulation of the forms of energy, providing the foundation for the *conservation of energy* principle presented in subsequent chapters.

3.1
Work

Work connotates an active or dynamic state, during which some mechanical effort has been exerted. This visualization is embodied in our definition of work:

Work: *Force times distance through which the force acts constantly.*

$$Wk = F\Delta x \tag{3-1}$$

In order to provide for cases where the force acting through a distance is not constant, we will prefer to consider a small amount of work, dWk, resulting from a force which, while varying over a finite distance Δx, is considered to be constant over a small distance dx. Then

$$dWk = Fdx \tag{3-2}$$

and using integral calculus we say that the work done over that finite distance Δx is

$$Wk = \int dWk = \int_{x_1}^{x_2} Fdx \tag{3-3}$$

where $\Delta x = x_2 - x_1$. If we plot force F vs. distance or displacement x we might have a graph which looks like that shown in figure 3–1. This graph (or curve) describes a force which varies with the distance. Using the concepts of calculus then, we can show that the area under the curve in a force-displacement graph is equal to the work Wk since the area under the curve is also equal to $\int Fdx$ between the two values x_1 and x_2.

Let us now look at two examples of work, one in which the force applied is constant during the motion and the other in which the force varies.

Example 3.1

Determine the work done in lifting a 60-lbf block vertically 6 feet.

Solution

The force required to support or move the weight is 60 lbf, a constant value, as indicated in figure 3–2. The distance is, of course, 6 feet and from our initial definition of work, equation (3–1), we have

$$\text{work} = 60 \text{ lbf} \times 6 \text{ ft}$$
$$= 360 \text{ ft-lbf} \qquad\qquad \textit{Answer}$$

Notice that the units on work are the same as those assigned to energy and that the potential energy of the weight has increased by an amount equal to the work done, 360 ft-lbf. Identifying the system in this problem as the 60-lb weight, we can define work in a slightly more general manner:

Work: *Energy in transition across the boundary of a system, which can always be identified with a mechanical force acting through a distance.*

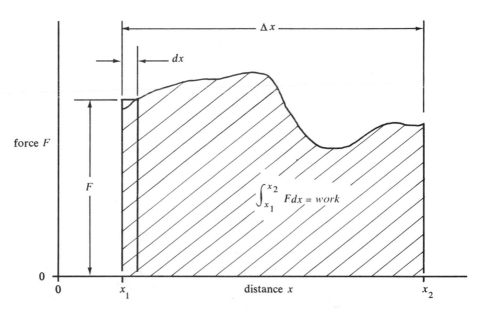

Figure 3-1   Typical force-distance relationship for a process involving work.

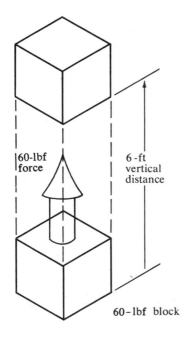

work = area

= 360 ft-lbf

distance x
feet

force-distance
relationship in
example 3.1

60-lbf
force

6-ft
vertical
distance

60-lbf block

Figure 3-2   Example 3.1.

We see then that work is the boundary effect on a system as opposed to the internal property of energy.

Example 3.2   Determine the work done in stretching a spring 3 inches from its free length of 12 inches if the spring has a modulus of 30 lbf/in deflection.

Solution   In this problem some terminology has been introduced which may not be familiar to you. First of all, notice that we are considering a spring which behaves much like the one used to support a weight in section 2.7; that is, if no force is applied to the spring it has a length called the *free length*. As some force is applied to the spring (or a weight hung from it) the spring will stretch some distance. As the force is increased, the spring continues to stretch further and the force per unit amount of stretch or deflection is called the *modulus* of the spring. If we plot on a sheet of graph paper what has just been stated we have a relation like that shown in figure 3–3. The slope of the line is the change in force divided by the corresponding change in deflection, Δx. Obviously, this problem represents a case of varying force with distance, and we therefore find the work done by referring to equation (3–3):

$$Wk = \int_{x_1}^{x_2} F dx$$

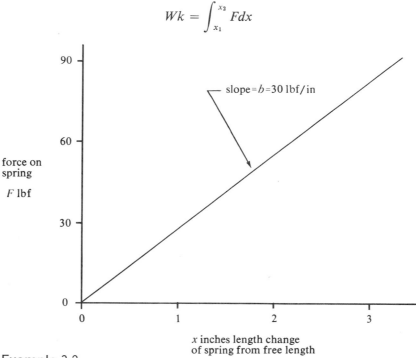

Figure 3–3   Example 3.2.

In this problem, the initial distance x_1 is zero and the final distance x_2 is 3 inches. From figure 3–3 we see

$$F = bx \quad \text{where} \quad b = 30 \text{ lbf/in}$$

and we substitute this relation into our work equation to obtain

$$Wk = b \int_{x_2}^{x_2} x dx$$
$$= \frac{1}{2}bx_2^2 - \frac{1}{2}bx_1^2$$
$$= \frac{1}{2}(30 \text{ lbf/in})(3 \text{ in})^2$$
$$= 135 \text{ in-lbf}$$
$$= 11.25 \text{ ft-lbf} \qquad \qquad Answer$$

This then is the work required to stretch the spring 3 inches. It corresponds to the area under the curve in figure 3–3 between $x = 0$ and $x = 3$ inches. We can say the work done to change the length of a spring which has a force-length relationship pictured in figure 3–3 is

$$Wk_{sp} = \frac{1}{2}bx^2 \qquad \qquad \textbf{(3–4)}$$

where x is the change in spring length from the free length of the spring.

Let us now look at a type of apparatus which we will see often in the remainder of this book. This apparatus includes a freely sliding frictionless piston of radius r, and a gas which pushes the piston out of the cylinder or which is compressed when the piston is retracted into the cylinder. This arrangement is shown in figure 3–4 and since the gas is not free to escape from the container, we will call this a *closed system*. (See section 3.8.) The apparatus we call a *piston-cylinder*. In this case let us assume that the gas contained in the cylinder is at some high pressure and subsequently moves the piston out of the cylinder. If the gas itself is to be considered as the system in this process, we see that work is being done by the system. In examples 3.1 and 3.2, work was done on the system. Obviously then, work is a directional phenomenon in the sense that it is energy crossing the boundary of the system into or out of the system. We can account for this direction by stating that *positive work is work done by the system and negative work is work done on the system*. We should note that work is not a vector. Although it is a product of two separate vectors (*force* and distance, or *displacement*), it does not have a vector direction. That is, the precise path followed by the work is not known to us, so we must conclude that work is a scalar product.

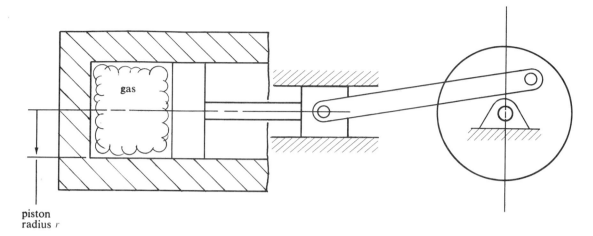

piston
radius r

Figure 3–4   Axial cross section of piston-cylinder device.

Now, we notice that as the gas in the piston-cylinder expands and pushes the piston out, the pressure will likely drop and since the force applied to the piston, F, is

$$F = p \text{ (Circular area of piston)} = p \times A \qquad (3\text{–}5)$$

the force will also drop. We, therefore, will have a case of varying force so that the work of the closed system piston-cylinder can be described with equation (3–2)

$$dWk_{cs} = Fdx$$

The subscript cs denotes *closed system* here. We know that the force, in terms of pressure, is given by equation (3–5) so that

$$dWk_{cs} = pAdx$$

Now, A is the circular area of the face of the piston so that Adx is a small change in the volume dV of the gas itself. We therefore write

$$dWk_{cs} = pdV \qquad (3\text{–}6)$$

which will be used many times. In order to find the work done during some finite change in volume of the gas system, let us plot pressure vs. volume for the case where the piston is extended out of a cylinder, thereby increasing the volume of gas from V_1 to V_2 as indicated in figure 3–5. From equation (3–3) we see that the work will be

$$Wk_{cs} = dWk_{cs} = \int_{V_1}^{V_2} pdV \qquad (3\text{–}7)$$

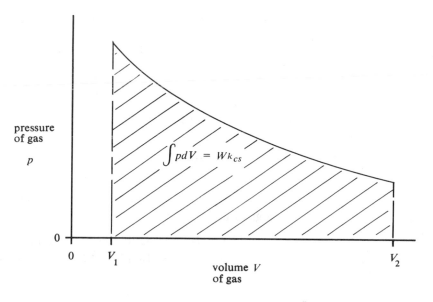

pressure
of gas

p

$\int pdV = Wk_{cs}$

0

0 V_1 volume V of gas V_2

Figure 3-5   Pressure-volume, *p-V*, diagram of closed thermodynamic system involved in a work process.

Note that the area under the curve in a *p-V* diagram is the work of a frictionless closed system Wk_{cs}. We will frequently say that the work of a closed system is $\int pdV$—implying the limits V_1 and V_2.

A relation which will be used frequently is the work per unit of system mass. This is, for the closed system

$$wk_{cs} = \frac{Wk_{cs}}{m} = \int \frac{pdV}{m}$$

or

$$wk_{cs} = \int pdv \qquad\qquad \textbf{(3–8)}$$

We will see that many specialized relationships for work come from either equation (3–7) or (3–8), among which is the following example of a process which has great utility.

Example 3.3   A 9-inch diameter piston-cylinder contains a gas which, under constant pressure, extends the piston 3 inches. Determine the work of this process if the gas pressure is 85 lbf/in² absolute.

Solution   In this case the pressure is constant so that equation (3–7) can be written

$$Wk_{cs} = p \int_{V_1}^{V_2} dV$$
$$= p(V_2 - V_1)$$
$$= p\Delta V \qquad\qquad (3\text{–}9)$$

The pressure is given as 85 psia and the change in volume can be calculated from

$$\Delta V = A \times 3 \text{ in}$$
$$= \pi r^2 \times 3 \text{ in}$$
$$= 3.14 \times 4.5^2 \times 3 \text{ in}^3 = 190.8 \text{ in}^3$$

which gives us

$$Wk_{cs} = 85 \text{ lbf/in}^2 \times 190.8 \text{ in}^3$$
$$= 16214.2 \text{ in-lbf} = 1351.2 \text{ ft-lbf} \qquad\qquad Answer$$

3.2
Heat

Heat is a word which probably has been more mistreated in technological language than any other single word. Following the manner of defining work in section 3.1, we will define heat:

Heat: *Energy in transition across the boundary of a system, which cannot be identified with a mechanical force acting through a distance.*

Heat occurs in a process when there is some temperature difference between the system and its surrounding. The direction of energy transition is always toward the area of lesser temperature. Heat will leave a system if it is hotter than its surroundings; if it is cooler, heat will enter the system. This energy transition will continue in the same direction until the system and its surroundings are separated or until thermal equilibrium is reached.

Heat will be identified by the symbol Q, and since it is energy in motion, the units can be ft-lbf. More commonly, however, heat is measured in Btus:

1 Btu: *The amount of heat to raise the temperature of 1 lbm of water 1° Fahrenheit when the water is at 39° Fahrenheit.*

The amount of heat moving in or out of the system can also be described by the amount per unit of system mass, which we will identify by q and measure in Btu/lbm. This term is related to Q by the equation $q = Q/m$.

The manner in which heat occurs in a process is called *heat transfer* and is synonymous with *heat*. That is, we will use the two terms to mean the same thing. Heat transfer is a branch of science which is well documented and has found many applications in technology. There are three forms of heat transfer: *conduction, convection,* and *radiation.* (They will be discussed individually in the following three sections.) Each of these represents a method of calculating Q, depending on the mode of heating, much like we calculated work in the previous section. There is one distinction, however, between the approach of the following three sections and the rest of this text where heat is calculated; here we must know the temperature of both the surroundings *and* the system, whereas in chapter 7 the property referred to as *entropy* will be introduced which will allow heat or heat transfer to be calculated from a knowledge of either the system *or* the surrounding temperature, but not necessarily both.

3.3 Conduction Heat Transfer

Suppose a pot of water is placed on a stove burner which has a temperature of between 700°R and 3000°R depending on the type of burner and the operating conditions. Obviously, the water will begin to warm up and, if left on the stove, will vaporize (boil).

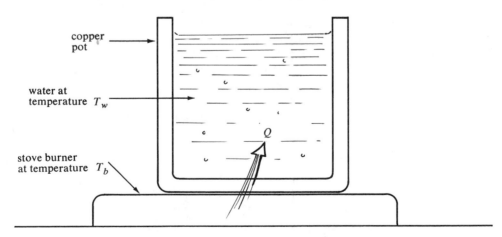

copper pot

water at temperature T_w

stove burner at temperature T_b

Q

Figure 3–6   Physical arrangement for example of heat transfer by conduction.

In the arrangement shown in figure 3–6, a change is certainly being effected on the water in the pot. We can easily identify this change as heat or a heat transfer process in which internal energy of the water is increased. Energy is being lost from the surface of the burner and, in the form of heat, is transferred or conducted through the copper pot into the water itself. The heat transfer involved here is called *conduction*, and the rate of conduction is given by Fourier's law of conduction:

$$\dot{Q} = \left(\frac{\kappa A \Delta T}{\Delta x}\right) \qquad (3\text{--}10)$$

where $\dot{Q}$ is a rate of heat transfer with respect to time, given by an amount of heat Q, divided by time τ, or more precisely

$$\dot{Q} = \frac{dQ}{d\tau} \qquad (3\text{--}11)$$

where $d\dot{Q}$ is a small amount of heat transferred during a small time period, $d\tau$. In equation (3–10), A is the surface area of the burner in contact with the corresponding area of the copper pot; ΔT is the temperature drop across the bottom of the pot which is $T_b - T_w$; Δx is the pot thickness; and κ is the thermal conductivity of the pot. In table 3–1 the thermal conductivity is listed for various types of material and temperatures. A more complete listing of the conductivity of materials can be found in various technical references, but note from table 3–1 that κ varies with temperature for all materials. Those materials listed as refractory type are characterized by their minimum dependence of κ on temperature, and are used in very high temperature applications such as furnaces and incinerators. Insulating materials have low thermal conductivities and are thus capable of retarding the heat transfer, while conducting materials have relatively high values of thermal conductivity for high heat transfer rates. Note from equations (3–10) and (3–11) that the material's conductivity determines the rate of heat transfer, but it does not determine the amount of heat transfer if the process can continue indefinitely.

Let us now look at the water that was placed on the stove burner.

Example 3.4

The thickness of the copper pot is assumed to be 0.060 inch and the water in the pot is at 75°F initially and 145°F after being on the stove 8 minutes. Determine the initial rate of heat transfer and the rate after 8 minutes. Assume the pot diameter to be 8 inches, and the surface temperature of the burner to be 1000°R.

Solution

The heat transfer rate can be found by using equation (3–10). The terms in this equation have the following values initially:

$$A = \frac{\pi D^2}{4} = \frac{(3.14)(8^2)}{4 \text{ in}^2}$$
$$= 50.24 \text{ in}^2$$

$\kappa = 220$ Btu/hr-ft-°F (approximately) where the copper pot temperature is assumed to be 60°R, from table 3–1

$\Delta T = 1000°R - (75 + 460)°R$

$= 465°R = 465°F$

$\Delta x = 0.060$ in

Using equation (3–10), we have

$$\dot{Q} = \frac{(220 \text{ Btu/hr-ft-°F})(50.24 \text{ in}^2)(465°F)}{(0.060 \text{ in})(12 \text{ in/ft})}$$

$= 7,138,000$ Btu/hr $= 7.138 \times 10^6$ Btu/hr

$= 1983$ Btu/s   *Answer*

After 8 minutes of heating the values are

$A = 50.24$ in²

$\kappa = 216$ approximately, from table 3–1, assuming an average pot temperature of 350°F

$\Delta T = 1000°R - (145° + 460°) = 395°F$

$\Delta x = 0.060$ in

which then yields, from equation (3–10)

$$\dot{Q} = \frac{(216)(50.24)(395)}{(0.060)(12)}$$

$= 5.95 \times 10^6$ Btu/hr

$= 1654$ Btu/s   *Answer*

From this example you can see that we have many variables in a heat transfer problem: temperature gradiants (differences); thermal conductivity (a constant which varies); and the rate of heat transfer. This term $\dot{Q}$, by the way, affects its own answer to the extent that it will contribute to the change in the temperature on either side of the conducting surface (the wall of the pot in the above problem). We are not concerned here with any further detailed analysis of conduction heat transfer, but let us close this section with another example.

Example 3.5

An iron pipe is used to convey steam from a steam turbine to a condenser where the steam will change from vapor to a liquid. During the process, the steam is found to be 240°F at a certain point in the length of the pipe. Assuming that the atmosphere surrounding the pipe is 88°F, determine the rate of heat loss per foot length of pipe if the pipe has an outside diameter of 4 inches and a wall thickness of $\frac{1}{8}$ inch. Figure 3–7 shows this configuration.

Table 3–1

Thermal Conductivity of Common Materials at Characteristic Temperatures

Material	Phase	Type of Thermal Conductor	Thermal Conductivity (κ, Kappa) Btu/ft-°F-hr						
			20°F	32°F	200°F	400°F	1000°F	1500°F	2000°F
Air	Gas	Insulation	0.0137	0.0139	0.0178	0.0219	. . . . .	. . . . . .	. . . . . .
Glass Wool	Solid	Insulation	0.0217	0.0231	0.0435	. . . .	. . . . .	. . . . .	. . . . . .
Rock Wool	Solid	Insulation	0.0321	0.0355	0.0373	. . . .	. . . . .	. . . . .	. . . . . .
Hydrogen	Gas	Insulation	. . . . . .	0.0990	0.1240	. . . .	. . . . .	. . . . .	. . . . . .
Helium	Gas	Insulation	. . . . . .	0.0820	0.0970	. . . .	. . . . .	. . . . .	. . . . . .
Pennsylvania Fire Brick	Solid	Refractory	. . . . . .	. . . .	. . . .	. . . .		0.71	0.76	0.84
Magnesite Brick	Solid	Refractory	. . . . . .	. . . .	. . . .	. . . .		2.90	2.32	2.14
Water	Liquid	Conducting	. . . . . .	0.320	0.392	. . . .	. . . . .	. . . . . .	. . . . . .
Aluminum	Solid	Conducting	. . . . . .	132.0	132.0	131.0	130.0	. . . . .	. . . . . .
Iron	Solid	Conducting	. . . . . .	43.0	40.0	36.0	25.0	. . . . .	. . . . . .
Silver	Solid	Conducting	. . . . . .	233.0	226.0	217.0	218.0	. . . . .	. . . . . .
Copper	Solid	Conducting	. . . . . .	220.0	218.0	215.0	214.0	. . . . .	. . . . . .

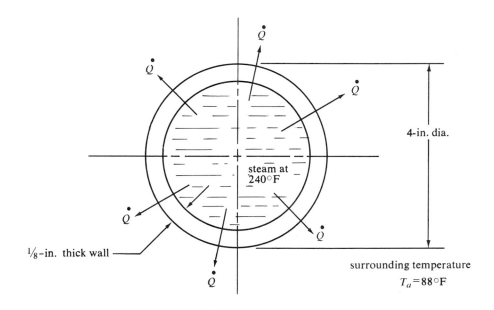

Figure 3–7   Radial cross section of steam pipe with conduction heat transfer to surroundings.

Solution

We can obtain the solution to this problem by applying equation (3–10). The thermal conductivity of the iron pipe is

$$\kappa = 40 \text{ Btu/hr-ft-°F} \quad \text{(approximately, from table 3–1)}$$

The area across which heat is transferred varies from inside to outside of the pipe, but we will assume the area to be determined from the outside of the pipe. Then

$$A = \pi \times \text{diameter} \times 1 \text{ ft length}$$
$$= (3.14)(4 \text{ in})(1 \text{ ft})(1/12 \text{ ft/in})$$
$$= 1.047 \text{ ft}^2$$

Also, the temperature drop across the pipe is

$$\Delta T = 240°F - 88°F = 152°F$$

Across the pipe thickness,

$$\Delta x = \frac{1}{8} \text{ in}$$
$$= \frac{1}{96} \text{ ft}$$

Now, $\dot{Q}$ is occurring in a radial manner as indicated in figure 3–7, but the total heat transfer rate around the pipe in a 1-foot section is given by:

$$\dot{Q} = \frac{(40 \text{ Btu/hr-ft-°F})(1.047 \text{ ft}^2)(152°F)}{(1/96 \text{ ft})}$$
$$= 6.111 \times 10^5 \text{ Btu/hr}$$
$$= 169.8 \text{ Btu/s} \qquad \qquad \textit{Answer}$$

This example is one of steady state, that is, at a point in space the properties remain the same with time, even though the material passing through (steam) will likely have its properties changed at some other point in space. A state where properties such as temperature change at a given point with time elapsing, is called a *transient condition*, which we have here avoided. Other examples of conduction heat transfer and a more complete description of this phenomenon can be found in the references listed in appendix C.

3.4 Convection Heat Transfer

In the previous section we considered two examples of conduction heat transfer through a container or a wall. In the first example, water was boiling in a pot, but a closer examination of this problem would reveal that the water at the bottom (closest to the source of energy or the stove burner) would boil first. In fact, although the water will not all be at the same temperature simultaneously, the temperature can be made more uniform by stirring the water. This type of heat transfer then is an example of *forced convection* and is the phenomenon involving conduction and flow of matter due to some outside agent (we stirred the water). Similarly, in example 3.5 the transfer of heat from the 240° steam through the wall of the pipe is another example of forced convection, since the steam was moving (mass transfer) and there was obviously a conduction of heat at the inside surface of the pipe. There is one other form of convection, called *free convection*, an example of which is the transfer of heat from the pipe to the surrounding air in example 3.5. The natural or free currents of air passing around the outside of the pipe cause a mass transfer and again, conduction at the outer surface of the pipe. If, however, someone were to operate a fan and force air around the pipe, rather than rely on nature, the phenomenon would be considered forced convection.

The rate of heat transfer, $\dot{Q}$, for convection heat transfer is found from the equation, known as Newton's law of heat transfer.

$$\dot{Q} = K_h A \Delta T \text{ Btu/hr} \qquad (3\text{--}12)$$

where K_h is the coefficient of heat transfer, a function of the type of materials involved, the magnitude of the motion of the material, and the temperature of the material; and where ΔT is the temperature difference between the temperatures of the two materials convecting heat across a surface area A.

The calculation of the coefficient of heat transfer K_h, is, of course, the important and critical problem in convection, and there are specialized methods required to calculate it. For a more complete and quantitative treatment of this type of heat transfer, you may want to refer to the heat transfer texts listed in appendix C.

3.5 Radiation Heat Transfer

Conduction and convection are two types of heat transfer which occur due to intimate contact between materials. Neither type can occur through a vacuum, but radiation, or radiant heat transfer, occurs in spite of matter. It is the type of heat or energy transfer which occurs between the earth and the sun, and without which people could not survive. Radiation is emitted from all matter, and the rate can be determined from the equation

$$\dot{Q} = \epsilon \sigma A T^4 \qquad (3\text{--}13)$$

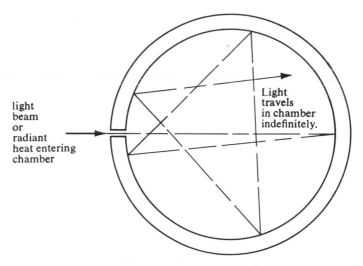

light beam or radiant heat entering chamber

Light travels in chamber indefinitely.

Figure 3-8   Cross section of black body and mechanism for complete absorption of radiant heat, $\epsilon = 1$.

where

$T =$ the temperature of the body emitting radiant heat

$A =$ the surface area

$\sigma = 0.174 \times 10^{-8}$ Btu/hr-ft^2°R^4 = Stefan-Boltzmann's constant

$\epsilon =$ emissivity, or coefficient of radiant heat transfer

For an object which has perfect emission and absorption, ϵ has a value of 1. A type of object or material having these characteristics is called a *black body* and can be approximated very nearly in lampblack, water, and and in hollow containers having only one opening through which radiation can be emitted or absorbed. An example of this last type is seen in figure 3–8, where we are considering a beam of light (one of the more common forms of radiation). Notice in the figure that the light beam enters the only opening to the chamber and subsequently reflects and re-reflects continuously. Conversely, this process could be considered in reverse and then all emission from the inner walls of the chamber would exit through the single opening, thus acting as a perfectly emitting body, or a black body.

Now let us consider the radiant interaction between two black bodies. Figure 3–9 shows two flat plates which are parallel and which have radiant heat transfer between them. The radiant heat transfer from plate (1) to plate (2) is represented by the term $\dot{Q}_{12}$ and, conversely, $\dot{Q}_{21}$ represents the heat transfer from plate (2) to plate (1). If we assume both

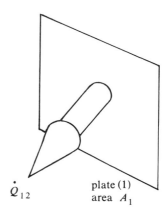

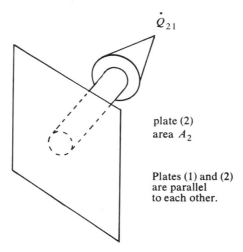

Figure 3–9   Mechanism of total radiant heat transfer between two plates.

plates are black body radiators, then we have $\dot{Q}_{12}$ = the heat emitted from plate (1) minus the heat absorbed by plate (1) from plate (2). Using equation (3–13) for a black body ($\epsilon = 1$), we obtain

$$\dot{Q}_{12} = \sigma(A_1T_1^4 - A_2T_2^4) \qquad \text{(3–14)}$$

Additionally, the term $\dot{Q}_{21}$ is given by

$$\dot{Q}_{21} = \sigma(A_2T_2^4 - A_1T_1^4) \qquad \text{(3–15)}$$

A comparison of equations (3–14) and (3–15) indicates that

$$\dot{Q}_{12} = -\dot{Q}_{12}$$

which seems reasonable for a condition of thermal equilibrium.

Example 3.6

Two black body plates are shown in figure 3–9. The plates are parallel and have areas of 3 square feet. The temperature of plate (1) is 300°F and 2000°F for plate (2). Find the net heat transfer and state whether it is directed toward plate (1) or plate (2).

Solution

The surface temperatures are

$$T_1 = 300°F + 460 = 760°R$$
$$T_2 = 2000°F + 460 = 2460°R$$

and from equation (3–14) we have

$$\dot{Q}_{12} = 0.174 \times 10^{-8} \text{ Btu/hr-ft}^2\text{-}°R^4 (3 \text{ ft}^2 \times 760^4 - 3 \text{ ft}^2 \times 2460^4)$$
$$= -1.894 \times 10^5 \text{ Btu/hr} \qquad \qquad Answer$$

In addition,

$$\dot{Q}_{21} = 1.894 \times 10^5 \text{ Btu/hr} \qquad \qquad Answer$$

We see that the net heat transfer is into plate (1) which is at the lower temperature; that is, there is more radiant energy transferred into plate (1) than is emitted by plate (1).

In this example problem many simplifying assumptions were made. Among these is the condition that heat transfer is only radiated between the two plates. In reality, however, each of the plates would probably radiate energy to other bodies or area and thereby alter the rates of heat transfer between the two. Also, it should be stressed that while heat transfer is conveniently categorized as *conduction, convection,* and *radiation,* any real heat transfer process is composed of all three of these types in varying amounts. A lengthier and more detailed treatment of radiant heat transfer is beyond the purposes of this book and the reader interested in further discussions of heat transfer is here referred to the references in appendix C.

3.6
Reversibility

In section 3.1, the work done by the gas contained in a piston-cylinder device was found to be given by $\int p\,dV$ or the area under a curve in a p-V diagram. What we want to know, however, is the amount of work available external to the piston-cylinder. This work will be equivalent to an external force F_x times a distance (figure 3–10). But the force F_x is given by

$$F_x = F_g - F_f - F_i$$

in figure 3–10. In all of this discussion, the piston is assumed to have no

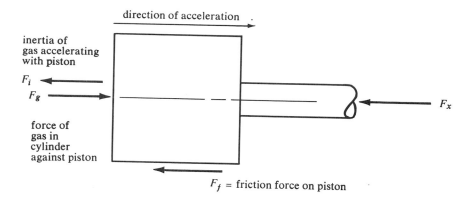

Figure 3–10

Forces acting on massless piston in actual piston-cylinder work process with friction and acceleration.

mass and therefore no inertia. The work transmitted outside the piston cylinder is then

$$Wk_x = \quad pdV - \text{friction work} - \text{inertia work} \qquad \textbf{(3–16)}$$

where *friction work* and *inertia work* are the efforts required to overcome friction of the piston and inertia of the gas, respectively. In reality, inertia is not overcome, but rather detracts from the gas pressure inside the piston. This is equivalent to the result of equation (3–16). Recall from physics mechanics that inertia (or inertia force) is given by the mass times the acceleration of the mass; and acceleration is the change of velocity per unit time. Inertia must then be overcome if the piston is to move, starting from rest, in a limited amount of time since there will always be some acceleration and consequently an inertia force of the gas which reduces the pressure acting on the piston.* With the intent of not treating all that is pertinent to the subject, friction and inertia will generally be ignored in this book as they are the two most mentioned items causing irreversibilities in processes. Now we can write equation (3–16) as

$$Wk_x = \int pdV - \text{irreversible work} \qquad \textbf{(3–17)}$$

---

*The student should observe that during deceleration of the piston, the inertia force will be opposite from that shown in figure 3–10, but the gas in the cylinder will not reflect this on the face of the piston. Consequently, inertia will detract from the effective gas pressure but will not, under reversed conditions, add to the pressure.

If the piston or mechanism is frictionless and a process is carried out very, very slowly so that the piston is always in a static or quasi-static condition then we can write

$$Wk_x = \int p\,dV \qquad\qquad (3\text{--}18)$$

since

irreversible work $= 0$

The process is then said to be *reversible* which means that work can be extracted from or put into the system, depending on the requirement. Of course, there are no actual reversible processes but many are close to this idealization and equation (3–18) is a good description for use in engineering work. Keep in mind that the term $\int p\,dV$ is an area under a curve in a *p-V* diagram and it can be found by calculus or simple geometry.

Now that we have considered *reversible* and *irreversible* as adjectives describing energy transfer processes (work) we should consider these words as they describe heat. Reversible work can be determined by neglecting irreversible work, but heat, by its very nature, is one-directional; that is, it always moves from a region of high temperature to a region of lower temperature. Reversible heat transfer is completely unreal at the macroscopic level, and a process which involves heat is irreversible unless the heat transfer is progressing at an infinitesimal rate or unless there is an infinitesimal temperature difference between the bodies involved in the heat transfer. Under these conditions we can have reversible heat transfer but, of course, it will take forever to observe the effects. We will, however, use the concept of reversible heat, and it will have the same connotation as reversible work to allow for simplified analysis of thermodynamic problems.

3.7 The Equivalence of Work and Heat

Work and heat have been defined as mutually exclusive phenomena; that is, if the energy transfer is heat, then it cannot be work, and vice versa. But heat and work are both energies in transition (or energy being transferred) so that work could ultimately affect a system exactly as if the process had involved heat instead of work. The reverse of this statement is not always true, as the second law of thermodynamics will later demonstrate; heat cannot always affect a system exactly like work. There is, however, an equivalence between the common unit of heat, the Btu, and the common unit describing work, the foot-pound force, which has been measured as

$$778.16 \text{ ft-lbf} = 1 \text{ Btu}$$

or

$$778.16 \text{ ft-lbf/Btu} = 1$$

Some authors treat the conversion factor 778.16 ft-lbf/Btu as an algebraic term and assign to it the letter *J*. In this text we will refrain from this practice and will consider the conversion from common heat units to common work units as equivalent to converting from feet to inches or any other unit conversion.

Note that when the above conversion factor is recalled we can place heat, work, and energy all in the same units of Btus or ft-lbfs, as we choose. Let us look at an example of this conversion.

Example 3.7

The internal energy of 3 lbm of air is 60 Btu. How much energy is this in ft-lbf?

Solution

This is strictly a unit conversion so that

$$U = 60 \text{ Btu} \times 778 \text{ ft-lbf/Btu}$$
$$= 46,680 \text{ ft-lbf} \qquad \qquad Answer$$

Obviously 1 Btu represents a much greater *amount* of energy than 1 ft-lbf and it is for this reason that the equivalence of the units of heat and work was not readily accepted by the scientific community when first proposed by James Joule in 1842.

**3.8
Types of
Systems**

When the concept of the system was introduced in chapter 2, it was emphasized that the identification of the system was a first step in the thermodynamic method of solving real problems. Here we will classify the system into one of three types, *open, closed,* or *isolated,* indicating that the *second step* in solving a problem is determining what type the identified system is. So that the learner can make the distinction we identify the three types of system:

Open System: *A system whose boundaries allow for mass transfer, heat transfer, and work. That is, the amount of mass and energy in an open system can change.*

Closed System: *A system whose boundaries allow for heat transfer and work, but not mass transfer. That is, the amount of mass of a closed system always remains the same, but the amount of energy can change.*

Isolated System: *A system whose boundaries prevent mass transfer, heat transfer, and work. That is, the amount of mass and energy of an isolated system remains the same.*

Certain peculiarities exist for each of the three systems and will be notated with appropriate subscripts (such as Wk_{cs} for work of a closed system, *cs*) when needed.

3.9
The Forms of Energy

During these past two chapters, *energy* has been a term which entered the discussion frequently. We will see in chapter 4 that conserving energy is the major task of thermodynamics, called the *conservation of energy* or the *first law of thermodynamics*. It involves, naturally, that ubiquitous property of the system, energy; in fact, other properties of the system (such as mass, pressure, volume, and temperature) will be measured so that the amount of energy can be subsequently determined. We know that in open or closed systems, the amount of energy can be increased or decreased; thus energy can be *static* (stationary) or *dynamic* (moving from one place to another).

Table 3–2

The Forms of Energy

Form	Type	Condition
Static energy	Potential Kinetic Internal (thermal) Electromagnetic Strain Chemical	System property
Dynamic energy, i.e., energy in transition	Work Heat Heat transfer	Not a system property. Dependent on process. Occurs only during a a process.

Table 3–2 concisely lists the forms in which energy exists and makes clear the distinction between static and dynamic energy. In studying the table, note that work and heat exist only when energy is dynamic, that is, only when energy is in a state of motion or transition. (See section 2.5.) As soon as energy becomes static, it changes form, and the forms of heat and work cease to exist. We will see the importance of this change in the next chapter.

Practice Problems

Problems marked with an asterisk * are generally more difficult and require the use of differential or integral calculus.

Section 3.1

3.1. A force of 20 lb is required to slide a 300-lbf box horizontally across a 20-ft long platform. What work is required?

3.2. In problem 3.1, what work is required to lift the boxes 20 ft vertically?

3.3. A 6-inch long spring having a modulus of 100 lbf/in is deflected an amount which requires 1250 in-lbf of work. Determine this deflection.

3.4. A spring, deflected 1 inch from its free length, is suddenly deflected 1 inch further. If the modulus of the spring is 140 lbf/in, what work is required to deflect the spring the second inch?

3.5. For a certain reversible process of a piston-cylinder, the graph pictured in figure 3-11 is obtained between the initial state, (1), and the final state, (2). Determine the work of this process.

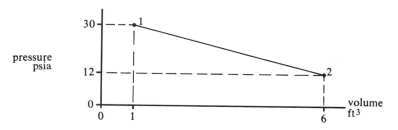

Figure 3–11   Problem 3.5.

3.6. A 3,000-lbm automobile is accelerated from rest to 60 mph in 10 seconds on a flat, straight highway.
  (a) How much work was done by the auto's engine in this process. (Hint: force = mass × acceleration and distance = $\frac{1}{2}a\tau^2$.)
  (b) If the acceleration to 60 mph required 15 seconds, what was the engine's work output?

3.7. For certain reversible processes of a closed system the pressure-volume relationships are given by the solid lines in figure 3–12. Find the work for these processes.

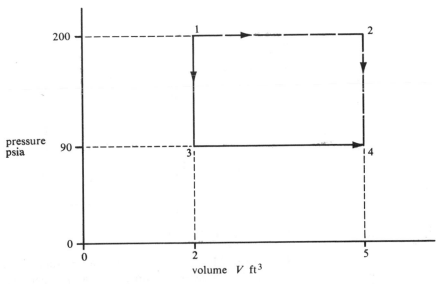

Figure 3-12   **Problems 3.7 and 3.8.**

3.8. In figure 3–12, the dashed lines represent a p-V relationship for a process which involves work. Find the work done.

3.9. Given the following relation between spring force and deflection, $F = kx^{3/2}$. Determine the work equation for this spring and the work to deflect the spring 2 inches from the free length, if $k = 60\ \text{lbf}/(\text{in})^{3/2}$.

3.10. A gas in a piston-cylinder obeys the relation $p = C/V$ where C is a constant. Derive a relationship for work in terms of C, V_1, and V_2.

3.11. For a process it is found that $pV^{1.5} = 32.2$ where p is in psia and V is in ft^3. Determine the work done if the volume increases from 2 to 3 ft^3.

Sections 3.2 and 3.3   **3.12.** Indicate which mode of heat transfer is primarily involved in the following situations:
(a) Frying eggs on a grill.
(b) Heating a room with a fireplace.
(c) Heating a room with an electric space heater.
(d) Broiling a steak.
(e) Welding with an acetylene torch.

3.13. A cryogenic sphere of 10-foot radius holds liquid oxygen at a temperature of $-325°$F. If the sphere is constructed of a $\frac{1}{2}$-inch thick steel plate, (assume steel and iron have the same conductivity), and the atmospheric temperature is 75°F, determine the heat transfer rate into the sphere. (Area $= 4\pi r^2$ for spheres).

3.14. For figure 3–13, the outside brick wall is at a temperature of 15°F at a certain time in severe winter weather. The inside wallboard, which is separated from the brick wall by a 6-inch air gap, is at 78°F. Determine, approximately, the heat rate per square foot area of wall for this condition and indicate the direction, whether from brick to wallboard or vice versa.

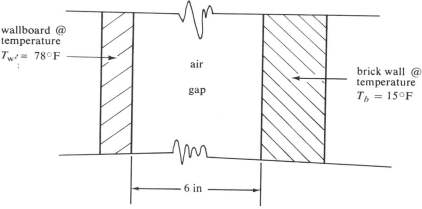

wallboard @
temperature
$T_w = 78°F$

air

gap

brick wall @
temperature
$T_b = 15°F$

6 in

Figure 3–13   **Problems 3.14 and 3.15.**

3.15. For problem 3.14, assume that during summer weather $T = 80°F$. Find the heat transfer rate through the air gap per unit area and the direction of it under this condition.

3.16. The rate of heat loss through a 1-ft² area of a furnace wall 16 inches thick is 500 Btu/hr. If the inside of the furnace is at 1800°F and the wall is assumed to be composed of fire brick, find the wall temperature of the outside surface.

3.17. The combustion chamber of an internal combustion engine is at a temperature of 2000°F when fuel is burned in this chamber. Assuming that the engine is made of iron and has an average thickness of 4 inches between the combustion chamber and the outside surface, determine the heat transfer rate per unit of wall area when combustion of fuel occurs. The air temperature around the engine is 80°F, but the outside surface of the engine is 100°F.

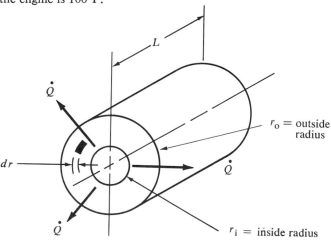

L

$\dot{Q}$

dr

$\dot{Q}$

$\dot{Q}$

r_o = outside radius

r_i = inside radius

Figure 3–14   **Problem 3.18.**

*3.18. Equation (3–10) can be written in differential form as $\dot{Q} = \kappa A\ dt/dx$. Noting that $A = 2\pi rL$ and $dx = dr$, in figure 3–14 derive a more accurate relation than equation (3–10) for calculating heat transfer rate per unit of length, $\dot{Q}/L$, through the tube if T_o = outside wall temperature and T_i = inside wall temperature.

*3.19. Using the result of problem 3.18, find the heat transfer rate per foot length through a copper pipe which has an outside diameter of 2 inches and an inside diameter of $1\frac{1}{2}$ inches. The pipe contains 180°F ammonia and is surrounded by 80°F air. Make any further assumptions you need.

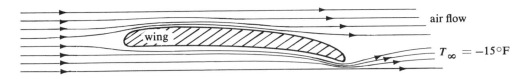

Figure 3–15   Problem 3.20.

Section 3.4

3.20. The cross section of a super-sonic aircraft wing is shown in figure 3–15, with the flow of air around the wing indicated by lines called "streamlines." The ambient atmospheric temperature of the undisturbed air, T_∞ is −15°F and the temperature of the surface of the wing, T_s, is found to be 900°F. Assuming the coefficient of heat transfer is 700 Btu/hr-ft²°F, find the rate of convection heat transfer per unit of wing area. (Note: answer should be in Btu/hr-ft².)

3.21. A 2-inch outside diameter boiler pipe containing water is heated by forcing hot air, at 800°F, around the pipe. If the coefficient of heat transfer is found to be 1000 Btu/hr-ft²-°F and the pipe temperature is 285°F, find the rate of heat transfer to the pipe per foot length of pipe. (Note: your answer should have units of Btu/hr-ft.)

Section 3.5

3.22. On a certain clear night, the temperature of the sky as seen on the earth is −55°F. If the temperature of the earth at the same time is 80°F, and both earth and sky are assumed to be black body radiators, find the net heat transfer rate to the sky per square foot of area.

3.23. A radiation pyrometer is a device utilizing radiant heat to measure temperature. Assume that the pyrometer has an area of 2 in² and a temperature of 70°F, when it is directed toward a fire having a temperature of 2000°F. Find the net heat rate toward the pyrometer, if black body radiation is assumed.

3.24. The surface of the sun seems to have a temperature of around 10,000°F. What would you guess the rate of heat emission from the sun per square foot area to be?

3.25. Assuming the surface of an electric stove heating element to be 100°F when operating, what would the rate of radiation heat transfer be from the element if it has an emissivity of 0.89? The element has an area of 28 in².

Section 3.7

3.26. Convert the following:
(a)   17 Btu/lbm to ft-lbf/lbm.
(b)   3350 ft-lbf to Btu.
(c)   2,000,000 in-ounces to Btu.

4

The First Law of Thermodynamics

In this chapter, the conservation principles of mass and energy will be introduced as the general vehicles for solving thermodynamic problems. Applying these concepts thus represents the third step in the thermodynamic method, after (1) identifying the system and its boundaries and (2) classifying the system as *open, closed,* or *isolated.* The *conservation of mass* will be presented for the general system, and for the *steady state* or *steady flow* condition.

The *conservation of energy,* or the first law of thermodynamics, will be stated and the equations representing this concept will be formulated for the system, with particular emphasis placed on the closed system. The isolated system will be discussed to illustrate the conversion of energy from one form to another when no work or heat is present.

Flow work and *enthalpy* will be introduced and will be used to formulate the equation of the first law of thermodynamics applied to the open system. Particular attention will be given to the steady flow energy equation, so common in engineering problems.

**4.1
Conservation
of Mass**

One of the most fundamental concepts of science is that mass is indestructible; that is, mass can neither be created nor destroyed. This is the principle known as the *conservation of mass* and for a closed or an isolated system we write that

$$\text{mass} = \text{constant} \qquad \text{(4–1)}$$

If the system is open so that mass can be transferred into or out of it, then the statement of the conservation of mass is written

$$m_{in} - m_{out} = \Delta m_{system} \qquad \text{(4–2)}$$

where m_{in} is the mass entering the system, m_{out} is the mass leaving the system, and Δm_{system} is the change in mass of the system. (See figure 4–1.)

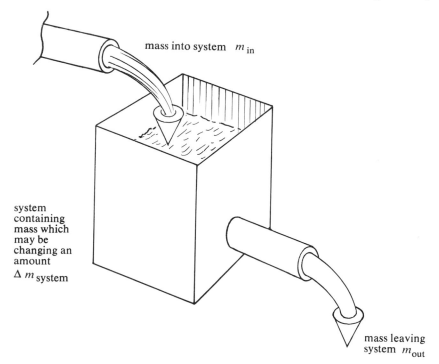

mass into system m_{in}

system
containing
mass which
may be
changing an
amount

Δm_{system}

mass leaving
system m_{out}

Figure 4–1   Conservation of mass.

The term Δm_{system} is positive if the system is gaining mass, and negative if it is losing mass. Equation (4–2) implies that all terms are related to some common time period, $\Delta\tau$, which can be divided into each of the terms in the equation, giving a new form of the mass balance written

$$\frac{m_{in}}{\Delta\tau} - \frac{m_{out}}{\Delta\tau} = \frac{m_{system}}{\Delta\tau} \qquad \text{(4–3)}$$

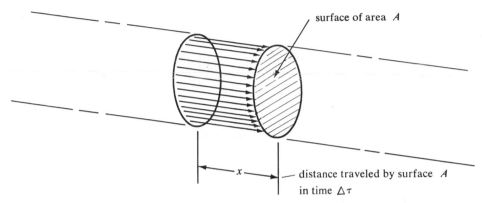

surface of area A

x

distance traveled by surface A
in time $\Delta \tau$

Figure 4–2   Illustration of mass flow.

Now, in mathematical language, as $\Delta \tau$ is made increasingly small, consequently making m_{in}, m_{out}, and Δm_{system} smaller since the time period is decreased, the terms each approach a differential so that we then write

$$\frac{dm_{\text{in}}}{d\tau} - \frac{dm_{\text{out}}}{d\tau} = \frac{dm_{\text{system}}}{d\tau} \qquad (4\text{–}4)$$

In this text, the differential with respect to time, $d/d\tau$, will be denoted by a "dot" so that equation (4–4) is then written

$$\dot{m}_{\text{in}} - \dot{m}_{\text{out}} = \dot{m}_{\text{system}} \qquad (4\text{–}5)$$

The two terms $\dot{m}_{\text{in}}$ and $\dot{m}_{\text{out}}$ are called the *mass flow rate into the system* and the *mass flow rate out of the system* respectively. The term $\dot{m}_{\text{system}}$ is the rate at which the system mass changes with respect to time. The mass flow rate is commonly found from the equation

$$\dot{m} = \rho A \bar{V} \qquad (4\text{–}6)$$

where A is the area across which mass is moving with velocity $\bar{V}$ and with density ρ.

To derive equation (4–6), we note that the amount of mass flowing across area A is equal to the volume of the mass times its density. We write this as

$$m = \rho V \qquad (4\text{–}7)$$

but the volume, V, is also equal to A times the distance through which the leading surface of the mass has traveled during some time period, $\Delta \tau$, as indicated in figure 4–2. If we denote this distance as x then

$$m = \rho A x \qquad (4\text{–}8)$$

and if this equation is divided by the time period, $\Delta \tau$, on both sides we have

$$\frac{m}{\Delta\tau} = \frac{\rho A x}{\Delta\tau} \tag{4-9}$$

In this last equation $x/\Delta\tau$ is just the velocity of the flowing mass, $\bar{V}$, which then gives us

$$\frac{m}{\Delta\tau} = \rho A \bar{V}$$

or, since we have denoted $m/\Delta\tau$ by the notation $\dot{m}$, then

$$\dot{m} = \rho A \bar{V} \tag{4-6}$$

Example 4.1

Water flows from a 1-inch diameter faucet with a velocity of 8.7 ft/s. Determine the mass flow rate of the water leaving the faucet.

Solution

The water is crossing the circular area of the faucet outlet, equal to pi (π) times the radius squared. Then the area A is calculated by

$$A = \pi\left(\frac{1}{2}\right)^2 \text{ in}^2$$

or

$$A = 0.7854 \text{ in}^2$$

The water is here assumed to be at 78°F where the density is approximately 62.4 lbm/ft³. The mass flow rate is then determined by using equation (4-6)

$$\dot{m} = \rho A \bar{V}$$

and substituting values into this equation gives us

$$\dot{m} = (62.4 \text{ lbm/ft}^3)(0.7854 \text{ in}^2)(8.7 \text{ ft/s})$$

To convert to consistent units, we must multiply by the factor 1/144 ft²/in² so that

$$\dot{m} = (62.4 \text{ lbm/ft}^3)(0.7854 \text{ in}^2)(8.7 \text{ ft/s})\left(\frac{1 \text{ ft}^2}{144 \text{ in}^2}\right)$$

Thus

$$\dot{m} = 2.96 \text{ lbm/s} \qquad\qquad \textit{Answer}$$

From this example we see that mass flow rate can have the units of lbm/s or any other unit equivalent to *mass per unit time*. In some instances the weight flow rate is specified or required. This quantity is defined as the weight per unit of time and is given by

$$\dot{W} = \frac{W}{\Delta\tau} \tag{4-10}$$

or in terms of mass flow rate and from equations (4–6) and (2–7)

$$\dot{W} = \frac{g}{g_c}\dot{m} \qquad (4\text{–}11)$$

Then

$$\dot{W} = \gamma A \bar{V} \qquad (4\text{–}12)$$

where γ is the specific weight as defined in section 2.8.

The volume flow rate, defined as the volume of material crossing an area per unit of time and written

$$\dot{V} = \frac{V}{\Delta\tau} \qquad (4\text{–}13)$$

is another term for describing flow rate. Since the volume V can be described by Ax we have, from equation (4–13), that

$$\dot{V} = A\bar{V} \qquad (4\text{–}14)$$

The units for weight flow rate can be lbf/s, lbf/hr, or other units of force per unit of time. The volume flow rate is described by units of ft³/s, ft³/min, gallons/hr, gallons/min (commonly written gpm), or any other compatible combination of volume per unit time.

Example 4.2

Given the situation of example 4.1, calculate the volume flow rate and weight flow rate, if the local gravitational acceleration is 31.0 ft/s².

Solution

The volume flow rate is calculated from the equation

$$\dot{V} = A\bar{V} \qquad (4\text{–}14)$$

where A was found to be 0.7854 in² and the velocity $\bar{V}$ was 8.7 ft/s. Then

$$\dot{V} = (0.7854 \text{ in}^2)(8.7 \text{ ft/s})\left(\frac{1 \text{ ft}^2}{144 \text{ in}^2}\right)$$

$$= 0.048 \text{ ft}^3/\text{s} \qquad \qquad Answer$$

Since the local gravitational acceleration was 31.0 ft/s², we use this information in equation (4–11)

$$\dot{W} = \frac{(31.0 \text{ ft/s}^2)(2.96 \text{ lbm/s})}{(32.2 \text{ ft-lbm/s}^2\text{-lbf})}$$

and

$$\dot{W} = 2.84 \text{ lbf/s} \qquad \qquad Answer$$

Recall the preceding discussions of the conservation of mass as described by equations (4–1) or (4–5) for closed and open systems respectively; there is no principle of conservation of weight or volume and caution must be used whenever reference is made to volume flow rate or weight flow rate—a conversion to mass flow rate is a safe approach.

Example 4.3   A cylindrical mixing tank having a diameter of 2 feet and containing 620 lbm of water is being filled from two water lines, one line delivering hot water at a rate of 0.7 lbm/s and a second line with $\frac{5}{8}$-inch diameter delivering cold water at 8 ft/s. If we assume the tank has an exit port of $\frac{3}{4}$-inch diameter from which mixed water discharges at 12 ft/s, determine the rate of change of water level in the tank and the mass of the water in the tank 10 seconds after flow begins. (See figure 4–3.)

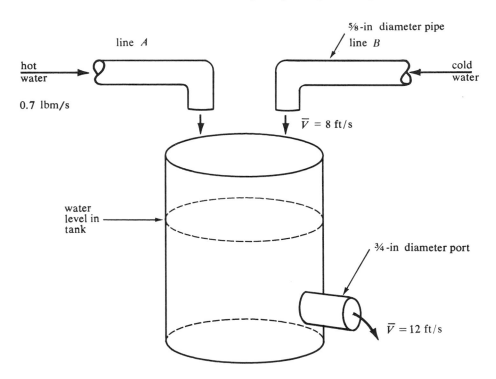

Figure 4–3   Mass flow balance. Example 4.3.

Solution   For determining the rate of change of water level, the rate of change of mass in the mixing tank must be found. Using the principles of conservation of mass, we write equation (4–5)

$$\dot{m}_{in} - \dot{m}_{out} = \dot{m}_{system}$$

where the system is the mixing tank. Now

$$\dot{m}_{in} = \text{mass flow from line } A + \text{mass flow from line } B$$

which gives us

$$\dot{m}_{in} = 0.7 \text{ lbm/s} + \left[\underbrace{\rho_B}_{\text{(density)}} \underbrace{A_B}_{\text{(area)}} \underbrace{\overline{V}_B}_{\text{(velocity)}} \right]$$

The velocity of line B, $\overline{V}_B$, is 8 ft/s; the area of B is $\pi \times (5/16)^2$ in². We assume the density to be 62.4 lbm/ft³. Then the mass flow from line B is

$$\dot{m}_B = (62.4 \text{ lbm/ft}^3)\left[\left(\pi \times \left(\frac{5}{16}\right)^2 \times \frac{1}{144} \text{ ft}^2 \right)\left((8 \text{ ft/s}) \right) \right]$$

and then

$$\dot{m}_B = 1.064 \text{ lbm/s}$$

so that

$$\dot{m}_{in} = 0.700 \text{ lbm/s} + 1.064 \text{ lbm/s}$$
$$= 1.764 \text{ lbm/s}$$

The mass flow out of the system is just that leaving the port of area $\pi \times (3/4)^2$ in². Assume the density of the mixed water is 62.0 lbm/ft³. We then have

$$\dot{m}_{out} = (62.0 \text{ lbm/ft}^3)\left[\pi \times \left(\frac{3}{4}\right)^2 \text{ in}^2 \right]\left(\frac{1}{144} \text{ ft}^2/\text{in}^2 \right)(12 \text{ ft/s})$$
$$= 9.13 \text{ lbm/s}$$

The mass rate of change for the system, $\dot{m}_{system}$, is then $1.764 - 9.13$ lbm/s or -7.366 lbm/s, from equation (4–5).

The rate of change of the water level in the mixing tank is $\dot{m}_{system}/pA_{tank}$,

$$\overline{V}_{level} = \frac{(1.764 - 9.13) \text{ lbm/s}}{(62.0 \text{ lbm/ft}^3)(\pi \times 1 \text{ ft}^2)}$$

or

$$\overline{V}_{level} = \frac{-7.366}{62 \times \pi \times 1 \text{ ft}^2} = -0.0378 \text{ ft/s} \qquad \textit{Answer}$$

The negative sign here means that the level is dropping and the mass of the system is decreasing.

Ten seconds after flow has commenced, the mass is determined from

$$\dot{m}_{system} = \frac{\Delta m_{system}}{\Delta \tau}$$

or, more clearly, using a little algebra

$$\Delta_T \dot{m}_{\text{system}} + m_{\text{system, initial}} = m_{\text{system}} \text{ @ } 10 \text{ s}$$

Then, since $\dot{m}_{\text{system}} = -7.366 \text{ lbm/s}$ initially from the above calculation, we have

$$-(7.366 \text{ lbm/s})(10 \text{ seconds}) + 620 \text{ lbm} = m_{\text{system}} \text{ @ } 10 \text{ seconds}$$

or

$$m_{\text{system}} \text{ @ } 10 \text{ seconds} = 546.34 \text{ lbm} \qquad \textit{Answer}$$

4.2 Steady Flow

As mass flows through a system there is often no loss or gain of mass in the system itself. This tells us then that

$$\dot{m}_{\text{in}} - \dot{m}_{\text{out}} = 0 \qquad \textbf{(4–15)}$$

since

$$\dot{m}_{\text{system}} = 0$$

This condition is called *steady flow* or *steady state* and is the rule rather than the exception in engineering and technological applications. Any engine producing power, refrigerator cooling foods, generator producing electric energy, or any device which is intended to perform for extended periods of time is in steady flow or at the very least, some of the components are in steady flow.

Example 4.4

A nozzle is commonly used to change the velocity of liquids or gases, by changing the cross-sectional area of the flow line. (See section 11.4 for a more complete description of nozzles.) Suppose we have a nozzle with air passing through so that within the nozzle no loss or accumulation of air occurs. The air is entering the nozzle with a velocity of 80 ft/s and a density of 0.08 lbm/ft³. The density of the air leaving is 0.068 lbm/ft³. The nozzle is circular in cross-sectional area and reduces evenly from an entrance diameter of 2 ft to an exit diameter of 1 ft. Determine the velocity of the air leaving the nozzle. (See figure 4–4.)

Solution

This problem is an example of steady flow, i.e.

$$\dot{m}_{\text{system}} = 0$$

so that we may use equation (4–15)

$$\dot{m}_{\text{in}} - \dot{m}_{\text{out}} = 0$$

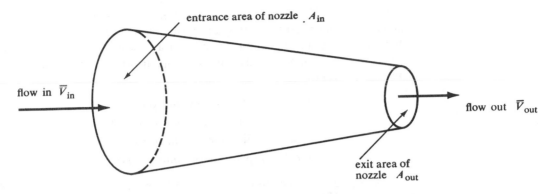

entrance area of nozzle A_{in}

flow in $\overline{V}_{in}$

flow out $\overline{V}_{out}$

exit area of
nozzle A_{out}

Figure 4–4   Nozzle flow. Example 4.4.

If we replace these two terms by using equation (4–6)

$$\rho_{in} A_{in} \overline{V}_{in} - \rho_{out} A_{out} \overline{V}_{out} = 0 \qquad (4\text{–}16)$$

we can substitute numbers into this equation. It is an important steady flow relative which we will use often. Into equation (4–16) we substitute values:

$$0 = (0.08 \text{ lbm/ft}^3)(\pi \times 1^2 \text{ft}^2)(80 \text{ ft/s}) - (0.068 \text{ lbm/ft}^3)\left[\pi \times \left(\frac{1}{2}\right)^2 \text{ ft}\right](\overline{V}_{out})$$

which can then be solved easily for

$$\begin{aligned}
\overline{V}_{out} &= \frac{(0.08 \text{ lbm/ft}^3)(3.14 \text{ ft}^2)(80 \text{ ft/s})}{(0.068 \text{ lbm/ft}^3)(3.14 \times \frac{1}{4} \text{ ft}^2)} \\
&= \frac{0.08 \times 4 \times 80 \text{ ft}}{0.068 \text{ s}} = 376.47 \text{ ft/s} \qquad \textit{Answer}
\end{aligned}$$

4.3 Conservation of Energy

As a system goes through a process, some properties of the system are altered and we have postulated that the mass is conserved. If the system is closed or isolated, then the system remains unchanged for any process; but if the system is open, the mass will change according to equation (4–5). Similarly, we now postulate that energy is conserved in any process of a system and we write

$$E_{in} - E_{out} = \Delta E_{system} \qquad (4\text{–}17)$$

where a positive energy change of the system implies an accumulation of energy in the system and a negative value implies loss of system energy. In terms of rates of change we say

$$\dot{E}_{in} - \dot{E}_{out} = \dot{E}_{system} \qquad (4\text{–}18)$$

in a manner similar to the conservation of mass in section 4.1.

In equation (4–17), E_{in} and E_{out} represent energy crossing the boundary of the system under scrutiny—they are terms of energy in transition. From chapter 3 we recall the definitions of heat and work, which are precisely the energy terms under discussion. We quite arbitrarily assign the following sign conventions:

> +*Heat*—implies energy *into* the system.
> +*Work*—implies energy *out of* the system.
> −*Heat*—implies energy *out of* the system.
> −*Work*—implies energy *into* the system.

and, using Q for heat and Wk for work, we get the following from equation (4–17):

$$Q - Wk = \Delta E_{\text{system}} \qquad \textbf{(4–19)}$$

Notice that we arrived at equation (4–19) by saying

$$E_{in} = +Q \quad \text{or} \quad -Wk$$

and

$$E_{out} = -Q \quad \text{or} \quad +Wk$$

With a positive heat transfer, we "heat up" the system; and with negative heat, we "cool down" the system. Similarly, work gained from a system is positive; work put into the system is negative.

Recalling from chapter 3 that a rate of heat transfer $\dot{Q}$ is a form of the rate of energy transition and defining here $\dot{Wk}$ as the rate of work or *power*, we then obtain the following from equation (4–19):

$$\dot{Q} - \dot{Wk} = \dot{E}_{\text{system}} \qquad \textbf{(4–20)}$$

The remainder of this text is devoted to clarification and examples of equations (4–19) and (4–20).

Remember that the energy being used in these equations is any of the forms of energy in the static condition referred to in chapter 3, that is, kinetic, potential, or internal energy. Other forms or adjectives are equally acceptable such as *strain*, *electromagnetic*, or *chemical*, but only the three common ones will generally be considered in this text.

Example 4.5

A fuel tank is filled with propane gas and heat is transferred from the surroundings at a rate of 30 Btu per hour. Determine the increase in energy of the propane gas over a 24-hour period. (See figure 4–5.)

Solution

We first identify the system to be analyzed—the propane gas inside the fuel tank. Assuming the tank is rigid and no gas or air is entering or

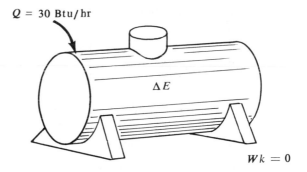

$Q = 30$ Btu/hr

ΔE

$Wk = 0$

Figure 4–5   Propane tank subject to heat transfer. Example 4.5.

leaving the tank, we have no mass change of the system and no reversible
work since there is no change in volume. We then write the first law as

$$\Delta E = Q - Wk$$

or

$$\Delta E = Q$$

since no work is present, that is,

$$Wk = 0$$

Since the rate of heat transfer was given, we can write

$$\dot{E} = \dot{Q}$$

so that, since $\dot{Q}$ is 30 Btu/hr,

$$\dot{E} = 30 \text{ Btu/hr}$$

The definition of the rate of energy change is

$$\dot{E} = \frac{\Delta E}{\Delta \tau}$$

where $\Delta \tau$ is the time during which the energy changes an amount ΔE. We
then have

$$\Delta E = (30 \text{ Btu/hr})\Delta \tau$$

and

$$\Delta \tau = 24 \text{ hr}$$

so

$$\Delta E = 720 \text{ Btu} \qquad\qquad\qquad Answer$$

Example 4.6   A piston-cylinder device does 7800 ft-lbf of work while losing 3.7 Btu of heat. What is the change of energy of the contents of the piston-cylinder?

Solution   Here we have a closed system where energy is conserved so we write equation (4–19)

$$\Delta E = Q - Wk$$

Since 7800 ft-lbf of work is done, work is a positive quantity in the equation. The amount of heat lost is a negative quantity, and substituting these two values into the equation we have

$$\Delta E = -3.7 \text{ Btu} - 7800 \text{ ft-lbf}$$

Since 778 ft-lbf is 1 Btu, we convert to get

$$\Delta E = -3.7 \text{ Btu} - \frac{7800}{778} \text{ Btu}$$

$$= -13.7 \text{ Btu} \qquad\qquad Answer$$

4.4
The First Law of Thermodynamics for a Closed System

The First Law of Thermodynamics: *Energy can be neither created nor destroyed but can only be converted to its various forms.*

When considering the conservation of energy, or the first law of thermodynamics, as it applies to a closed system we can write the energy balance in the following ways:

$$\Delta E = Q - Wk_{cs}$$

or

$$m\Delta e = mq - mwk_{cs} \qquad\qquad (4\text{–}21)$$

or

$$\Delta e = q - wk_{cs} \qquad\qquad (4\text{–}22)$$

In these equations the terms q and wk_{cs} are heat per unit mass of the system and work per unit mass of the system, respectively. The system mass m is usually expressed in lbm or grams. If the process is *reversible* we can invoke equation (3–7)

$$Wk_{cs} = \int_{V_2}^{V_1} p\,dV$$

and then have

$$\Delta E = Q - \int_{V_2}^{V_1} p\,dV \qquad\qquad (4\text{–}23)$$

or

$$\Delta e = q - \int_{v_2}^{v_1} p \, dv \qquad \text{(4–24)}$$

We can also consider time as a variable so that equation (4–21) becomes

$$\dot{E} = \dot{Q} - \dot{W}k_{cs} \qquad \text{(4–25)}$$

Using the same terms as for equation (4–21) we have

$$\frac{d}{dt}me = m\frac{d}{dt}e + e\frac{d}{dt}m \qquad \text{(4–26)}$$

but

$$\frac{dm}{dt} = 0 = \dot{m} \text{ (closed system)} \qquad \text{(4–27)}$$

so that

$$\dot{E} = \frac{d}{dt}me = m\dot{e} \qquad \text{(4–28)}$$

Similarly

$$\dot{Q} = m\dot{q} \qquad \text{(4–29)}$$

and

$$\dot{W}k_{cs} = m\dot{w}k_{cs} \qquad \text{(4–30)}$$

which then gives us, for the first law of thermodynamics applied to closed systems,

$$\dot{e} = \dot{q} - \dot{w}k_{cs}$$

or

$$m\dot{e} = \dot{Q} - \dot{W}k_{cs} \qquad \text{(4–31)}$$

The closed system is the classic visualization of the thermodynamic system. All we have considered up to this point is clearly accountable, and the primary task of applying any of the above equations is to define the considered system. We have at this time one form of equations (4–23) or (4–24) which can be solved arithmetically. This is the *constant pressure process* considered in example 3.3 where we arrived at equation (3–9).

$$Wk_{cs} = p\Delta V$$
$$= p(V_{final} - V_{initial})$$

For *constant pressure reversible process* only, equation (4–21) becomes

$$\Delta E = Q - p(V_{final} - V_{initial}) \qquad \text{(4–32)}$$

4.5
The First Law of Thermo-dynamics for an Isolated System

Considerations of isolated systems eliminate any heat or work or mass transfer. All we can say about an isolated system then is

$$\Delta E = 0$$

Since $\Delta m = 0$ for isolated systems and $E = me$ then

$$m\Delta e = 0 \quad \text{or} \quad \Delta e = 0 \qquad (4\text{--}33)$$

for a system having mass m and energy e. All that happens here is that energy is converted from one form to another, and interestingly there is no indication outside of the isolated system as to what is transpiring inside. Some consider our universe as an isolated system, although it has not been established as such and although no actual isolated systems have been identified in the strictest sense.

Example 4.7

In figure 4–6, a closed chamber having walls which do not allow heat transfer (called *adiabatic walls*) contains dust composed of 20 lbm of particles. The total kinetic energy of the particles as they move about the chamber is 78 ft-lbf. Determine the change in energy as the dust settles to the bottom if the chamber is 10 ft high.

Solution

Since the chamber does not allow heat transfer ($Q = 0$) and since it is closed off, it cannot convey mechanical work ($Wk = 0$); thus we write

$$\Delta E = 0$$

or

$$\Delta KE + \Delta PE + \Delta U = 0$$

The initial kinetic energy is 78 ft-lbf and the final is zero. Then

$$\Delta KE = KE_{final} - KE_{initial}$$
$$= 0 - 78 \text{ ft-lbf}$$

We assume the dust is initially dispersed uniformly throughout the chamber, and we visualize the total initial potential energy as the mass (20 lbm) times the *average* height of the particles (5 ft). Then

$$PE_{initial} = \frac{g}{g_c} Mz_{\text{average}} = \left(\frac{32.2 \text{ ft/s}^2}{32.2 \text{ ft-lbm/s}^2\text{-lbf}}\right)(20 \text{ lbm})(5 \text{ ft})$$
$$= 100 \text{ ft-lbf}$$

where we assume $g = 32.2 \text{ ft/s}^2$ locally. The final potential energy is zero since we used the bottom of the chamber as the level where $z = 0$. This gives us for our process of dust settling on the bottom

$$\Delta PE = PE_{final} - PE_{initial} = 0 - 100 \text{ ft-lbf}$$
$$= -100 \text{ ft-lbf}$$

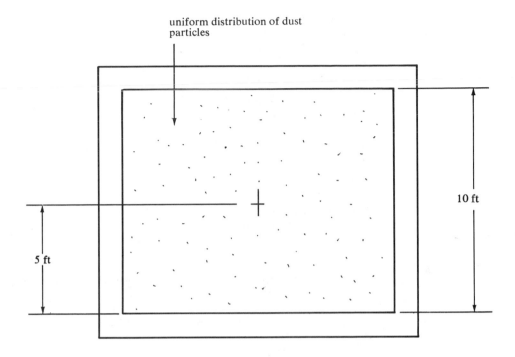

uniform distribution of dust
particles

10 ft

5 ft

Figure 4–6 Adiabatic chamber containing dust. Example 4.7.

and

$$\Delta U = -\Delta KE - \Delta PE$$

or

$$\Delta U = U_{\text{final}} - U_{\text{initial}} = -(-78 \text{ ft-lbf}) - (-100 \text{ ft-lbf})$$
$$= 178 \text{ ft-lbf}$$

If we arbitrarily say

$$U_{\text{initial}} = 0$$

then

$$U_{\text{final}} = 178 \text{ ft-lbf}$$

In this example we used the term *adiabatic wall* to denote no heat transfer. Throughout the text we will quite often refer to an *adiabatic process* which is a process where there is no heat transfer (Q or $\dot{Q}$ is zero), whether the process applies to open, closed, or isolated systems. Example 4.7 is an adiabatic process.

4.6
Flow Work and
Enthalpy

Thermodynamics has been shown to apply quite generally to closed and isolated systems. No mass transfer is involved and only energy transfer in the form of work and heat are of concern. But most of the systems encountered in technology are open systems, and we should hope to use the thermodynamic approach for these as well. Let us then look carefully at the unique mechanism of the open system—mass moving across the system boundary. In figure 4–7, we have defined an open system and for simplistic purposes provided only two inlet pipes, A and B, through which mass may enter or leave the system. Now consider some mass moving into the system on the left boundary of the system in figure 4–7. In order that the mass fully enter the system, it must move the distance x_A along pipe A. But for it to move this distance *someone or something external must exert a force*, F_A, through the distance x_A. We then say there is flow work required to "flow the mass" into the system:

$$\text{Flow work} = F_A x_A$$

Also, F_A is equal to the pressure p_A of the system, times the cross-sectional area of the pipe, A. Then

$$\text{Flow work at } A = p_A A x_A = p_A V_A \tag{4–34}$$

since

$$Ax_A = \text{Volume of the mass entering at } A$$
$$= V_A$$

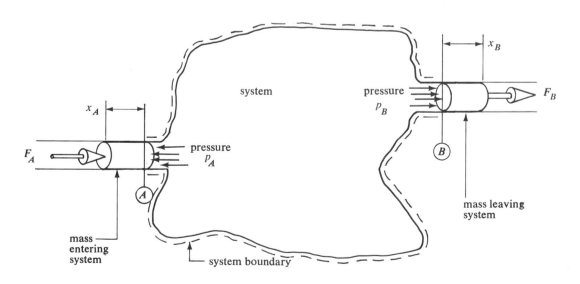

Figure 4–7   Open system.

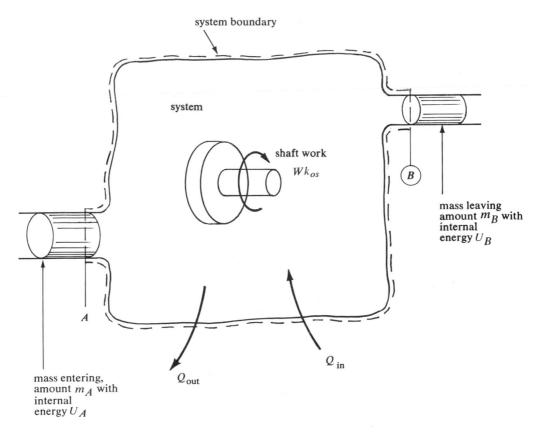

system boundary

system

shaft work
Wk_{os}

B

mass leaving
amount m_B with
internal
energy U_B

A

Q_{out}

Q_{in}

mass entering,
amount m_A with
internal
energy U_A

Figure 4–8 Open system.

The flow work per lbm of entering mass at *A* is

$$p_A \frac{V_A}{M_A} = p_A v_A \qquad (4\text{–}35)$$

At pipe *B* we consider that mass is leaving the system so that the mass will move out of its own, propelled by a back pressure p_B, and, associated with it, the flow work out of the system. At *B*, the flow work per pound-mass is $p_B v_B$ in a completely analagous development as for pipe *A*. Keep in mind though, that while work (or flow work) was required external to this system at *A* to shove the mass in, work is done by the system in ejecting mass at *B*.

Now let us look at the energy balance or first law for this system. In figure 4–8 is our open system with some mechanical additions to account for the fact that work and heat may occur. We write equation (4–21)

$$\Delta E = Q - Wk$$

or, since in this illustration

$$Wk = \text{shaft work} - \text{flow work at } A + \text{flow work at } B$$

we have, for our energy balance

$$\Delta E = Q - Wk_{os} - \text{flow work at } B + \text{flow work at } A \quad \text{(4–36)}$$

The flow work at A was *into* the system, or done *on* the system, and therefore negative. The flow work at B was *out* and therefore positive. Upon substituting into the energy balance both signs change and therefore the equation (4–36) results.

The term Wk_{os} is the open system work, here given as the shaft work. Using equation (4–35) for flow work, we then substitute these into equation (4–36) to obtain

$$\Delta E = Q - Wk_{os} - p_B v_B m_B + p_A v_A m_A \quad \text{(4–37)}$$

If we break up the energy terms on the left of this equation into their identifiable types we have

$$\Delta E = \Delta KE + \Delta PE + \Delta U \quad \text{(4–38)}$$

but

$$\Delta E = \Delta E_{\text{system}} - E_A + E_B$$

The terms E_A and E_B are $KE + PE + U$ evaluated at A and B respectively. Then, equation (4–37) becomes

$$\Delta E_{\text{system}} - KE_A - PE_A - U_A + KE_B + PE_B + U_B$$
$$= Q - Wk_{os} - m_B p_B v_B + m_A p_A v_A \quad \text{(4–39)}$$

We can write U_B as $m_B u_B$ and U_A as $m_A u_A$. Then if we take the flow work terms to the left side of the equation we get

$$\Delta E_{\text{system}} - KE_A - PE_A - m_A u_A - m_A p_A v_A + KE_B + PE_B + m_B u_B +$$
$$m_B p_B v_B = Q - Wk_{os}$$

In this equation we could algebraically combine the terms $u + pv$, namely

$$m_A(u_A + p_A v_A) \quad \text{and} \quad m_B(u_B + p_B v_B)$$

The quantities inside the brackets here are defined as the enthalpy and are denoted by h. Then $h_A = u_A + p_A v_A$, and $h_B = u_B + p_B v_B$, or generally

$$h = u + pv \quad \text{(4–40)}$$

Enthalpy is specified by units of energy per unit mass, and if it is multiplied by the mass, the resulting property is called *total enthalpy, H.*

We have

$$H = mh$$
$$= mu + mpv$$
$$= U + pV \qquad\qquad (4\text{-}41)$$

Total enthalpy is specified by energy units. Remember that enthalpy and total enthalpy are merely mathematical combinations which may or may not have physical significance in a given problem.

Example 4.8

An air pump shown in figure 4–9 takes in air at 14.7 psia, 78°F, and with a density of 0.075 lbm/ft³ at station (1). The pump compresses the air flowing to the tank so that at station (2) the density is found to be 0.050 lbm/ft³ and the pressure is 100 psia. Determine the flow work and enthalpy of the air at stations (1) and (2), if the internal energy of the air is 60 Btu/lbm at (1) and 180 Btu/lbm at (2).

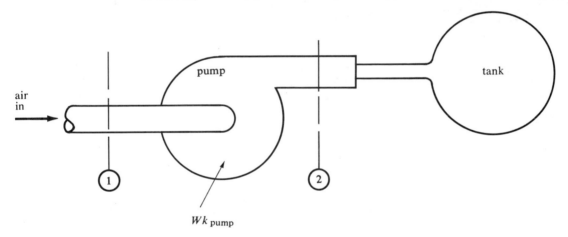

Figure 4–9   Air pump. Example 4.8.

Solution

The flow work at station (1) is negative, and is equal to $-(p_1v_1)$ or $-(p_1/\rho_1)$ so that flow work at (1) = -14.7 lb/in² × 1 ft³/0.075 lbm. Converting to compatible units, we get

-14.7 lb/in² × 144 in²/ft² × 1 ft³/0.075 lbm
$$= 28{,}224 \text{ ft-lbf/lbm} \quad \textit{Answer}$$

At station (2)

flow work = $+100$ lb/in² × 144 in²/ft² × 1 ft³/0.050 lbm
$$= 288{,}000 \text{ ft-lbf/lbm} \qquad\qquad \textit{Answer}$$

To calculate the enthalpy we use equation (4–40)

$$h = u + pv$$

At station (1)

$$h_1 = u_1 + p_1v_1$$

or

$$h_1 = 60 \text{ Btu/lbm} + 28{,}224 \text{ ft-lbf/lbm}$$
$$= \left(60 + \frac{28{,}224}{778}\right) \text{Btu/lbm}$$
$$= 96.3 \text{ Btu/lbm} \qquad\qquad\qquad Answer$$

At station (2)

$$h_2 = 180 \text{ Btu/lbm} + \frac{288{,}000}{778} \text{Btu/lbm}$$
$$= 550.2 \text{ Btu/lbm} \qquad\qquad\qquad Answer$$

4.7
The First Law of Thermo-dynamics for an Open System

One way to consider the energy balance or first law of thermodynamics is as the bookkeeping rule of balancing the "energy budget" of a system. All energy must be accounted for in a given process as is done in the closed and isolated systems. Here let us take an accounting of the energy of an open system process—very generally and intuitively—in the same manner as in section 4.6. We could generalize the system as anything, such as a pump, fan, radiator, cooling tower, boiler, turbine, or a biological system. In figure 4–10 is shown a general open system which allows heat transfer, work, and mass flow into and out of the region. Assuming mass enters at station (1) and leaves at station (2), then from $\Delta E = Q - Wk$ we write

$$\text{KE}_2 + \text{PE}_2 + H_2 - \text{KE}_1 - \text{PE}_1 - H_1 + \Delta\text{KE}_s + \Delta\text{PE}_s + \Delta U_s$$
$$= Q_{\text{in}} - Q_{\text{out}} - Wk - Wk_s \quad \textbf{(4–42)}$$

Combining Wk and Wk_s into the term called Wk_{os}, and combining Q_{in} and Q_{out} into Q and merely recalling the convention of signs for heat and work (section 4.3) then

$$\text{KE}_2 + \text{PE}_2 + H_2 - \text{KE}_1 - \text{PE}_1 - H_1 + \Delta\text{KE}_s + \Delta\text{PE}_s + \Delta U_s$$
$$= Q - Wk_{os} \quad \textbf{(4–43)}$$

This equation is generally too cumbersome to use, so if we neglect kinetic and potential energy of the system, we have

$$\text{KE}_2 + \text{PE}_2 + H_2 - \text{KE}_1 - \text{PE}_1 - H_1 + \Delta U_s = Q - Wk_{os} \quad \textbf{(4–44)}$$

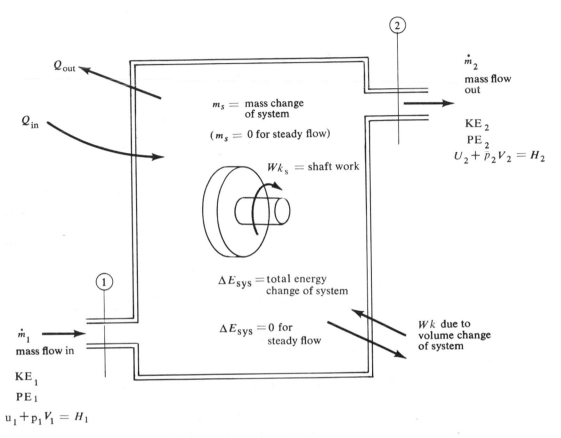

Figure 4–10   General open system.

Even this equation can be extremely difficult to use under completely general conditions of variable flow, mass change of system, or variations in heat or work. For most common engineering problems, the assumption of steady flow is sufficiently general. Steady flow means that the mass flow rates in and out are constant in time and that there is no change in system mass with time. This is equivalent then to writing

$$\text{system mass} = \text{constant} \tag{4–45}$$

and

$$m_2 = m_1 = \text{constant} \tag{4–46}$$

The mass flow rates in and out are the same from conservation of mass. If steady state is assumed (the state of the system remains constant in time) then

$$\Delta E_s = \Delta U_s = 0$$

and the energies evaluated at stations (1) and (2) are constant in time. The steady flow-steady state energy equation (hereafter referred to as the *steady flow energy equation*) is then written

$$KE_2 + PE_2 + H_2 - KE_1 - PE_1 - H_1 = Q - Wk_{os} \quad \textbf{(4-47)}$$

or

$$\dot{m}_2(ke_2 + pe_2 + h_2) - \dot{m}_1(ke_1 + pe_1 + h_1) = \dot{m}q - \dot{m}wk_{os} \quad \textbf{(4-48)}$$

and, if $\dot{m} = \dot{m}_1 = \dot{m}_2$, equation (4-48) can be reduced to

$$\Delta ke + \Delta pe + \Delta h = q - wk_{os} \quad \textbf{(4-49)}$$

Expanding this equation gives us a well-known form of the first law of thermodynamics applied to the open system under steady flow conditions,

$$\frac{\bar{V}_2^2 - \bar{V}_1^2}{2g_c} + \frac{g(z_2 - z_1)}{g_c} + h_2 - h_1 = q - wk_{os} \quad \textbf{(4-50)}$$

Quite frequently power or rates of heat transfer are specified. The steady flow equation for rates, equivalent to equation (4-50), is equation (4-48) here expanded and rewritten as

$$\dot{m}\left(\frac{\bar{V}_2^2 - \bar{V}_1^2}{2g_c} + \frac{g(z_2 - z_1)}{g_c} + h_2 - h_1 \right) = \dot{m}q - \dot{m}wk_{os} \quad \textbf{(4-51)}$$

All terms on the left in equation (4-51) should be easily identifiable with some physical quantity. On the right side of the equation, heat or heat transfer rates can be calculated from those equations discussed in chapter 3, but Wk_{os} is without a defining equation. We say for a closed system and a reversible process that

$$Wk_{cs} = \int pdV$$

or Wk_{cs} is the area under a curve in a *p-V* diagram.

Here we note that since Wk_{os} is the same *work* we have referred to *minus* the flow work, then we can write

$$Wk_{os} = Wk_{cs} - \Delta pV \quad \textbf{(4-52)}$$

where ΔpV is the representation of the two flow work terms, namely,

$$\Delta pV = p_2V_2 - p_1V_1$$

Mathematically this is the same as $\int_{p_1v_1}^{p_2v_2} d(pv)$ and since for reversible processes

$$Wk_{cs} = \int_{V_1}^{V_2} pdV$$

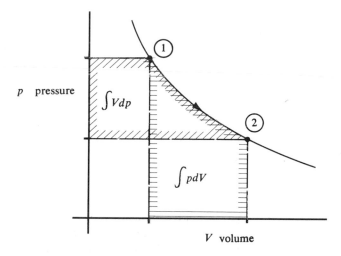

Figure 4-11   Graphical comparison of *pdV* and *Vdp* terms.

Then from equation (4–52)

$$Wk_{os} = \int_{V_1}^{V_2} pdV - \int_{p_1V_1}^{p_2V_2} d(pV)$$

But, recalling that the derivative of a product such as *pV*, written $d(pV)$ is $pdV + Vdp$, then

$$-Vdp = pdV - dpV$$

We can then see that

$$Wk_{os} = -\int Vdp \qquad \textbf{(4–53)}$$

Equation (4–53) is true for a reversible steady flow process without kinetic or potential energy changes and is shown geometrically in figure 4–11. We have stated that $Wk_{cs\ rev}$ is equivalent to the area under a curve in a *p-V* diagram; the $Wk_{os\ rev}$ is equal to the area to the left of the curve in the *p-V* diagram as can be seen in figure 4–11. We will use equation (4–53) as the general definition for open system work when kinetic, potential and chemical energy changes are neglected.

Per unit mass the reversible work of the open system is

$$wk_{os} = -\int vdp \qquad \textbf{(4–54)}$$

and the power for an open system under steady flow conditions is

$$\dot{W}k_{os} = -\dot{m}\int vdp \qquad \textbf{(4–55)}$$

If kinetic or potential energy changes are not negligible, then the work must be gotten from

$$wk_{os} = -\int vdp - \Delta ke - \Delta pe \qquad \text{(4–56)}$$

Example 4.9

A steam turbine running under reversible steady flow conditions takes in steam at 200 psia and exhausts it at 15 psia. Assuming the steam has a specific volume of 4.0 ft³/lbm at the inlet and the pressure-volume relation is

$$p = 228.48 - 7.12v$$

where p is in psia units and v is in ft³/lbm units, determine the work done per lbm of steam flowing through the turbine. Neglect kinetic and potential energy changes of the steam.

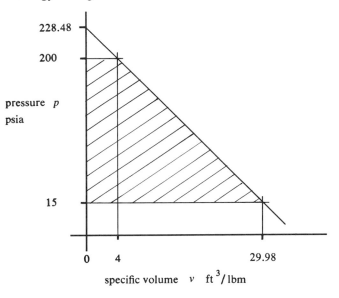

Figure 4–12   *p-v* diagram for process of example 4.9.

Solution

If we construct the *p-v* diagram for the expanding steam, we have the curve shown in the graph of figure 4–12. The work done per unit mass is the shaded area of the figure, or $wk_{os} = -\int vdp$. This area is a rectangle and a triangle, i.e.

$$wk_{os} = (200 - 15) \text{ lbf/in}^2 \times 144 \text{ in}^2/\text{ft}^2 \times 4 \text{ ft}^3/\text{lbm}$$

$$+ (200 - 15)(144) \text{ lbf-ft} \times \frac{1}{2} \times (29.98 - 4.00) \text{ ft}^3/\text{lbm}$$

$$= 106{,}560 \text{ ft-lbf/lbm} + 346{,}054 \text{ ft-lbf/lbm}$$

or

$$wk_{os} = 452,614 \text{ ft-lbf/lbm}$$

or

$$wk_{os} = 581.8 \text{ Btu/lbm} \qquad \qquad \text{\textit{Answer}}$$

We may also calculate this answer by using equation (4–54).

We have from the given pressure-volume relation that

$$v = (p - 288.48)\frac{\text{lbm}^2 \text{ ft}^3}{7.12 \text{ lb/in}^2 \times \text{lbm}}$$

If we substitute this into equation (4–54) we get

$$\begin{aligned} wk_{os} &= \int_{p_1}^{p_2} (p - 228.48)\frac{1}{7.12}\, dp \\ &= \frac{1}{7.12}\left(\frac{1}{2}p^2 - 228.48p\right)_{p_1}^{p_2} \\ &= \frac{1}{7.12}\left(\frac{1}{2}(15^2 - 200^2) - 228.48(15 - 200)\right) \\ &= \frac{1}{7.12}\left(-19887.5 + 42268.8\right) \\ &= \frac{144 \text{ in}^2/\text{ft}^2}{7.12 \text{ lbf lbm/in}^2 \text{ ft}^3}(22381.3 \text{ lbf}^2/\text{in}^4) \\ &= 452,656 \text{ ft-lbf/lbm} = 581.8 \text{ Btu/lbm} \qquad \text{\textit{Answer}} \end{aligned}$$

This answer agrees with the geometric one within an accuracy of 1%.

Example 4.10

A pump shown in figure 4–13 delivers 30 gallons per minute of water at 78°F and at 10 psig through a $\frac{1}{2}$-inch diameter pipe. Assuming that the water is at 14.7 psia and has negligible velocity at the inlet station (1) and there is no heat transfer or friction in the system, determine the power required by the pump $\dot{W}k_{\text{pump}}$.

Solution

This is obviously an open system and is assumed to be under steady flow-steady state conditions. The boundary is indicated by a dotted line in figure 4–13. We can then apply equation (4–51) to our system.

$$\dot{m}\left(\frac{V_2^2 - V_1^2}{2g_c} + \frac{g(z_2 - z_1)}{g_c} + h_2 - h_1\right) = \dot{Q} - \dot{W}k_{os}$$

Since no heat is transferred, $\dot{Q}$ is zero and V_1 is approximately zero by the problem statement. The elevation of the water passing through is increased by 75 ft so that $z_2 - z_1 = 75$ ft.

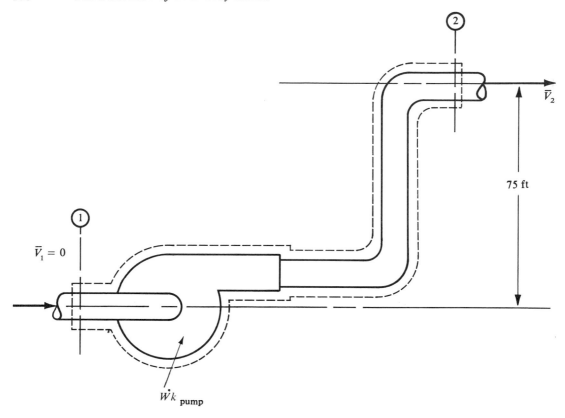

Figure 4–13   Example 4.10

We assume there is no internal energy change of the water so that

$$u_2 = u_1$$

and therefore

$$h_2 - h_1 = u_2 - u_1 + p_2 v_2 - p_1 v_1 = p_2 v_2 - p_1 v_1$$

and, for water, we can safely assume that the specific volume does not change; that is, if we assume the density of water to be 62.4 lbm/ft³, then $v = 1/62.4 = 0.016$ ft³/lbm. The mass flow rate is

$$\dot{m} = \dot{V}\rho$$

and

$$\dot{V} = 30 \text{ gallons/min} = 1 \text{ ft}^3/7.48 \text{ gal} \times 30 \text{ gal/min}$$
$$= 4.01 \text{ ft}^3/\text{min} = 0.067 \text{ ft}^3/\text{s}$$

so that

$$\dot{m} = 0.067 \text{ ft}^3/\text{s} \times 62.4 \text{ lbm/ft}^3$$
$$= 4.18 \text{ lbm/s}$$

The velocity at station (2) is

$$\bar{V}_2 = \frac{\dot{V}}{A_2}$$

where A_2 the cross-sectional area at (2) is

$$\pi \times \left(\frac{1}{2} \text{ in}\right)^2 \times \frac{1}{4} = 0.196 \text{ in}^2$$
$$= 0.0014 \text{ ft}^2$$

Then the velocity $\bar{V}_2$ is calculated

$$\bar{V}_2 = \frac{0.067 \text{ ft}^3/\text{s}}{0.0014 \text{ ft}^2} = 47.86 \text{ ft/s}$$

Substituting values into the energy equation (4–51) gives

$$4.18 \text{ lbm/s}\bigg(\frac{(47.86 \text{ ft/s})^2}{2 \times 32.2 \text{ ft-lbm/s}^2\text{-lbf}}$$
$$+ \frac{32.2 \text{ ft/s}^2}{32.2 \text{ ft-lbm/s}^3\text{-lbf}}(75 \text{ ft})$$
$$+ (10 + 14.7) \text{ lbf/in}^2 \times (0.016 \text{ ft}^3/\text{lbm}) \times (144 \text{ in}^2/\text{ft}^2)$$
$$- (14.7 \text{ lbf/in}^2)(144 \text{ in/ft}^2)(0.016 \text{ ft}^3/\text{lbm})\bigg) = -\dot{W}k_{os}$$

Calculating these:

4.18 lbm/s(35.6 ft-lbf/lbm + 75.0 ft-lbf/lbm + 56.9 ft-lbf/lbm
$$- 33.9 \text{ ft-lbf/lbm}) = -\dot{W}k_{os}$$

yielding

$$\dot{W}k_{os} = 558.5 \text{ ft-lbf/s}$$

There are 550 ft-lbf/s in 1 horsepower (hp) so that

$$Wk_{os} = \frac{558.5}{550} \text{ hp}$$
$$= 1.015 \text{ hp} \qquad\qquad\qquad \textit{Answer}$$

**4.8
Summary**

In thermodynamics, the conservation principles are at the foundation of all applications. The first conservation law considered is that of mass,

$$m_{\text{in}} - m_{\text{out}} = \Delta m_{\text{system}} \qquad\qquad (4\text{–}2)$$

or

$$\dot{m}_{in} - \dot{m}_{out} = \dot{m}_{system} \qquad (4\text{--}5)$$

for steady flow we have

$$\dot{m}_{in} = \dot{m}_{out} \qquad (4\text{--}15)$$

or

$$\rho A \bar{V} = \text{constant} \qquad (4\text{--}16)$$

The conservation of energy, here considered the first law of thermo-dynamics, is the principle upon which this text will be primarily built. For the system we have

$$\Delta E = Q - Wk \qquad (4\text{--}19)$$

or

$$\dot{E} = \dot{Q} - \dot{W}k \qquad (4\text{--}20)$$

and for isolated systems these reduce to

$$\Delta E = 0 = \dot{E}$$

For the closed system

$$\Delta E = Q - Wk_{cs} \qquad (4\text{--}21)$$

where, for reversible processes we can use

$$Wk_{cs} = \int pdv$$

The closed system energy balance per unit mass is

$$\Delta e = q - wk_{cs}$$

For the open system, flow work is identified as the energy associated with motion of mass across system boundaries. This term is evaluated from the product pV, or pv per unit mass, and together with internal energy is called *enthalpy H*, i.e.

$$H = U + pV \qquad (4\text{--}41)$$

or

$$h = u + pv \qquad (4\text{--}40)$$

The open system conservation of energy for steady flow condition is, per unit mass,

$$\frac{V_2^2}{2g_c} - \frac{V_1^2}{2g_c} + \frac{g}{g_c}(z_2 - z_1) + h_2 - h_1 = q - wk_{os}$$

where wk_{os} is the work of the open system and is evaluated from the equation

$$wk_{os} = -\int vdp - \Delta ke - \Delta pe \qquad (4\text{--}56)$$

Practice Problems

Sections 4.1 and 4.2

4.1. Water having a density of 62.5 lbm/ft³ is flowing steadily with a velocity of 10 ft/s through a round pipe. There is a restriction within the pipe where the diameter is one-half ($\frac{1}{2}$) the normal pipe diameter. Determine the water velocity at the restriction.

4.2. A 6-inch diameter pipe is being used to convey 600 lb of water per minute. Assuming the water has a density of 62.4 lbm/ft³ determine the velocity of the water.

4.3. Methyl alcohol at 68°F flows through a plastic tube, and there are required 2 lbm/s delivered. If the plastic tube limits the velocity of methyl alcohol to 15 ft/s or less, what is the minimum diameter of the tubing acceptable?

4.4. Shown in figure 4–14 is a converging-diverging circular nozzle through which benzene is flowing. Assume the temperature of the benzene remains at 68°F as it passes through the nozzle. If there are 60 lbm/s flowing through the nozzle at station A, determine
  (a) Mass flow at the nozzle throat, station B, and at the exit, station C.
  (b) Velocity of benzene at station B.
  (c) Velocity of benzene at station C.

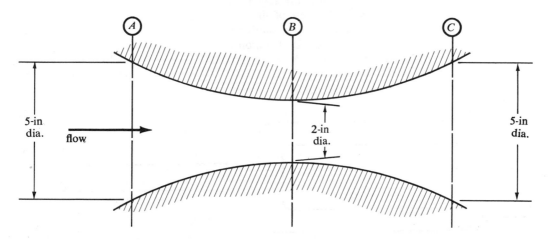

Figure 4–14   Problem 4.4.

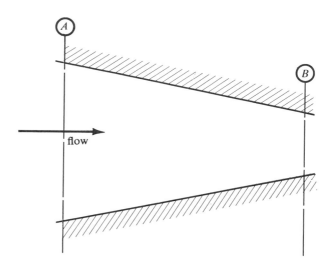

Figure 4–15   Problem 4.5.

4.5. Air flows through a converging nozzle shown in figure 4–15 with a velocity of 800 ft/s at station A. The density of the air is found to be 0.03 lbm/ft³ at A and 0.07 lbm/ft³ at B. If the area of the nozzle at A is 2 ft² and at B is 1 ft² determine
(a)   Mass flow rate of the air through the nozzle.
(b)   Velocity of the air at station B.

4.6. Air flows through the converging-diverging nozzle of problem 4.4. The density of the air is found to be 0.045 lbm/ft³ at station A, 0.060 lbm/ft³ at station B, and 0.050 lbm/ft³ at station C. If the velocity of air entering at station A is 400 ft/s, determine
(a)   Mass flow rate of air.
(b)   Velocity of air at station B.
(c)   Velocity of air at station C.

4.7. Water at 68°F is delivered through a pipe at 300 gallons/min (gpm). If the pipe has an inside diameter of $1\frac{1}{2}$ inches determine the specific kinetic energy of the water.

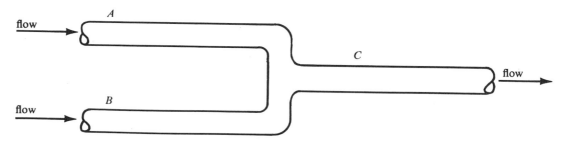

Figure 4–16   Problem 4.8.

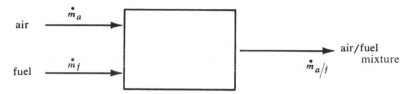

Figure 4–17   Problem 4.9.

4.8. Pipes A and B are joined together to supply pipe C as shown in figure 4–16. Pipe A is required to convey 400 gpm of water and pipe B, 700 gpm. If the pipes restrict the maximum velocities to 20 ft/s, determine the diameters of pipes A, B, and C.

4.9. Shown in figure 4–17 is a system diagram for a simple carburetor used to mix air and fuel. Under ideal conditions assume the carburetor mixes 0.04 lbm of fuel for every lbm of air where the density of air is 0.08 lbm/ft³. Assume the fuel has a density of 60 lbm/ft³ and the required air fuel mixture is 2 lbm/min. Determine the mass flow rates of fuel and air.

4.10. Water flows through ¾-inch I.D. boiler tubes. Before being heated the water in the pipes has a density of 62.5 lbm/ft³ and a velocity of 10 ft/s. What would you expect the velocity of the water to be downstream, after it has been heated such that the water density is 61.8 lbm/ft³?

4.11. In figure 4–18, air flows through a 1-inch diameter duct which is being heated from below. If the entering air at station A, has a density of 0.08 lbm/ft³ and has a density of 0.04 lbm/ft³ at B, find the entering air velocity if it is 100 ft/s at station B.

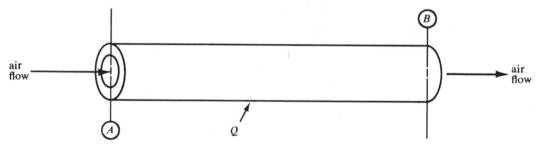

Figure 4–18   Problem 4.11.

4.12. In a combustor of a jet engine 30,000 ft³/min of air having a density of 0.06 lbm/ft³ enters through a cross-sectional area of 1 ft². In the combustor 0.02 lbm of fuel is mixed and burns with every 1 lbm of air. If the burned gases exit through a 1-ft² area, determine their velocities. Assume the burned gases have a density of 0.01 lbm/ft³.

4.13. 80,000 lbm/hr of steam is exhausted from a steam turbine at a nuclear power station. This steam, flowing through a 2-inch diameter pipe with a specific volume of 250 ft³/lbm enters a closed condenser where the steam cools and becomes liquid water with a density of 62.0 lbm/ft³. This water then flows at 80 ft/s through a pipe back to the nuclear reactor where it is boiled and becomes steam for another passage through the steam turbine. What diameter of pipe would you select for conveying the water from the condenser?

4.14. A refrigeration system rated at 60 tons uses 260 lbm/min of Freon-12. During the flow cycle of freon through the refrigerator, it is required to lose pressure at a certain point called the *expansion valve*. If the velocity through this valve is restricted to values of 100 ft/s or less, what diameter tube should be used if the freon has a density of 78 lbm/ft³?

4.15. A sprinkler consisting of an open tank 2 by 2 by 2 feet discharges 2 lbm/s of water at 78°F through holes in the tank bottom. If the tank is half full at a certain time, how great a flow rate must be provided from an external faucet to fill the tank to the top with water in 2 minutes if it is assumed the discharge remains constant at 2 lbm/s?

4.16. Balloons are filled with air having a specific volume of 12 ft³/lbm. If the filled balloons have a volume of 0.5 ft³ and the air is available at 0.01 lbm/s, determine the time required to fill each balloon.

4.17. Into a mixing chamber flows 30 lbm/s of water at 185°F, 20 lbm/s of water at 55°F and 36 lbm/min of methyl alcohol (figure 4–19). If 50.2 lbm/s of mixed solution is flowing out, determine the rate of accumulation or decline of mass within the mixing chamber.

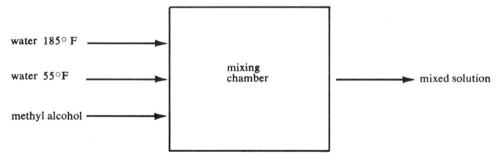

Figure 4–19   Problem 4.17.

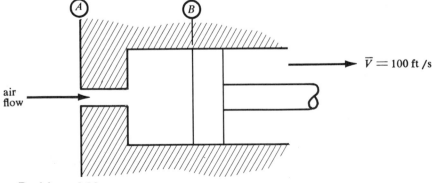

Figure 4–20   Problem 4.18.

4.18. In figure 4–20, a 3-inch diameter piston travels outward at 100 ft/s at a given instant. If air at 14.7 psia and 78°F is flowing in through a 1-in² port at A, find the velocity of air through the part required to keep the cylinder at a uniform density.

Sections 4.3 and 4.4 **4.19.** Gases enclosed in a frictionless piston-cylinder increase their internal energy by 30 Btu when 40 Btu of heat are added. Determine the work of the piston-cylinder and indicate whether it is output or input to the piston-cylinder device. Assume there are no kinetic or potential energy changes.

For problems 4.20 through 4.33 assume the data apply to closed reversible systems with no kinetic or potential energy changes.

4.20. For the process where $\Delta U = -20$ Btu, $Q = -20$ Btu, find Wk_{cs}.

4.21. For the process where $Wk_{cs} = -200$ ft-lbf, $Q = 0$, find the change in internal energy.

4.22. For the process where $wk_{cs} = 778$ ft-lbf/lbm and $\Delta u = 0.75$ Btu/lbm, determine the heat transferred per lbm.

4.23. For the process where $q = 62.5$ Btu/lbm and $wk_{cs} = 60$ Btu/lbm, determine the total energy change if the system mass is 2 lbm.

4.24. For the process involving 0.01 lbm of air, find the internal energy change per lbm if $Wk_{cs} = 15,560$ ft-lbf and $Q = -10$ Btu.

4.25. For a process involving no heat transfer, find the output if the internal energy change is -16.8 Btu.

4.26. For the process where $q = 8$ Btu/lbm and $wk_{cs} = 6224$ ft-lbf/lbm, find the energy change.

4.27. During a reversible process, 10 horsepower are being delivered external to the system. If the system energy is changing by -10 Btu/s, determine the rate of heat transfer.

4.28. Into a closed boiler containing 1000 lbm of water there are transferred 1500 Btu/s of heat. Determine the rate of change of energy of the water and the specific energy change; that is, find $\dot{U}$ and $\dot{u}$. Also determine the power involved in the process.

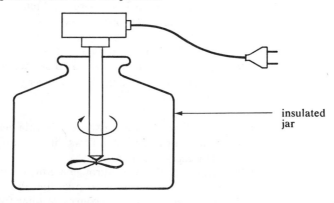

insulated jar

Figure 4–21   Problems 4.31 and 4.32.

4.29. A battery supplies 100 watt-hours of electric energy. If 10 Btu are lost by the battery during this process, what was the total decline in energy of the battery?

Section 4.4

4.30. One lbm of air at 78°F and 75 psia is contained in a piston-cylinder device. During a reversible process where the volume of the cylinder increases by 2 ft³ due to the piston moving out, the pressure of the air remains constant. Determine the work produced during this process.

4.31. Three lbm of methanol are contained in a closed, perfectly insulated jar (no heat transfer allowed across boundary). A paddle is inserted through the top as shown in figure 4–21 and, driven by an electric motor, agitates the methanol so that its internal energy increases by 24 Btu. Determine
 (a) Wk_{cs}.
 (b) Irreversible work Wk_{irr} and reversible work, $Wk_{cs\ rev}$.
 (c) Q.
 (d) wk_{cs}, wk_{irr}, and q.

4.32. In figure 4–21, the electric motor supplies 20 Btu of paddle work (or irreversible work) to stir up 3 lbm of methanol contained in the insulated jar. Assuming the jar is not perfectly insulated so that 1 Btu of heat is transferred out during the process, find the change in energy of the methanol.

4.33. A heat engine drives an electric generator, thereby producing 100,000 watts of power. If this process is completely reversible and the heat engine has no change in its internal energy, determine the rate of heat transfer required to the heat engine.

Section 4.5

4.34. Two 1-lbm balls, each with 15 ft-lbf of kinetic energy in the positions shown in figure 4–22, constitute an isolated system along with the insulated container. The balls bounce around inside the box and finally come to rest at the bottom. By what amount has the internal energy increased in the system during this dissipation process?

4.35. An isolated system composed of 30 lbm and with a total energy of 3000 Btu exists. What is its energy and mass
 (a) 2 hours after observing it?
 (b) 2 years after observing it?
 (c) When do you observe a change in its mass and/or energy?

4.36. An isolated system is composed of grain in a box as shown in figure 4–23. By what amount is the internal energy capable of increasing if the grain density is 38 lbm/ft³?

Section 4.6

4.37. Twenty lbm/s at a pressure of 200 psia, temperature of 500°F and specific volume of 2.72 ft³/lbm flows through a pipe. Determine its flow work or flow energy per lbm.

4.38. Water at 78°F flows through a 2-inch diameter pipe with a velocity of 30 ft/s. If the pressure is 60 psia, find the flow work of the water per lbm.

4.39. A fuel pump is used to convey gasoline from a tank to a mixing chamber. At the entrance to the pump, the gasoline has a pressure of 14 psia and a

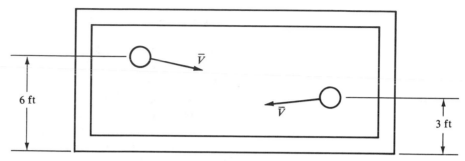

Figure 4–22   Problem 4.34.

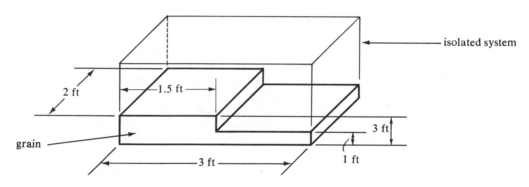

Figure 4–23   Problem 4.36.

density of 42 lbm/ft³. As the gasoline leaves the pump, it has a pressure of 14.8 psia and its density is 42 lbm/ft³. If 0.1 lbm/s of fuel is pumped, determine the rate of change in total flow work of the gasoline per unit of time due to the pump.

4.40. Calculate the rate of flow energy, $(\dot{pV}) = \dfrac{d}{dt}pV$, for
(a) Problem 4.37.
(b) Problem 4.38.

For problems 4.41 through 4.46, calculate flow energy per lbm, enthalpy, and total enthalpy.

4.41. Two lbm of ammonia at 30 psia, 9.492 ft³/lbm and internal energy of 1130.2 Btu.

4.42. Seven lbm of steam having a pressure of 20 psia, specific volume of 26.946 ft³/lbm, and specific internal energy of 1163.3 Btu/lbm.

4.43. Seventy-five lbm of air at 14.7 psia, density of 0.08 lbm/ft³ and internal energy of 85 Btu/lbm.

4.44. One lbm of nitrogen at 20 psia, 120 Btu/lbm of internal energy, and density of 0.1 lbm/ft³.

4.45. One lbm of nitrogen at 200 psia, 1000 Btu/lbm of internal energy, and 0.1 lbm/ft³ density.

4.46. Ten lbm of Freon-22, used as a refrigeration medium, under pressure of 112 psia with a specific volume of 0.7 ft³/lbm and internal energy of 122.93 Btu/lbm.

Section 4.7

4.47. Steam having a specific enthalpy of 70 Btu/lbm and pressure of 1 psia enters an adiabatic pump. Upon leaving the pump the steam is at a pressure of 250 psia and has a specific enthalpy of 70.75 Btu/lbm and no kinetic or potential energy changes occur through the pump. Determine
(a)   Pump work, wk_{os}, per lbm of steam.
(b)   Average density of steam, assuming the pump is reversible in nature.

4.48. A compressor used in a gas turbine engine increases the pressure of 3,000 lbm/min of air from 15 psia to 150 psia without changing the kinetic or potential energy of the air. If the enthalpy of entering air is 118 Btu/lbm and of the compressed air is 230 btu/lbm, determine the power required to drive the compressor in horsepower if it is assumed to be reversible and adiabatic.

4.49. Steam with an enthalpy of 1275 Btu/lbm enters a nozzle at station A (figure 4–24) with a velocity of 50 ft/s. The exit area of the nozzle at station B is $\frac{1}{3}$ the size of the inlet area at A, and the sides are assumed to be adiabatic, allowing no heat transfer. Determine the enthalpy of the steam leaving the nozzle if the steam is incompressible.

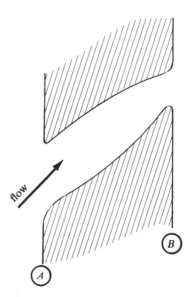

Figure 4–24   Problem 4.49.

4.50. A steam turbine shown in figure 4–25 produces 290 Btu per lbm of steam supplied to it while it loses 8 Btu/lbm of heat. If the entering steam has an enthalpy value of 1530 Btu/lbm and there is negligible kinetic energy loss through the turbine, determine the enthalpy of exiting steam.

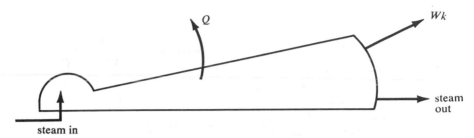

Figure 4–25   Problem 4.50.

4.51. A simplified single-cylinder internal combustion spark ignition engine which produces 20 horsepower is shown in figure 4–26. It loses 40 Btu/min in radiated and conducted heat while using 1.64 lbm/min of fuel and air mixture which is assumed to have an enthalpy of 1140 Btu/lbm of mixture. Determine the specific enthalpy of the exhaust gases, assuming the engine is frictionless.

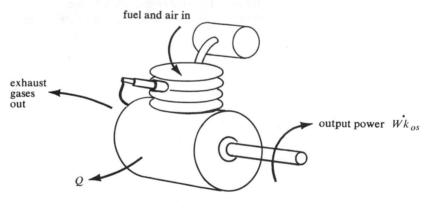

Figure 4–26   Problem 4.51.

4.52. In figure 4–27, a fan driven by a ¼-hp electric motor moves 40 lbm/min of air. Assuming this process is done without heat transfer and the fan is able to divert the flow of air in a uniform direction, determine the average velocity V_{av} of air leaving the fan if it has negligible velocity before passing around the fan. (Note: Assume the air has no enthalpy change.)

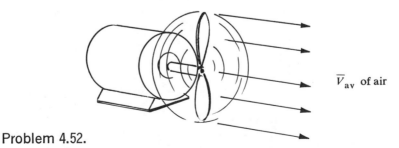

Figure 4–27   Problem 4.52.

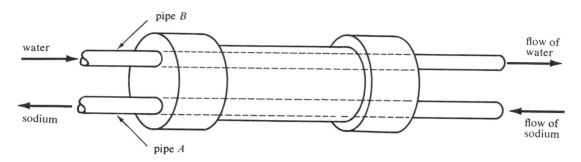

water

sodium

pipe B

pipe A

flow of water

flow of sodium

Figure 4–28   Problem 4.53.

4.53. A parallel flow heat exchange transfers energy from one stream to another. In figure 4–28, 2 lbm/s of sodium flows through pipe *A* and 10 lbm/s of water flows through *B*. If the sodium loses 85 Btu/lbm during its passage through the exchanger, what is the change in specific enthalpy of the water if 2 Btu/s of heat are lost to the surroundings?

5

Equations of State
and Calorimetry

The concepts of energy exchanges (heat and work) and the conservation of energy during such exchanges or processes were presented in the previous two chapters. In the discussions of these concepts, it was inferred that the description of the system or its state was sufficiently complete. In this chapter, elementary attempts at satisfactorily describing the state of the system are presented. These involve the use of *equations of state* to relate known system properties to the unknown properties at a given location and time.

The equations of state relating pressure, volume (or density), and temperature for gases will be introduced with particular attention being given to the relationship resulting from the assumption of a *perfect gas*. Solids and liquids will be treated very lightly due to the difficulty of the mathematics required to relate satisfactorily their properties at a given state.

Joule's experiment, which proved that the internal energy of a perfect gas is related only to the temperature, will be presented. Then the *caloric equations of state* will be introduced for describing gases, liquids, and solids. These equations relate the temperature to the internal energy through the concept of specific heats.

Calorimetry, or the methods for measuring internal energy and enthalpy, will be presented, precipitating an introduction to the common thermodynamics tables (a condensed collection of which appears in the text appendix) and their uses in engineering problem solving.

**5.1
Equations of
State — The
Perfect Gas**

There are many properties, such as pressure, volume, temperature, mass, density, energy, and shape, which help to describe the thermodynamic system. In order to define the system state and make the first law of thermodynamics usable for processes, we need a complete description of the system, which means we need an infinite number of data points to identify fully all the system properties. We need this description of the system state immediately when the process begins and at least once more at the conclusion of the process — then we can apply the concepts of the first law. Fortunately, we are assuming homogeneous and simple systems which means that only a few property data need to be determined to represent the system state.

In addition, some of the properties of the system, such as mass, density, and volume, are related at given states so that somewhat fewer properties need to be measured in the physical situation. Here we would like to develop some other relations which can still further reduce the number of properties which need to be measured to determine the system state.

Gases, such as air, have long been observed to change density as the pressure changed. Leonardo da Vinci remarked in the late 1400s that "Air does not resist unless it grows denser." This was an observation that the density of air increases as the pressure increases, which is exactly the relation commonly referred to as *Boyle's law,* namely

$$\frac{p}{\rho} = \text{constant} = B \qquad (5\text{--}1)$$

Boyle's law is correct only for light or thin gases and for those cases where temperature remains constant. Equation (5–1) is generally written as

$$pv = B \qquad (5\text{--}2)$$

In conjunction with this law, another relationship called *Charles' law* has been found to describe the volume and temperature of a gas at constant pressure. This is

$$V = CT \qquad (5\text{--}3)$$

where C is another constant. It can be shown that during an arbitrary process of a gas which obeys Boyle's and Charles' laws independently, the gas also obeys the relationship

$$pv = RT \qquad (5\text{--}4)$$

We can also write this relationship, without any loss of generality as

$$pV = mRT \qquad (5\text{--}5)$$

and call this the *perfect gas law*. Equations (5–4) and (5–5) are also referred to as the *perfect gas equations* since any gas obeying them is defined as a perfect or an ideal gas. Note that Boyle's and Charles' laws are special relations resulting from restrictions imposed on equations (5–4) and (5–5).

In these equations, the constant R is denoted as the gas constant, values for which are tabulated in the appendix table B.4. From this table it can also be seen that the gas constant is related to the molecular weight of a gas, MW, through the relationship

$$R = \frac{1544 \text{ ft-lbf/lbm-mole}}{\text{MW}} \qquad (5\text{--}6)$$

where the constant 1544 ft-lbf/lbm-mole is called the *universal gas constant, R_u*. In the CGS or metric system R_u is 8.3143 joules/g-mole°K. Values for molecular weights of all the natural elements are listed in appendix table B.18.

The perfect gas relationship is the most common and easily used equation of state, but caution must be exercised in its application. Materials which obey the perfect gas equations have the following qualitative characteristics:

(a) Sufficiently rarefied so that no attractive or repulsive forces exist between any of the atoms.

(b) Sufficiently dense so that the material can be considered a continuous, uniform medium without voids or vacuums.

(c) Atoms with perfectly elastic collisions (reversible collisions) between themselves and the container enclosing the material.

These assumptions are quite difficult to check, but it is generally helpful to "have a feel" for these characteristics. Those materials which obey the perfect gas law include air, most light gases, and free electrons in a solid, electrically conducting material. Obviously, liquids, solids, and dense gases such as moist steam, do not obey the perfect gas law and the re-lationships (5–4) or (5–5) should not be used for defining these materials.

One of the earliest attempts at deriving a more general equation of state was van der Waal's relationship

$$\left(p + \frac{a}{v^2}\right)(v - b) = RT \qquad (5–7)$$

where a and b are constants. This relationship was introduced to describe real gases better than the perfect gas law does, but it still has many short-comings. A better general approach to describing gases by an algebraic relationship is the *virial equation of state:*

$$\frac{pv}{RT} = 1 + \frac{b}{v} + \frac{0.625\ b^2}{v^2} + \frac{0.2896\ b^3}{v^3} + \ldots \qquad (5–8)$$

While equations (5–7) and (5–8) represent more precise relationships between gas properties, the perfect gas relationship is still much more widely used due to its relative simplicity.

The equations of state for solids are generally in the realm of pure research, and any reference to an equation for solids would be misleading in this text. For liquids and dense gases, one of the equations of state used with some success is written

$$\frac{pV}{RT} = \frac{2(V/N)^{1/3}}{2(V/N)^{1/3-2} - 2d} \qquad (5–9)$$

where N is the number of molecules in the volume V and d is the molecular diameter. Again, this equation is not as simple as the perfect gas law, and consequently, is generally used only if tables of properties are not available.

Example 5.1

A perfect gas at 10 psia and 40°F is enclosed in a rigid tank of 3 ft³. This gas is heated to 540°F and 20 psia at which point its density is 0.1 lbm/ft³. Determine the mass and the gas constant for this material.

Solution

Since the gas is contained in a container of constant volume (3 ft³), we can use the relation between density, volume, and mass to determine the mass. Thus,

$$\rho = \frac{m}{V}$$

and from which we have

$$m = \rho V$$

Substituting values into this equation yields

$$m = (0.1 \text{ lbm/ft}^3)(3 \text{ ft}^3)$$

or

$$m = 0.3 \text{ lbm} \qquad\qquad \textit{Answer}$$

The gas constant can now be found from the perfect gas equation (5–5)

$$pV = mRT$$

and the gas constant can be determined at either state. Let us first calculate it at the state with a pressure of 10 psia and 30°F. We can rearrange equation (5–5) to read

$$R = \frac{pV}{mT}$$

and substituting values into this equation obtain

$$R = \frac{(10 \text{ lbf/in}^2)(3 \text{ ft}^3)(144 \text{ in}^2/\text{ft}^2)}{(0.3 \text{ lbm})(40 + 460°R)}$$

or

$$R = 28.8 \text{ ft-lbf/lbm}°R \qquad\qquad \textit{Answer}$$

At the second state, as a check on this last answer, we get

$$R = \frac{(20)(3)(144)}{(0.3)(540 + 460)}$$
$$= 28.8 \text{ ft-lbf/lbm}°R$$

and this agrees with the result calculated at the first state, as it should.

**5.2
Joule's
Experiment**

Before we proceed with a detailed investigation of how internal energy is related to the common properties of temperature, pressure, and volume, let us here see how a simple experiment can indicate some powerful conclusions regarding internal energy and its relation to other properties. This experiment, named after James Joule who first performed it in 1843, involves the use of the apparatus shown in figure 5–1. Tanks *A* and *B* are submerged in a water bath at room temperature. Tank *A* is filled with air at some high pressure, say 300 psia, while tank *B* is empty or near minus (−) 14.7 psig. The two tanks are connected by a valve which is opened quickly. The temperature is recorded before and after the valve is opened and is found to be the same. During this process, the air expanded to fill both tanks, the pressure reached an equilibrium in

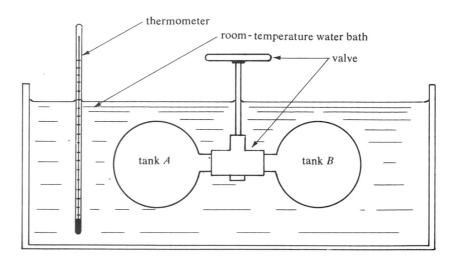

Figure 5-1   Apparatus for Joule's experiment.

A and B, and the volume of air increased from that in tank A initially to that in both tanks after expanding. The temperature remained constant, however; and since no work or heat was done across the boundary of the two tanks A and B, the internal energy remained fixed or constant — this from the first law of thermodynamics, i.e.

$$\Delta U = Q - Wk$$

but

$$Q = 0 \quad \text{and} \quad Wk = 0$$

so that

$$\Delta U = 0$$

From these observations it is seen that internal energy of the air cannot be a function of pressure or volume of the air, and consequently, must be a function of temperature only. This is true because pressure and volume changed, while temperature and internal energy remained unchanged during the expansion. We then write for air,

$$U = f(T) \tag{5–10}$$

and this has been found true for all gases which can be considered as perfect gases. The significance here is that the internal energy of a perfect gas is *only* a function of temperature and this result is a helpful simplification in studying gases.

5.3
The Specific
Heats

In section 5.2 was introduced a result which is very helpful; that is, internal energy is only a function of temperature for perfect gases. This result must be expanded to be useful. For instance, there are many functions which satisfy equation (5–10), a few of which are

$$U = CT + B \qquad\qquad (5\text{–}11)$$

$$U = CT^2 + B \qquad\qquad (5\text{–}12)$$

and

$$U = CT^3 + BT \qquad\qquad (5\text{–}13)$$

where C and B are constants. Keep in mind that these three equations are attempts at describing internal energy for perfect gases. Quite generally, for any gas or liquid, we define the change in internal energy per unit temperature change during a constant volume process as the total specific heat at constant volume C_v and write

$$C_v = \left(\frac{dU}{dT}\right)_{v=\text{constant}} \qquad\qquad (5\text{–}14)$$

The notation here of bracketing the differential and subscripting with "v = constant" indicates that the differential is only with respect to temperature, and not volume. Alternately the specific heat at constant volume, c_v, is related to the total, C_v, by

$$mc_v = C_v$$

or

$$c_v = \frac{C_v}{m} \qquad\qquad (5\text{–}15)$$

and then

$$c_v = \left(\frac{du}{dT}\right)_{v=\text{constant}}$$

The units for total specific heat are easily seen to be energy per unit temperature, that is Btu/°R, calorie/°K, or other similar units. The units for specific heat at constant volume are then in terms of energy per unit mass per unit temperature.

For perfect gases, U is only a function of temperature and in this case, equation (5–14) becomes

$$C_v = \frac{dU}{dT} \qquad\qquad (5\text{–}16)$$

or, for changes in internal energy,

$$dU = C_v dT \qquad\qquad (5\text{–}17)$$

where now no restriction is placed on the differential regarding constant volumes. Equation (5–17) is correct for gases or liquids which do not behave as perfect gases, but correct only if the process during which the internal energy and temperature changes occur is one which has no volume change. Again, for perfect gases, no such restriction is placed on equation (5–17).

Now let us integrate equation (5–17) to arrive at a *caloric equation of state;* first we will assume C_v is constant and then

$$\int dU = C_v \int dT$$

from which we obtain the caloric equation of state

$$U = C_v T + U_0 \tag{5–18}$$

where U_0 is a constant of integration. Obviously, equation (5–11) is a good description of internal energy of a perfect gas with constant C_v since it agrees with equation (5–18), provided that C and C_v are identified as one and the same. The value of U_0 cannot be determined but is arbitrarily assigned some value, generally zero. For U_0 assigned the value of zero, internal energy has the value of zero when absolute temperature is zero. This may or may not be true physically, but the absolute value is not as important as is the change of value in internal energy. This point can easily be checked by noting that in the statements of the first law of thermodynamics, only changes in energy were considered, not single absolute values.

From equation (5–17) or (5–18) the change in internal energy for constant specific heats is

$$\Delta U = C_v \Delta T = m c_v \Delta T$$

and

$$\Delta u = c_v \Delta T \tag{5–19}$$

For any substance, we define now the total specific heat at constant pressure, C_p, as the change in enthalpy per unit change in temperature when pressure is kept constant, and write

$$C_p = \left(\frac{dH}{dT}\right)_{p=\text{constant}} \tag{5–20}$$

The specific heat at constant pressure c_p is related to C_p by

$$m c_p = C_p \tag{5–21}$$

or

$$c_p = \frac{C_p}{m} = \left(\frac{dh}{dT}\right)_{p=\text{constant}} \tag{5–21}$$

Recalling the definition of enthalpy,

$$H = U + pV$$

we see that for a perfect gas, U is a function of temperature only and pV is equivalent to mRT which is also only a temperature function. For a perfect gas, then, enthalpy is a function of temperature only and equation (5–20) can be written

$$C_p = \frac{dH}{dT} \tag{5–22}$$

or

$$dH = C_p dT \tag{5–23}$$

From equation (5–23) we can get another caloric equation of state by assuming C_p is constant. Then

$$\int dH = C_p \int dT \tag{5–24}$$

and from this we obtain

$$H = C_p T + H_0 \tag{5–25}$$

Here H_0 denotes an integration constant in the same manner as U_0 for equation (5–18). We see then that like internal energy, enthalpy cannot be determined absolutely; only changes from one state to another can be measured with some precision. The change in enthalpy for constant specific heats is obtained from integrating equation (5–24) between limits, so that

$$\Delta H = C_p \Delta T = m c_p \Delta T \tag{5–26}$$

and

$$\Delta h = c_p \Delta T$$

From equation (5–22) substituting the terms $U + pV$ for H, we get

$$C_p = \frac{dU}{dT} + \frac{dpV}{dT}$$

and since $pV = mRT$ for perfect gases

$$C_p = \frac{dU}{dT} + mR \frac{dT}{dT}$$

or

$$C_p = \frac{dU}{dT} + mR \qquad (5\text{–}27)$$

Now, from equation (5–16), the first term on the right side of equation (5–27) is identified as C_v or mc_v so that

$$C_p = mc_v + mR \qquad (5\text{–}28)$$

and consequently, since $C_P = mc_p$, we obtain

$$c_p = c_v + R \qquad (5\text{–}29)$$

The equations (5–28) and (5–29) are true only for perfect gases but represent useful tools in problem solving.

Another notation which will be used in subsequent analyses is the ratio of specific heats, k, defined as

$$k = \frac{C_p}{C_v} = \frac{c_p}{c_v} \qquad (5\text{–}30)$$

Under most conditions, k has a value near 1.4 and unless it is stated otherwise the student can safely use this value for a perfect gas. Values of k for various gases are listed in table B.4 of the appendix.

5.4 Measurement of Internal Energy

We have no energy meters available with which to measure internal energy of a system — but we have thermometers, and armed with them and the caloric equations of state as given by (5–18) and (5–25), we can obtain internal energy and enthalpy data for perfect gases. We do, however, need to know values for the specific heats, and the following example should give you the basis for determining specific heats at constant volume.

Example 5.2

In figure 5–2, air at 70°F is contained in a rigid tank which is placed on an electric burner. The sides and top of the tank are insulated while the bottom allows free heat transfer. A thermometer is inserted into this device to record the temperature while the air is heated. After 20 minutes, it is found that the air temperature inside the tank is 270°F and the amount of electric energy required to heat up the air was 1.0 watt-hour. Determine C_v and c_v for the air. Assume the mass of air contained is 0.1 lbm.

Solution

We can assume air behaves as a perfect gas over the range of 70°F to 270°F and even if it does not, the process is one of constant volume so

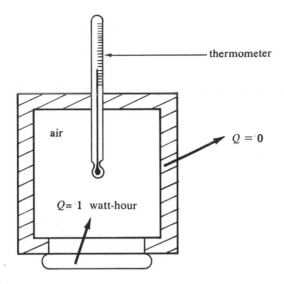

air

Q = 0

Q= 1 watt-hour

thermometer

Figure 5–2   Example 5.2.

that we can use

$$dU = C_v dT \qquad (5\text{–}17)$$

and assume C_v will be constant. Then we can use the caloric equation of state (5–18), calling state (2) the final and state (1) the initial, we have

$$U_2 = C_v T_2 + U_0$$
$$U_1 = C_v T_1 + U_0$$

and the change of energy is

$$U_2 - U_1 = C_v(T_2 - T_1) = \Delta U$$

This result agrees with a result obtainable from equation (5–19) for changes in internal energy.

Stipulating the system as the air inside the tank, we see that no work was done; but heat was transferred into this system, the amount of which was 1.0 watt-hour. We assumed no heat losses during the process so that, using the first law of thermodynamics

$$\Delta U = Q - Wk = Q$$

since $Wk = 0$. Then

$$\Delta U = 1.0 \text{ watt-hour}$$

Also, we know that

$$\Delta U = U_2 - U_1 = C_v(T_2 - T_1)$$

We can see that the temperature T_2 is 270°F + 460 or 730°R, while T_1 is 70°F + 460 or 530°R. Notice, however, that temperature changes are the same for absolute or thermometric scales, that is, Rankine and Fahrenheit are equivalent when considering changes in temperature. Then

$$\Delta U = C_v(730 - 530)$$
$$1.0 \text{ watt-hour} = C_v(200°R)$$

and

$$C_v = 1.000 \text{ watt-hour}/200°R$$
$$= 0.005 \text{ watt-hour}/°R$$

Since there are 3.414 Btu per watt-hour we have

$$C_v = 0.005 \text{ watt-hour}/°R \times 3.414 \text{ Btu/watt-hours}$$

or

$$C_v = 0.01707 \text{ Btu}/°R \qquad\qquad\qquad Answer$$

The specific heat at constant volume, c_v, is determined from

$$c_v = \frac{C_v}{m}$$

from which

$$c_v = \frac{0.01707 \text{ Btu}/°R}{0.1 \text{ lbm}}$$

and then

$$c_v = 0.1707 \text{ Btu/lbm°R} \qquad\qquad\qquad Answer$$

Table B.4 of the appendix lists values for specific heats and our answer agrees within 0.0011 Btu/lbm°R for that of air.

Calorimetry: *The science and technology concerned with precisely measuring energy and enthalpy.*

From it the values of specific heats listed in table B.4 were determined in much the same manner as described by example 5.2. In addition, some gases, as well as liquids, have been found to have specific heat values which vary with the temperature of the substance — not an unexpected turn of events. In table B.13 are equations which show that specific heats at constant pressure, c_p, are functions of temperature, and utilizing the relation for perfect gases

$$c_p - c_v = R \qquad\qquad\qquad (5-27)$$

we can see that c_v is also a function of temperature for those gases listed in the table. For variable specific heats we must use, provided we still have a perfect gas,

$$dU = C_v dT \qquad (5\text{-}17)$$

in order that a change in internal energy may be found. To obtain this change in energy, the equation must be integrated so that we have

$$\int_{U_1}^{U_2} dU = U_2 - U_1 = \Delta U = \int_{T_1}^{T_2} C_v dT \qquad (5\text{-}31)$$

or, per pound-mass,

$$\int du = u_2 - u_1 = \Delta u = \int c_v dT \qquad (5\text{-}32)$$

Either of these two equations can be used for determining internal energy data for gases and liquids deviating from a perfect gas. From tests using empirical methods of calorimetry, temperature and power are recorded. By assigning a value of zero for u_1 in equation (5–32) when T is zero degrees absolute, u_2 is identified then as the power required to elevate the material temperature above zero. In measuring the power data, we bring into use much sophisticated instrumentation — suffice it here to say that from an apparatus as described in example 5.2, watt-meters could measure the power. From other methods of supplying heat to the container of test substances, such as Bunsen burners, heat pipes, or heat exchangers, the power data can also be determined; and from some of the later application chapters we will see how this is done.

Air is not precisely a perfect gas, nor does it have a constant value of its specific heat at constant volume. In table B.6 are listed data of internal energy as functions of temperature for air — data which were determined empirically in a manner just discussed. Table B.12 contains internal energy-temperature data for other gases. In those thermodynamic tables where internal energy data are missing but enthalpy is listed, use of the equation

$$u = h - pv$$

can by made to obtain the data.

Example 5.3   Determine the change in internal energy of 2 lbm of oxygen gas as it changes temperature from 70°F to 90°F.

Solution   We assume that oxygen is a perfect gas and also that the specific heats are constant over the specific temperature range. Then we have

$$\Delta U = mc_v \Delta T \qquad (5\text{-}19)$$

for which we identify the following terms

$$m = 2 \text{ lbm}$$
$$\Delta T = 90°F - 70°F = 20°F = 20°R$$

and

$$c_v = 0.157 \text{ Btu/lbm°R (from table B.4 of the appendix)}$$

Substituting these values into the caloric equation of state (5-17) yields

$$\Delta U = (2 \text{ lbm})(0.157 \text{ Btu/lbm°R})(20°R)$$
$$= 6.28 \text{ Btu} \qquad \qquad Answer$$

and we have determined the change in internal energy.

5.5 Measurement of Enthalpy

We determine enthalpy by using equation (5-23), provided the substance is a perfect gas. Even then, we must have a value for c_p in order that we obtain the desired values. The following example should provide the background for understanding the origin of values for c_p.

Example 5.4

Twelve grams of air are contained in a frictionless piston-cylinder device. The piston weighs 2 newtons and the walls of the cylinder and the piston are perfectly insulated (adiabatic surfaces) allowing no heat transfer. This device is shown in figure 5-3 where a paddle, driven by an electric motor, is inserted through the bottom. The paddle shaft is frictionless and, after the electric motor has expended 155.4 watt-seconds of energy to drive the paddle, the air temperature in the cylinder is found to have increased from 10°C to 80°C. Determine the total specific heat at constant pressure, C_p, and the specific heat at constant pressure, c_p.

Solution

Here we have an addition of energy to our system (the air inside the cylinder) which is completely irreversible. The paddle wheel stirred up the air in 155.4 watt-seconds, and thereby increased the air temperature through internal friction of the air itself. In example 5.2, the addition of energy to the air could have been made in this same way rather than by heat transfer. The heat for this process is zero so we write the first law

$$\Delta U = Q - Wk = Wk$$

The work term is composed of a reversible and an irreversible part. The irreversible work is, as we have mentioned, equal to 155.4 watt-seconds,

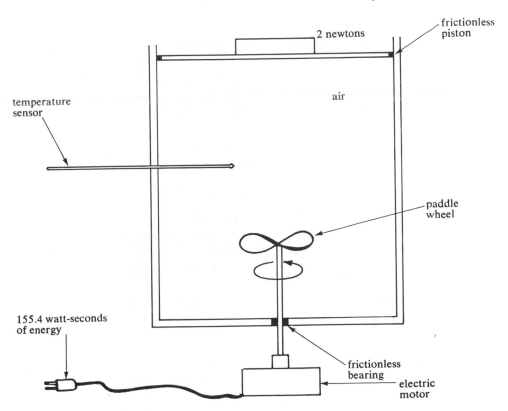

frictionless
piston

2 newtons

air

temperature
sensor

paddle
wheel

155.4 watt-seconds
of energy

frictionless
bearing

electric
motor

Figure 5–3   Device for determining specific heat values.

while the reversible work is the motion of the piston. Since the air is
elevated in temperature, it also will tend to increase pressure; but the
piston will be raised instead, to increase the volume of air. The process
here is conducted at constant pressure since the piston weight is constant
as it moves up. Since we have a frictionless piston, the reversible work
done by the system is given by

$$Wk_{cs \text{ rev}} = \int p dV = p \int dV$$

and then

$$Wk_{cs \text{ rev}} = p(V_2 - V_1)$$

The first law then becomes

$$U_2 - U_1 = -(-155.4 \text{ watt-seconds}) - p(V_2 - V_1)$$

or

$$U_2 - U_1 + pV_2 - pV_1 = 155.4 \text{ watt-seconds}$$

Now, using the definitions for enthalpy, we see the left side here is $H_2 - H_1$ and then using equation (5-24), we see that

$$\int dH = H_2 - H_1 = \int C_p dT$$

We assume that air is a perfect gas and that the specific heat is a constant. This gives us

$$H_2 - H_1 = C_p(T_2 - T_1)$$

Now, we equate this last result with the previous equation,

$$C_p(T_2 - T_1) = 155.4 \text{ watt-seconds}$$

and substituting values, obtain

$$C_p = \frac{155.4 \text{ watt-seconds}}{(80 + 273 - 10 - 273)} °K$$
$$= \frac{155.4 \text{ watt-seconds}}{70°K}$$
$$= 2.22 \text{ watt-seconds}/°K \qquad \qquad \textit{Answer}$$

The specific heat at constant volume is

$$c_p = \frac{C_p}{m}$$
$$= \frac{2.22 \text{ watt-seconds}/°K}{12 \text{ g}}$$
$$= 0.185 \frac{\text{watt-second}}{\text{g}°K}$$

or, since a watt-second is a *joule*,

$$c_p = 0.185 \text{ joule/g}°K \qquad \qquad \textit{Answers}$$

This result can easily be converted to English engineering units. From table B.14 we have

$$1054 \text{ joules} = 1 \text{ Btu}$$
$$454 \text{ grams} = 1 \text{ lbm}$$

and, from chapter 2 we saw,

$$\frac{9}{5}°K = °R$$

Then

$$c_p = 0.185 \frac{\text{joule}}{\text{g}°K} \times \frac{1 \text{ Btu}}{1054 \text{ joules}} \times \frac{454 \text{ g}}{\text{lbm}} \times \frac{9°K}{5°R}$$
$$= 0.24 \text{ Btu/lbm}°R \qquad \qquad \textit{Answer}$$

which agrees with the specific heat listed in table B.4.

The determination of C_p and C_v for various materials is pursued in the same manner. The primary reason is that it is a means of obtaining enthalpy data for use in engineering and technology problem solving. Gases which behave as perfect gases, or reasonable facsimilies thereof, but which have variable specific heats, can have enthalpy property changes determined from the equations

$$\Delta H = \int C_p dT \tag{5-33}$$

$$\Delta h = \int c_p dT \tag{5-34}$$

with the value of C_p obtained from tables such as B.4 or B.13.

Example 5.5

Determine the change in enthalpy per lbm of methane gas as it goes through a temperature change from 600°F to 200°F.

Solution

Assuming that methane gas behaves as a perfect gas with constant values for the specific heats over the temperature range, we have

$$\Delta h = c_p \Delta T \tag{5-24}$$

for which

$$\Delta T = 200°F - 600°F$$
$$= -400°F$$

and

$$c_p = 0.5318 \text{ Btu/lbm°R (from table B.4 of the appendix)}$$

Substituting values into the above equation gives us

$$\Delta h = (0.5318 \text{ Btu/lbm°R})(-400°R)$$
$$= -212.72 \text{ Btu/lbm} \qquad \text{\textit{Answer}}$$

Example 5.6

Determine the change in enthalpy per lbm for carbon monoxide (CO) as the gas is cooled from 1500°F to 500°F, assuming the gas does not have a constant specific heat.

Solution

We will assume the gas, CO, is a perfect gas so that

$$\Delta h = \int c_p dT \tag{5-34}$$

and we must find the relation c_p has to temperature. From table B.13 we can see that there is a choice of relations for CO, namely

$$c_p = \left(9.46 - \frac{3.29(10^3)}{T} + \frac{1.07(10^6)}{T^2}\right) \text{ Btu/lbm mole}°\text{R}$$

or

$$c_p = [a + b(10^{-3})T + c(10^{-6})T^2 + d(10^{-9})T^3] \text{ cal/g-mole}°\text{K}$$

where a, b, c, and d can take on two different sets of values. Let us use the second equation with the following set of constants:

$$a = 6.480$$
$$b = 1.566$$
$$c = -0.2387$$
$$d = 0$$

Then

$$c_p = (6.48 + 1.566\ T \times 10^{-3} - 0.2387\ T^2 \times 10^{-6}) \text{ cal/g-mole}°\text{K}$$

and the initial and final temperatures are

$$T_1 = 1500°\text{F} + 460° = 1960°\text{R}$$
$$= \frac{5}{9} \times 1960°\text{R} = 1089°\text{K}$$

and

$$T_2 = 500°\text{F} + 460 = 960°\text{R}$$
$$= \frac{5}{9} \times 960 = 533°\text{K}$$

respectively. We now integrate equation (5–34),

$$\Delta h = \int_{1089°\text{K}}^{533°\text{K}} (6.48 + 1.566\ T \times 10^{-3} - 0.2387\ T^2 \times 10^{-6})dT$$

$$= \left(6.48\ T + 0.783\ T^2 \times 10^{-3} - 0.0796\ T^3 \times 10^{-6}\right)_{1089°\text{K}}^{533°\text{K}}$$

$$= [6.48(533 - 1089) + 0.783 \times 10^{-3}(533^2 - 1089^2) - 0.0796$$
$$\times 10^{-6}\ (533^3 - 1089^3)]\ \text{cal/g-mole}$$
$$= \{-3603 - 706 + 91\}\ \text{cal/g-mole}$$
$$= -4218\ \text{cal/g-mole}$$

Per gram, the change in enthalpy is found by using the molecular weight of CO, 28 g/g-mole,

$$\Delta h = -4218\ \text{cal/g-mole} \times \frac{1}{28}\ \text{g/g-mole CO}$$
$$= -150.6\ \text{cal/g} \qquad\qquad \textit{Answer}$$

This answer could now easily be converted to Btu/lbm if it is desired.

**5.6
The Thermo-
dynamic Tables**

For perfect gases we can use the equations (5–17) and (5–23) in integrated form,

$$\int dU = \int C_v dT = \Delta U$$

$$\int dH = \int C_p dT = \Delta H$$

and with constant values C_v and U_p, obtain equations (5–19) and (5–26)

$$\Delta U = C_v \int dT = C_v \Delta T$$

$$\Delta H = C_p \int dT = C_p \Delta T$$

to determine changes in internal energy and enthalpy. All we need here are temperature data and numerical values for the specific heats. These equations are also reasonably accurate for nonperfect gases, particularly if we utilize variable specific heats and integrate equations (5–17) and (5–23) with the according complication as shown in example 5.6. However, for the vast bulk of materials subject to engineering usage — liquids, dense gases, and solids — we have tables of thermodynamic properties at our disposal with which to determine enthalpy and internal energy. Tabular values of these properties are also recommended over caloric equations of state like (5–19) and (5–26) for gases approximating perfect gases if the desired results must be accurate to more than 5%.

Tables B.1, B.2, B.3, B.6, B.8 B.9, B.10, B.11, B.12, and B.13 in the appendix list enthalpy and internal energy value for various substances at representative temperatures and pressure, and using interpolation of data we can solve all the problems presented in this text. More extensive data of thermodynamics properties are published in various literature, and much effort is being expended in calorimetry to compile more precise and expanded data treating more and more materials. The following example should be helpful if you are not familiar with *interpolation*, a technique giving detailed scope to tabular data.

Example 5.7

Determine the enthalpy of saturated mercury vapor at 65 psia.

Solution

From table B.10, which lists the properties of mercury vapor, we can see that the enthalpy of saturated vapor, h_g, is 148.6 Btu/lbm at 60 psia and 148.9 Btu/lbm at 70 psia. The differences are, for enthalpy

$$148.9 - 148.6 = 0.3 \text{ Btu/lbm}$$

and for pressure

$$70 - 60 = 10 \text{ psia}$$

For a pressure of 65 psia, we can write

$$\frac{65 \text{ psia} - 60 \text{ psia}}{10 \text{ psia}} = \frac{h - 148.6 \text{ Btu/lbm}}{0.3 \text{ Btu/lbm}}$$

and then

$$\frac{5}{10} \times 0.3 \text{ Btu/lbm} = (h - 148.6) \text{ Btu/lbm}$$

Solving this then gives us h, the enthalpy of saturated mercury vapor at 65 psia, or

$$\frac{5}{10} \times 0.3 \text{ Btu/lbm} + 148.6 \text{ Btu/lbm} = h$$

and

$$h = 148.75 \text{ Btu/lbm}. \qquad \textit{Answer}$$

We will have more examples of the use of the tables of thermodynamic properties in succeeding chapters.

5.7 Summary

In this chapter we have attempted to relate the properties at given states through algebraic relationships, called *equations of state* and *caloric equations of state*. The most elementary and usable material to be studied is the perfect gas, defined by the equations of state,

$$pv = RT \qquad (5\text{--}4)$$
$$pV = mRT \qquad (5\text{--}5)$$

There are other equations, such as van der Waal's equation and the virial equation, which are attempts to relate properties for dense gases and liquids, but which will not generally be used here.

The caloric equations of state for perfect gases with variable specific heats are

$$\Delta U = \int C_v dT \qquad (5\text{--}31)$$

or

$$\Delta u = \int c_v dT \qquad (5\text{--}32)$$

and

$$\Delta H = \int C_p dT \tag{5-33}$$

or

$$\Delta h = \int c_p dT \tag{5-34}$$

which follow from definitions of the specific heats. The total specific heats, C_v and C_p, were defined for any material as

$$C_v = \left(\frac{dU}{dT}\right)_{v=\text{constant}} \tag{5-14}$$

and

$$C_p = \left(\frac{dH}{dT}\right)_{p=\text{constant}} \tag{5-20}$$

From Joule's experiment it was concluded that for perfect gases, internal energy and enthalpy are functions of temperature only. Equations (5–14) and (5–20) then reduce to (5–31) and (5–33) respectively.

The specific heats, more widely used than the total specific heats are given by

$$c_v = \frac{C_v}{m} = \left(\frac{du}{dt}\right)_{v=\text{constant}}$$

and

$$c_p = \frac{C_p}{m} = \left(\frac{dh}{dT}\right)_{p=\text{constant}} \tag{5-21}$$

for any substance. For perfect gases, the following two useful relations hold:

$$C_p - C_v = mR \tag{5-28}$$

and

$$c_p - c_v = R \tag{5-29}$$

Equations (5–31), (5–32), (5–33), and (5–34) are used, along with the assumption of constant values for C_v and C_p, to give us

$$\Delta U = C_v \Delta T \tag{5-19}$$
$$\Delta H = C_p \Delta T \tag{5-26}$$
$$\Delta u = c_v \Delta T \tag{5-19}$$
$$\Delta h = c_p \Delta T \tag{5-26}$$

These specific equations allow us to calculate changes in internal energy and enthalpy for *perfect gases* with constant specific heats. Tables of thermodynamic properties developed from calorimetric techniques are used to obtain enthalpy and internal energy of nonperfect gases and liquids.

Practice Problems

Problems designated with an asterisk * preceding the number should only be attempted by those having a background which includes the knowledge of integral calculus.

Given following conditions, determine the unknown property of a perfect gas, for problems 5.1–5.12.

Section 5.1

5.1. Pressure is 20 psig, volume is 3 cubic feet, mass is 1.5 lbm, and the gas constant is 30 ft-lbf/lbm °R.

5.2. Density is 0.1 lbm/ft³, the gas constant is 45 ft-lbf/lbm°R, and the temperature is 180°F.

5.3. Oxygen gas at 200 psia and 800°F.

5.4. A gas whose molecular weight is 13.5 g/g-mole and which is at 300,000 dynes/cm² gage pressure and 800°K. (R_u = 8.3143 joules/g-mole°K)

5.5. Two hundred seventy grams of argon gas at a pressure of 2800 dynes/mm² (absolute), and a volume of 1300 liters.

5.6. A gas having a specific volume of 7 ft³/lbm, a pressure of 120 psia and a temperature of 1000°F.

5.7. A gas contained in a 3-liter container, at a pressure of 3 bars (3×10^6 dynes/cm²), temperature of 700°C and with a mass of 0.66 gram.

5.8. Determine the gage pressure if the atmospheric pressure is 14.7 psia, the gas constant is 96 ft-lbf/lbm°R, the temperature is 700°R, and the specific volume is 10 ft³/lbm.

5.9. Carbon dioxide gas (CO_2) at a pressure of 15 psia and 90°F.

5.10. Seventy lbm of a gas are contained in a rigid container at 200 psia and 80°F. The gas is then expanded to fill a 2000-ft³ volume at a pressure of 20 psia and a temperature of 70°F. Determine the volume of the rigid container.

5.11. During a constant pressure process, 0.05 lbm of hydrogen gas increases in temperature from 70°F to 200°F. Determine the final volume of the gas if its density is initially 0.09 lbm/ft³.

5.12. A constant temperature process is executed during which the pressure of an ideal gas increases by a ratio of 10 to 1. If the gas was initially enclosd in 2 liters, determine the final volume.

5.13. Determine the temperature of carbon monoxide gas at 20 psia and 6 ft³/lbm using van der Waal's equation of state and using the values

$$a = 375 \text{ at-ft}^6/\text{mole}^2$$
$$b = 0.63 \text{ ft}^3/\text{mole}$$

Compare your answer to that derived from using the perfect gas law.

5.14. Determine the increase in internal energy of 2 lbm of argon gas if its temperature increases from 85°F to 255°F.

5.15. Calculate the change in internal energy per lbm of sulfur dioxide as its temperature increases 50°R.

5.16. Six hundred thirty grams of propane gas are cooled from 38°C to 13°C. Determine the total internal energy change and the specific internal energy change.

5.17. Helium gas is cooled from 80°C to 20°C in 60 minutes. Determine the internal energy change and the rate of change of energy per gram.

5.18. Neon gas exhibits a change in temperature of 1000°C. Determine its change in internal energy per gram.

5.19. Three lbm of a perfect gas, having constant specific heat, can absorb 70 Btu of heat while increasing temperature by 80°R. Determine C_v and c_v of the gas.

*5.20. The specific heat, c_v, of a certain gas obeys the relation,

$$c_v = (0.25) + 0.01\ T)\ \text{Btu/lbm°R}$$

Determine the change in internal energy of this gas as its temperature increases from 200°F to 300°F.

*5.21. Determine, within 1.1% error, the change of internal energy of oxygen gas between 1000°R and 2000°R. (Hint: Use table B.13 and equation $c_v = c_p - \text{R}$.)

*5.22. Determine the change in internal energy of ammonia (NH_3) as its temperature increases from 700°K to 800°K. (Hint: Use table B.13 and equation $c_v = c_p - \text{R}$.)

5.23. A perfect gas increases temperature from 70°C to 130°C. Determine the change in enthalpy if the specific heat, c_p, is found to be 0.2 cal/g°K.

5.24. Determine the increase in total enthalpy of 100 lbm of sulfur dioxide gas (SO_2) between 100°F and 250°F.

5.25. Determine the enthalpy change of helium as it is cooled from 120°F to 60°F.

*5.26. Assuming variable specific heats, determine the enthalpy change of propane gas (C_3H_8) as it is heated from 85°F to 200°F. Check your answer against the answer you get by assuming constant specific heats.

5.27. A perfect gas is known to have a gas constant value of 78.5 ft-lbf/lbm°R and a ratio of specific heats, k, of 1.28. Determine c_p and c_v.

5.28. Determine the values of k and c_p of a perfect gas which has c_v of 0.225 Btu/lbm°R and a gas constant of 66 ft-lbf/lbm °R.

Use the tables in the appendix for problems 5.29–5.36.

5.29. Determine the enthalpy of dry air at 600°R.

5.30. Determine the enthalpy of dry air at 1150°R and 500°R.

5.31. Determine the internal energy for the following gases at 3400°R: O_2, N_2, CO, and H_2O.

5.32. Determine enthalpy and internal energy of saturated mercury vapor at 50 psia.

5.33. Determine the enthalpy and internal energy of saturated freon (F-12) vapor at 120°F.

5.34. Determine the enthalpy and internal energy of saturated freon (F–22) liquid at 160°F.

5.35. Determine the enthalpy of superheated steam at 200 psia and 1000°F.

5.36. Determine the enthalpy of saturated steam vapor at 14.696 psia.

6

Processes

In chapter 5, those equations relating system properties at one particular state were introduced. Here we introduce equations called *process equations*, which determine and relate a system's properties at various states during a process. The advantage of an accurately descriptive process equation is that it gives additional information necessary to determine the amount of work done during a reversible process. The reversible process is given heaviest emphasis in this text, but some attempt is made to show how irreversibilities might affect predicted properties and work.

Tables which list the various forms of the relationships resulting from process equations are presented at the end of the chapter. These tabulations can circumvent much of the mathematical manipulations which students may find too difficult, and they should be useful references for subsequent problem solving.

6.1
The Constant Pressure Process

The most common process occurring in our mechanized, technological society is probably the constant pressure process — any closed system executing a change in volume against the atmosphere is involved in a constant pressure process. There are, of course, many other situations which will produce this type of process; we will now investigate a general one.

For a closed reversible system, the work is calculated from $\int p dV$. In particular, if the pressure is constant, we have

$$Wk_{cs} = p \int_{V_1}^{V_2} dV = p(V_2 - V_1)$$

or

$$Wk_{cs} = p(\Delta V) = p(V_2 - V_1) \tag{6-1}$$

Here V_2 is the final volume of the closed system, and V_1 is the initial volume. This result has been reached previously, but is repeated here for completeness. It is correct for any material — liquid, imperfect gas, or perfect gas.

For perfect gases we have other results which are quite useful. For the constant pressure process we have

$$p_1 = p_2 = \text{constant} \tag{6-2}$$

where the subscripts (1) and (2) represent initial and final states. For the perfect gas this implies

$$\frac{V_1}{T_1} = \frac{V_2}{T_2}$$

or

$$\frac{V_1}{V_2} = \frac{T_1}{T_2} \tag{6-3}$$

Thus, from this result and given enough data we can calculate properties at different states. This relation is true for either open or closed systems.

For the open system, we have indicated that reversible work can be found from the equation

$$Wk_{os} = -\int V dp$$

During the constant pressure process this integral vanishes, i.e.

$$dp = 0$$

so that

$$Wk_{os} = 0$$

and for an open system, the only contributions to work when pressure remains constant must be in the form of kinetic or potential energy changes. Frequently, the constant pressure process is referred to as the *isobaric process*.

Example 6.1

In figure 6–1, a perfect gas flows through a chamber at constant pressure, with no change in velocity. If the density of the gas increases by a factor of 2.5, and the inlet temperature is 1200°F, determine the work and the exhaust temperature.

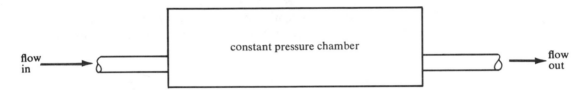

constant pressure chamber

flow in

flow out

Figure 6–1 **Example 6.1.**

Solution

Since we are here concerned with a system through which gas flows, that is, an open system, we can directly find the work done. We have, for this isobaric process

$$Wk_{os} = 0$$

The exhaust temperature we seek from equation (6–3)

$$\frac{V_1}{V_2} = \frac{T_1}{T_2}$$

and, using the relationship $\rho = m/V$, we obtain

$$\frac{\rho_2}{\rho_1} = \frac{T_1}{T_2} \qquad (6\text{–}4)$$

We then have

$$\frac{\rho_2}{\rho_1} = 2.5 = \frac{T_1}{T_2} = \frac{(1200 + 460)°\text{R}}{T_2}$$

yielding, for the exhaust temperature

$$T_2 = \frac{1660°\text{R}}{2.5} = 664°\text{R} \qquad \qquad Answer$$

6.2
The Constant
Volume
Process

The *constant volume process*, commonly called the *isometric process*, is approximated quite frequently in engineering situations. Even solid materials can be conveniently considered if we assume that the materials retain their initial volume throughout a twisting, stretching, or shearing process. This then is the meaning of a constant volume process — no volume change.

For a closed, simple system the work done during an isometric process is zero; at least the reversible work due to volume change is zero since

$$Wk_{cs} = \int p dV$$

and obviously here $dV = 0$, so that $Wk_{cs} = 0$.

We can, of course, have irreversible work or reversible work manifested by changes in kinetic, potential, or strain energy. (See example 6.3.) Reversible work in the open system, neglecting kinetic or potential energy changes, is

$$Wk_{os} = -\int_{p_1}^{p_2} V dp = -V \int_{p_1}^{p_2} dp = -V(p_2 - p_1) \qquad \textbf{(6–5)}$$

or, per lbm of material

$$wk_{os} = -v(p_2 - p_1) \qquad \textbf{(6–6)}$$

For perfect gases subjected to isometric processes, whether the system is open or closed, we have

$$\frac{p_1}{p_2} = \frac{T_1}{T_2} \qquad \textbf{(6–7)}$$

Remember that while they appear very similar, equations (6–3) and (6–7) are in no way related to each other — they are each descriptive of entirely different processes.

Example 6.2

A water pump, shown in figure 6–2, operating under steady flow conditions, moves water at 78°F from a region of low pressure (15 psia) to a region of high pressure (250 psia). Determine the work done per pound-mass of water and the power required if the water is supplied at 20 lbm/s.

Solution

The system involved here is a water pump whose boundaries cross an inlet and an exhaust water pipe. Since we are here treating water, we assume it is incompressible and therefore of constant volume. In this circumstance

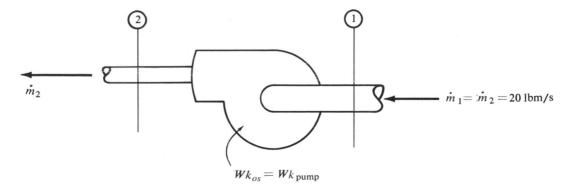

Figure 6-2   Water pump of example 6.2.

$$wk_{os} = -\int vdp$$

$$= -v\int dp$$

$$= -v\Delta p = -v(p_2 - p_1)$$

so that, using a value of 0.016 ft³/lbm for the specific volume of water at 78°F,

$$wk_{os} = -(0.016 \text{ ft}^3/\text{lbm})(250 - 15 \text{ lbf/in}^2)(144 \text{ in}^2/\text{ft}^2)$$
$$= -541.4 \text{ ft-lbf/lbm} \qquad \qquad \textit{Answer}$$

Now, the power is determined from

$$\dot{W}k_{os} = \dot{m}wk_{os}$$

and

$$\dot{W}k_{os} = 20 \text{ lbm/s } (-541.4 \text{ ft-lbf/lbm})$$
$$= -10829 \text{ ft-lbf/s} = -19.7 \text{ hp} \qquad \textit{Answer}$$

For the irreversible case, this power requirement must be increased; but we could not determine exactly how much by using only thermodynamics.

Example 6.3   A cylindrical bar 10 inches long and 1 inch in diameter is subjected to an axial load of 3000 lbf, as shown in figure 6–3. The bar is composed of steel having a modulus of elasticity of 30×10^6 lbf/in² where the modulus of elasticity, Y, is defined as

$$Y = \frac{\text{stress}}{\text{strain}} \qquad \qquad \textbf{(6–8)}$$

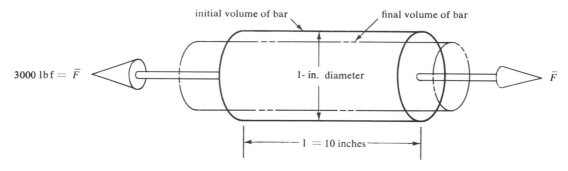

initial volume of bar   final volume of bar

$3000 \; \text{lbf} = \bar{F}$ 1- in. diameter $\bar{F}$

l = 10 inches

Figure 6–3   Strain energy application. Example 6.3.

The stress and strain are

$$\text{stress} = \frac{\text{force}}{\text{area}} = \frac{F}{A}$$

and

$$\text{strain} = \xi = \frac{\Delta l}{l}$$

where Δl is the change in length of a given length l bar. Determine the work done for this process, if the bar does not change volume when stretched and if no heat is transferred.

Solution   This problem has been included in this text to demonstrate the scope of the thermodynamic approach in analyzing a diverse set of technical problems. We have here a constant volume process for a closed system, but the system is not simple. We are assuming the material is perfectly elastic (which means it will return to its initial shape when external forces are removed) and, therefore, reversible. However, we have interactions between the constituent atoms and our first law becomes

$$\Delta E = Q - Wk_{cs}$$

We assume no heat transfer or changes in kinetic and potential energy or enthalpy, but we must account for strain energy, SE. Strain energy results from the interaction of the atoms or molecules. We define the change in strain energy by

$$\Delta\text{SE} = \int V \times \frac{F}{A} \times d\left(\frac{\Delta l}{l}\right) = \int V \times \frac{F}{A} \times d\xi \qquad \textbf{(6–9)}$$

Note that $F/A = Y\xi$ from equation (6–8). Then the strain energy change is

$$\Delta\text{SE} = \int VY\xi d\xi \qquad \textbf{(6–10)}$$

From our first law equation applied to the bar, since $Q = 0$,

$$\Delta SE = -Wk_{os} \tag{6-11}$$

Since we seek the work required to stress the bar under 3000 lbf load, we set $\xi_1 = 0$ and from the definition of the modulus of elasticity, $Y = \dfrac{F/A}{\xi}$, obtain

$$\xi_2 = \frac{3000 \text{ lbf}}{YA}$$

$$= \frac{3000 \text{ lbf}}{(\pi \times \tfrac{1}{4}) \text{ in}^2 (30 \times 10^6 \text{ lbf/in}^2)}$$
$$= 1.27 \times 10^{-4} \text{ in/in}$$
$$= 1.27 \times 10^{-4} = \xi_2$$

Then

$$Wk_{cs} = -\Delta SE = -\int_{\xi_1}^{\xi_2} VY\xi \, d\xi$$

$$= \left[-VY\left(\frac{1}{2}\xi^2\right) \right]_{\xi_1}^{\xi_2}$$

$$= -VY\left(\frac{1}{2}\right)(\xi_2^2 - \xi_1^2)$$

We now substitute numbers into this equation. First, the volume is just the volume of the cylindrical bar, i.e.

$$V = (10 \text{ in})\left(\pi \times \frac{1}{4} \text{ in}^2\right)$$

and then

$$Wk_{cs} = -\left(\frac{10\pi}{4} \text{ in}^3\right)(30 \times 10^6 \text{ lbf/in}^2)\left(\frac{1}{2}\right)[(1.27 \times 10^{-4})^2 - 0]$$

or

$$Wk_{cs} = 1.93 \text{ in-lbf} \qquad\qquad \textit{Answer}$$

This represents the energy expanded in applying 3000 lbf to the described bar and allowing equilibrium to be slowly achieved. In equilibrium the bar will have been stretched 1.27×10^{-4} inches/inch of bar length or a total of $1.27 \times 10 \times 10^{-4}$ or 1.27×10^{-3} inches.

Notice that for a modulus of elasticity, Y, which remains constant during a stress process the work can be written

$$Wk_{cs} = \frac{1}{2}VY(\xi_2^2 - \xi_1^2) \tag{6-12}$$

also. If the initial strain, ξ_1, is zero then

$$Wk_{cs} = \frac{1}{2}VY\xi_2^2 \qquad (6\text{–}13)$$

This relationship holds true for those materials having no prestress.

6.3 The Constant Temperature Process

We now treat the *constant temperature* or *isothermal* process which has many applications. This type of process will frequently involve heat transfer and work, as well as energy changes. If the system undergoing an isothermal process can be described as a perfect gas, we will see that the enthalpy and internal energy changes are zero.

Let us first consider the closed, reversible, isothermal system. Assuming the system is composed of a perfect gas we have

$$pV = mRT = \text{constant} = C \qquad (6\text{–}14)$$

or

$$p_1V_1 = p_2V_2 = C \qquad (6\text{–}15)$$

As a consequence of this, let us consider the reversible work for the closed system. Here

$$Wk_{cs} = \int p\,dV$$

and from equation (6–14)

$$Wk_{cs} = \int \frac{C\,dV}{V} \qquad (6\text{–}16)$$

We will integrate this between the limits of V_1 and V_2, i.e.

$$Wk_{cs} = C\int_{V_1}^{V_2} \frac{dV}{V}$$

The integration produces the result

$$Wk_{cs} = \left(C \ln V\right)_{V_1}^{V_2} = C(\ln V_2 - \ln V_1)$$

or

$$Wk_{cs} = C \ln \frac{V_2}{V_1} \qquad (6\text{–}17)$$

where the constant C has the value calculated from equation (6–15). Remember that equation (6–17) is correct for perfect gases only and represents the area under the curve

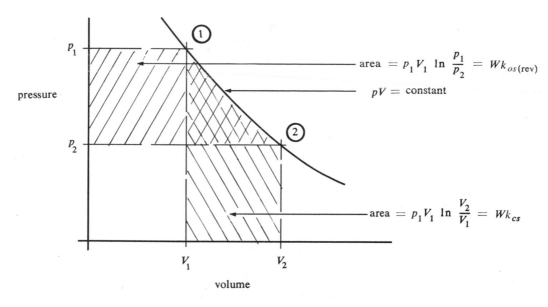

area $= p_1 V_1 \ln \dfrac{p_1}{p_2} = Wk_{os(rev)}$

$pV = \text{constant}$

area $= p_1 V_1 \ln \dfrac{V_2}{V_1} = Wk_{cs}$

Figure 6-4 p-V diagram for isothermal process with a perfect gas.

$$pV = \text{constant}$$

on a p-V diagram, as indicated in figure 6–4.

For closed systems without strain, kinetic, or potential energy, the internal energy change is zero for isothermal processes and, from the first law

$$\Delta U = 0 = Q - Wk_{cs}$$

We then have

$$Q = Wk_{cs} = C \ln \frac{V_2}{V_1}$$

In addition, by using equation (6–15) we have

$$\frac{p_1}{p_2} = \frac{V_2}{V_1}$$

so that

$$\ln \left(\frac{V_2}{V_1} \right) = \ln \left(\frac{p_1}{p_2} \right)$$

and substituting this result into equation (6–17) yields

$$Wk_{cs} = C \ln \left(\frac{p_1}{p_2} \right) \tag{6–18}$$

Let us now consider the isothermal process in an open system. Again, as for the closed system, we assume a perfect gas and obtain

$$p_1V_1 = p_2V_2 = \text{constant} = C$$

and observe that the reversible work is given by

$$Wk_{os} = -\int V dp$$

$$= -C\int_{p_1}^{p_2} \frac{dp}{p} \tag{6-19}$$

From this we get

$$Wk_{os} = -C \ln \left(\frac{p_2}{p_1}\right) \tag{6-20}$$

or, since $\ln (p_2/p_1) = -\ln (p_1/p_2)$,

$$Wk_{os} = C \ln \left(\frac{p_1}{p_2}\right) \tag{6-21}$$

which is the same result we had for the closed system with a perfect gas. We could write this last equation as

$$Wk_{os} = C \ln \left(\frac{V_2}{V_1}\right) \tag{6-17}$$

as well.

For those gases not obeying the perfect gas relation, the above results must be altered, and generally with an increase in algebraic difficulty and complication. At this point we will not consider isothermal processes for other than perfect gases.

Example 6.4

During the compression of 0.01 lbm of air in a cylinder (see figure 6–5), heat is transferred through the cylinder walls to keep the air at a constant temperature. The air pressure increases from 15 psia to 150 psia after the air is fully compressed. The initial specific volume of the air is 7.4 ft³/lbm. Determine the operating air temperature, the change in internal energy and in enthalpy, the work done, and the heat transferred during this process.

Solution

This is an isothermal process and we will assume it to be reversible as well. If the air is behaving as a perfect gas, which we assume, then the operating temperature can be found from

$$T = \frac{pV}{mR} = \frac{pv}{R}$$

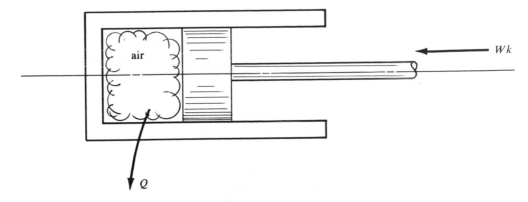

Figure 6–5   Isothermal process of a piston-cylinder device.

or, initially

$$T_1 = \frac{p_1 v_1}{R} = \frac{(15 \text{ lbf/in}^2)(7.4 \text{ ft}^3/\text{lbm})(144 \text{ in}^2/\text{ft}^2)}{53.3 \text{ ft-lbf/lbm}^\circ\text{R}}$$

so that

$$T_1 = 300^\circ\text{R}$$

and then

$$T_2 = 300^\circ\text{R}$$

The change in internal energy is

$$\Delta U = m c_v \Delta T = 0$$

and for the enthalpy change we have

$$\Delta H = m c_p \Delta T = 0$$

The work done is reversible and this we obtain from equation (6–17),

$$Wk_{cs} = C \ln \left(\frac{V_2}{V_1}\right)$$

or, more conveniently, from equation (6–18)

$$Wk_{cs} = C \ln \left(\frac{p_1}{p_2}\right)$$

The constant is determined first:

$$\begin{aligned}
C &= p_1 V_1 = p_1 m v_1 \\
&= (15 \text{ lbf/in}^2)(0.01 \text{ lbm})(7.4 \text{ ft}^3/\text{lbm})(144 \text{ in}^2/\text{ft}^2) \\
&= 159.8 \text{ ft-lbf}
\end{aligned}$$

Then

$$Wk_{cs} = (159.8 \text{ ft-lbf}) \ln \frac{15}{150}$$

$$= (159.8)\left(-\ln \frac{150}{15}\right)$$

$$= 159.8 \, (-\ln 10)$$

$$= -368 \text{ ft-lbf} \qquad\qquad Answer$$

The heat transferred is equal to the work done so that

$$Q = -368 \text{ ft-lbf} \qquad\qquad Answer$$

and Q is, as the sign indicates, removed from the system. For the irreversible isothermal process, the internal energy change can still be zero; but the work and heat increase in absolute values; that is, more work is required and more heat transfer is demanded to retain constant temperature.

6.4
The Adiabatic
Process

During the preceding sections we have considered processes where an important property is fixed or constant, that is, constant pressure, volume, or temperature. Here we will consider the condition when heat transfer is zero — called the *adiabatic process*. While no real process is completely adiabatic, there are physical conditions when it is approximated, such as a well-insulated system, or a rapidly occurring process.

The assumption of a perfect gas system has proved to be profitable before, so we will assume this condition again. Let us then write the first law for the adiabatic process with a perfect gas.

$$\Delta U = Q - Wk_{cs} \qquad\qquad (6\text{--}22)$$

but $Q = 0$ so

$$\Delta U = m\int c_v dT = -Wk_{cs} \qquad\qquad (6\text{--}23)$$

Now let us assume the process is reversible as well. Then the work term can by replaced by $\int p dV$ and we get

$$m\int c_v dT = -\int p dV \qquad\qquad (6\text{--}24)$$

For the perfect gas, $p = mRT/V$ so that

$$m\int c_v dT = -\int \frac{mRTdV}{V}$$

and dividing both sides by mT yields

$$\int c_v \frac{dT}{T} = -\int \frac{RdV}{V} \tag{6-25}$$

We now assume that c_v is a constant and recall that R can be replaced by $c_p - c_v$. Then

$$c_v \int \frac{dT}{T} = -(c_p - c_v) \int \frac{dV}{V} \qquad \frac{c_v - c_p}{c_v}$$

or

$$\int \frac{dT}{T} = -\left(\frac{c_p}{c_v} - 1\right) \int \frac{dV}{V} = (1 - k) \int \frac{dV}{V} \tag{6-26}$$

Integrating both sides of this equation gives us

$$\ln T = (1 - k) \ln V + \text{constant}$$

or, since $T = pV/mR$,

$$\ln \frac{pV}{mR} = (1 - k) \ln V + \text{constant}$$

This can be written

$$\ln p + \ln V = \ln V - k \ln V + C_0 \tag{6-27}$$

where C_0 is a new constant containing all the other constants. Now we have, from equation (6–27)

$$\ln p + k \ln V = C_0 \tag{6-28}$$

or

$$pV^k = C \tag{6-29}$$

The curve of this is plotted in figure 6–6. In this curve, C is a newer constant to account for the antilog operation between equations (6–28) and (6–29). For reversible adiabatic processes, equation (6–29) is correct if the system involved is composed of a perfect gas. Let us see what this has done for us.

The reversible work of a closed system indicated as the area under a curve described by $pV^k = $ constant in a p-V diagram is

$$Wk_{cs} = pdV$$

From equation (6–29) we write

$$p = \frac{C}{V^k}$$

and then

$$Wk_{cs} = \int_{V_1}^{V_2} \frac{CdV}{V^k} \tag{6-30}$$

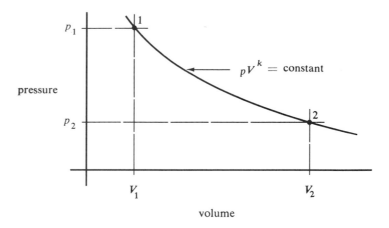

Figure 6-6   *p-V* diagram for the reversible adiabatic process with a perfect gas.

Integrating between limits of V_1 and V_2 yields

$$Wk_{cs} = C\left(\frac{1}{1-k}\right)(V_2^{1-k} - V_1^{1-k}) \qquad (6\text{--}31)$$

This can be condensed somewhat by substituting equation (6–29),

$$C = p_1V_1^k = p_2V_2^k$$

into equation (6–31)

$$Wk_{cs} = \frac{1}{1-k}(p_2V_2^kV_2^{1-k} - p_1V_1^kV_1^{1-k})$$

to obtain

$$Wk_{cs} = \frac{1}{1-k}(p_2V_2 - p_1V_1) \qquad (6\text{--}32)$$

Since this is an equation for perfect gases it could easily be written as

$$Wk_{cs} = \frac{mR}{1-k}(T_2 - T_1) \qquad (6\text{--}33)$$

It should not be surprising that, for the reversible adiabatic process, the change in internal energy and enthalpy are found from the equations

$$\Delta U = mc_v\Delta T$$

and

$$\Delta H = mc_p\Delta T$$

assuming constant specific heats for the perfect gas.

For open systems, the reversible work is gotten from the relation

$$Wk_{os} = \frac{k}{1-k}(p_2V_2 - p_1V_1) \qquad (6\text{–}34)$$

which can easily be derived in the manner of getting equation (6–32). This is left as an exercise for the student. (See problem 6.35.) Also, we can revise equation (6–34) to read

$$Wk_{os} = \frac{(mRk)}{1-k}(T_2 - T_1) \qquad (6\text{–}35)$$

by a simple algebraic manipulation, assuming we still have a perfect gas.

Example 6.5

A gas turbine receives air at 2000°K, allows reversible adiabatic expansion of the air past the turbine buckets, and exhausts this air at 2 lbm/s to the atmosphere at 1000°K. Determine the heat transferred, the work produced per lbm of air, and the power produced if there are negligible kinetic and potential energy changes for the air.

Solution

Our system is here specified as the gas turbine which is, to be brief, a device converting energy from a flowing stream (such as air) to a rotational energy of a wheel or shaft. The system is obviously an open one as indicated in figure 6–7, so we write the first law as

$$\Delta H = Q - Wk_{os}$$

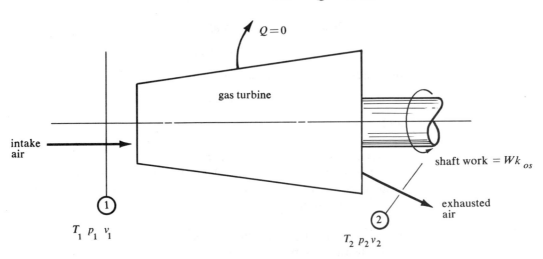

Figure 6–7   Sketch of system for adiabatic expansion of air in a gas turbine.

In terms of rates per unit time, this becomes

$$\dot{H} = \dot{Q} - \dot{W}k_{os} \qquad (6\text{-}36)$$

which can be rewritten as

$$\dot{m}\Delta h = \dot{Q} - \dot{m}wk_{os} \qquad (6\text{-}37)$$

The enthalpy change, Δh, is obtained from

$$\Delta h = c_p \Delta T$$

or

$$\Delta h = h_2 - h_1 = c_p(T_2 - T_1)$$

after assuming the air is a perfect gas with constant specific heat. From table B.4 of the appendix we find c_p to be 0.24 Btu/lbm°R or,

$$c_p = 0.24 \text{ Btu/lbm°R} \times \frac{9 \text{ °R}}{5 \text{ °K}} = 0.432 \text{ Btu/lbm °K}$$

Then

$$\Delta h = 0.432 \text{ Btu/lbm°K} \times (1000°K - 2000°K) = -432 \text{ Btu/lbm}$$

and since the process is adiabatic, $\dot{Q} = 0$. Then equation (6-37) can be written

$$\dot{m}(-432 \text{ Btu/lbm}) = -\dot{m}wk_{os} \qquad (6\text{-}38)$$

from which directly we have

$$wk_{os} = 432 \text{ Btu/lbm air} \qquad \textit{Answer}$$

The mass flow rate, $\dot{m}$, is given as 2 lbm/s so that

$$\dot{W}k_{os} = 2(432) \text{ Btu/s} = 864 \text{ Btu/s}$$

or, converting to horsepower units

$$\dot{W}k_{os} = 864 \text{ Btu/s} \times 1.41 \text{ hp-s/Btu} = 1218 \text{ hp} \qquad \textit{Answer}$$

and we have our solution.

Irreversibilities included in this process will reduce the actual output power by reducing the enthalpy change or producing strain energy of the system. In the irreversible adiabatic case then we expect deformation or wear in the system.

Before leaving the reversible adiabatic process of perfect gases let us return to the equation

$$pV^k = C \qquad (6\text{-}29)$$

and seek relations between p, V, and T. From this equation we can easily write

$$p_1 V_1^k = p_2 V_2^k$$

or

$$\frac{p_1}{p_2} = \left(\frac{V_2}{V_1}\right)^k \tag{6–39}$$

By using the perfect gas relation we have $p_1 = mRT_1/V_1$ and $p_2 = mRT_2/V_2$ which yields for equation (6–39) the result

$$\frac{mRT_1/V_1}{mRT_2/V_2} = \left(\frac{V_2}{V_1}\right)^k$$

and

$$\frac{T_1}{T_1} = \left(\frac{V_1}{V_2}\right)\left(\frac{V_2}{V_1}\right)^k$$

which reduces then to

$$\frac{T_1}{T_2} = \left(\frac{V_2}{V_1}\right)^{k-1} \tag{6–40}$$

Also we could revise this last equation to read

$$\frac{V_2}{V_1} = \left(\frac{T_1}{T_2}\right)^{1/k-1} \tag{6–41}$$

or

$$\left(\frac{V_2}{V_1}\right)^k = \left(\frac{T_1}{T_2}\right)^{k/k-1} \tag{6–42}$$

which, substituted back into equation (6–39) yields

$$\left(\frac{p_1}{p_2}\right) = \left(\frac{T_1}{T_2}\right)^{k/k-1} \tag{6–42}$$

We will need these relations in succeeding problem solutions.

6.5 The Polytropic Process

In the discussions of the processes considered in this chapter, we have always arrived at an equation relating pressure to volume — an equation which can be obtained from the general equation

$$pV^n = C \tag{6–43}$$

which is called the *polytropic equation*. For the isobaric process we set n equal to zero in equation (6–43) and find

$$pV^0 = p = c$$

Similarly, for the isometric process we revise (6–43) to read

$$p^{1/n}V = C^{1/n} = C' \text{ (a new constant)}$$

and set n equal to infinity. Then

$$p^0V = V = C'$$

as we required for this process.

For the constant temperature process of a perfect gas we set n equal to unity (1) and obtain from the polytropic equation

$$pV^1 = pV = C$$

For the reversible adiabatic process we arrived at equation (6–29) which is identical to the polytropic equation but with $k = n$.

The most general process then we call the *polytropic process*, which is any process of a system which can be described by equation (6–43), where n can take any value desired. We therefore consider n an empirical constant (it could also be a very complicated variable if desired!) which fits equation (6–43) to actual data.

Table 6–1 summarizes the previous discussion, and in figure 6–8 are plotted the curves characterizing the five common processes considered here.

Table 6–1
The Polytropic Exponent

Processes	Value of n in polytropic equation $pV^n = C$
Isobaric	0
Isometric	∞
Isothermal	1
Reversible Adiabatic	k
Polytropic	n

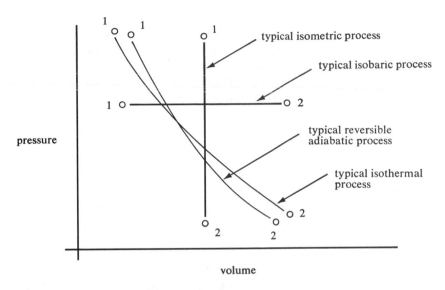

Figure 6–8   The various polytropic processes, shown on the *p-V* plane, for perfect gases.

For the closed system, the reversible work of a polytropic process can readily be shown to be

$$Wk_{cs} = \frac{1}{1-n}(p_2V_2 - p_1V_1) \qquad (6\text{--}44)$$

by a derivation much like that for the work of a reversible adiabatic process. The open system reversible work, likewise can be obtained from the equation

$$Wk_{os} = \frac{n}{1-n}(p_2V_2 - p_1V_1) \qquad (6\text{--}45)$$

which is closely related to the corresponding reversible adiabatic case. Similarly, the relation between pressure, volume, and temperature at various states along a polytropic process are given by

$$\frac{p_1}{p_2} = \left(\frac{V_2}{V_1}\right)^n = \left(\frac{v_2}{v_1}\right)^n \qquad (6\text{--}46)$$

$$\frac{T_1}{T_2} = \left(\frac{V_2}{V_1}\right)^{n-1} = \left(\frac{v_2}{v_1}\right)^{n-1} \qquad (6\text{--}47)$$

and

$$\frac{p_1}{p_2} = \left(\frac{T_1}{T_2}\right)^{n/n-1} \qquad (6\text{--}48)$$

Example 6.6   For the gas turbine of example 6.5, assume the expansion of air is reversible and polytropic, with n having a value of 1.5. Assume the inlet air temperature is 2000°K, the pressure ratio is the same as the reversible adiabatic process, and the mass flow of air is 2 lbm/s. Under these conditions determine the work produced per pound-mass of air, the power produced, and the heat transfer rate.

Solution   First, we observe that the system is the gas turbine and is an open system. The work per pound-mass, reversible, is

$$wk_{os} = \frac{n}{1-n}(p_2 v_2 - p_1 v_1) \tag{6-49}$$

and assuming a perfect gas, this becomes

$$wk_{os} = \frac{nR}{1-n}(T_2 - T_1) \tag{6-50}$$

Now we must determine the exhaust temperature T_2 in order to evaluate the work. Since the pressure ratios are the same for the reversible adiabatic process of example 6.5 and this example, we will determine this ratio from equation (6–42)

$$\frac{p_1}{p_2} = \left(\frac{T_1}{T_2}\right)^{k/k-1}$$

which yields

$$\frac{p_1}{p_2} = \left(\frac{2000°K}{1000°K}\right)^{1.4/0.4} = 2^{3.5} = 11.3$$

Here we used a value of 1000°K for T_2 from example 6.5 to obtain the pressure ratio. We will find the actual temperature T_2 a little later. The pressure ratio is normally defined as the final (exhaust) over the initial (inlet) pressure. The pressure ratio for our polytropic process then is

$$\frac{p_2}{p_1} = \frac{1}{11.3} = 0.0884$$

and, from equation (6–48)

$$\frac{p_2}{p_1} = \left(\frac{T_2}{T_1}\right)^{n/n-1}$$

for the polytropic process. We solve this for T_2,

$$T_2 = T_1(p_2/p_1)^{(n-1)/n}$$

Substituting values into this relationship then gives us

$$T_2 = 2000°K \ (0.078)^{0.5/1.5} = 891°K$$

and we can now calculate the work done.

$$wk_{os} = \frac{1.5}{1 - 1.5} \, 53.3 \text{ ft-lbf/lbm}^\circ\text{R} \times \left(\frac{9^\circ\text{R}}{5^\circ\text{R}}\right)(891^\circ\text{K} - 2000^\circ\text{K})$$
$$= 319,200 \text{ ft-lbf/lbm}$$

The power produced is $\dot{m}wk_{os}$ so that

$$\dot{m}wk_{os} = \dot{W}k_{os} = 2 \text{ lbm/s} \times 319,200 \text{ ft-lbf/lbm} = 638,400 \text{ ft-lbf/s}$$
$$= \frac{1 \text{ hp}}{550 \text{ ft-lbf/s}} \times 638,400 \text{ ft-lbf/s}$$

and then

$$\dot{W}k_{os} = 1161 \text{ hp} \qquad\qquad Answer$$

The enthalpy change, assuming air retains a constant specific heat cp of 0.24 Btu/lbm°R over the temperature variation, is

$$\Delta h = c_p \Delta T = 0.24 \text{ Btu/lbm}^\circ\text{R} \times \frac{9^\circ\text{R}}{5^\circ\text{K}} \times (891^\circ\text{K} - 2000^\circ\text{K})$$
$$= -479 \text{ Btu/lbm}$$

The first law is now written as

$$\dot{m}\Delta h = \dot{Q} - \dot{W}k_{os}$$

since we are neglecting kinetic and potential energy changes. From this equation we get

$$\dot{Q} = \dot{m}\Delta h + \dot{W}k_{os}$$
$$= 2 \text{ lbm/s} \times (-479 \text{ Btu/lbm}) + 1161 \text{ hp}$$
$$= 1.4 \frac{\text{hp}}{\text{Btu/s}}(-958 \text{ Btu/s}) + 1161 \text{ hp}$$
$$= -180.2 \text{ hp} = -252.3 \text{ Btu/s} \qquad\qquad Answer$$

In order for this process to continue in steady state or steady flow, there must be transferred from the turbine 71.4 Btu/s of heat. If this heat is not removed, the predicted 358 hp in power cannot be extracted without altering the air flow or pressure and temperature ratios — namely increasing the temperature inside the turbine.

6.6
Summary

We have considered the most common processes encountered in thermodynamics or energy transfer problems. The numerous equations accumulated during these last two chapters are listed in tables 6–2 and 6–3 to provide ready references. Note that table 6–2 lists the general

Table 6–2
General Process
(Excluding Processes Involving Phase Change)

Term	Reversible Isobaric	Reversible Isometric	Reversible Isothermal	Reversible Adiabatic	Reversible Polytropic
p-V relation	$p = c$	$V = c$	$T = c$	$s = c$	$pV^n = c$
$*Wk_{cs}$	$p(V_2 - V_1)$	0	$\int pdV$	$\int pdV$	$\frac{1}{1-n}(p_2V_2 - p_1V_1)$
$*Wk_{os}$	0	$V(p_1 - p_2)$	$-\int Vdp$	$-\int Vdp$	$\frac{n}{1-n}(p_2V_2 - p_1V_1)$
Q_{os}	$m\Delta h$	$m\Delta u$	$-\int Vdp$	0	$m\Delta h + Wk_{os}$
Q_{cs}	$m\Delta h$	$m\Delta u$	$\int pdV$	0	$m\Delta U + Wk_{cs}$
Δh	$\int c_p dT$	$\int c_p dT$	0	$\int c_p dT$	$\int c_p dT$
Δu	$\int c_v dT$	$\int c_v dT$	0	$\int c_v dT$	$\int c_v dT$
Δs	$\int \frac{dh}{T}$	$\int \frac{du}{T}$	$\frac{q}{T}$	0	$\int \frac{dq}{T}$

*Assuming there are no kinetic, potential, or strain energy changes.

equations which are applicable to any material but which are frequently too cumbersome; table 6–3 lists the specific equations pertinent to perfect gases only. In table 6–3 are listed only the reversible cases and in addition a new property, *entropy*, is tabulated as a change in its value, Δs. We will speak more of entropy in the next chapter but its formulas are presented here for completeness.

Practice Problems

Problems designated with an asterisk * preceding the number should only be attempted by those having a background which includes the knowledge of integral calculus.

Section 6.1

6.1. Determine the work of a process when the volume changes from 3 ft^3 to 5 ft^3. The pressure remains constant with a value of 70 psia.

6.2. A piston-cylinder contains 0.003 lbm of air at 18 psia and 100°F. The piston is then compressed at constant pressure so that the volume of air

Table 6–3

Perfect Gas Process

Term	Reversible Isobaric	Reversible Isometric	Reversible Isothermal	Reversible Adiabatic	Reversible Polytropic
p-v relation	$p = c$	$V = c$	$T = c$	$pV^k = c$	$pV^n = c$
p-v-T relations	$\dfrac{V_1}{V_2} = \dfrac{T_1}{T_2}$	$\dfrac{p_1}{p_2} = \dfrac{T_1}{T_2}$	$\dfrac{p_1}{p_2} = \dfrac{V_2}{V_1}$	$\dfrac{p_1}{p_2} = \left(\dfrac{V_2}{V_1}\right)^k$ $\dfrac{p_1}{p_2} = \left(\dfrac{T_1}{T_2}\right)^{k/k-1}$	$\dfrac{p_1}{p_2} = \left(\dfrac{V_2}{V_1}\right)^n$ $\dfrac{p_1}{p_2} = \left(\dfrac{T_1}{T_2}\right)^{n/n-1}$
*Wk_{cs}	$p(V_2 - V_1)$	0	$p_1 V_1 \ln\left(\dfrac{V_2}{V_1}\right)$	$\dfrac{1}{1 - k}(p_2 V_2 - p_1 V_1)$	$\dfrac{1}{1 - n}(p_2 V_2 - p_1 V_1)$
*Wk_{os}	0	$V(p_1 - p_2)$	$p_1 V_1 \ln\left(\dfrac{V_2}{V_1}\right)$	$\dfrac{k}{1 - k}(p_2 V_2 - p_1 V_1)$	$\dfrac{n}{1 - n}(p_2 V_2 - p_1 V_1)$
Q_{os}	$m\Delta h$	$m\Delta u$	$p_1 V_1 \ln (V_2/V_1)$	0	$m\Delta h + Wk_{os}$
Q_{cs}	$m\Delta h$	$m\Delta u$	$p_1 V_1 \ln (V_2/V_1)$	0	$m\Delta u + Wk_{cs}$
Δh	$c_p\Delta T$	$c_p\Delta T$	0	$c_p\Delta T$	$c_p\Delta T$
Δu	$c_v\Delta T$	$c_v\Delta T$	0	$c_v\Delta T$	$c_v\Delta T$
Δs	$c_p \ln\left(\dfrac{T_2}{T_1}\right)$	$c_v \ln\left(\dfrac{T_2}{T_1}\right)$	$R \ln (V_2/V_1)$	0	$c_v \ln (T_2/T_1)$ $+ R \ln (V_2/V_1)$

*Assuming there are no kinetic, potential, or strain energy changes.

is halved. Determine the final temperature, final volume, and work required for this process.

6.3. A water pump transfers an incompressible liquid without kinetic or potential energy changes. If the pressure also remains constant, what work is required to drive the pump? Assume the process to be reversible.

6.4. Air contained in a frictionless piston-cylinder expands from 1 in³ to 10 in³. The mass of air is 0.02 lbm at a temperature of 140°F, while the air has a volume of 1 in³. If the pressure remains constant during the process, calculate the final air temperature and the work.

6.5. Ammonia gas flows through a chamber at constant pressure. The specific volume of the entering ammonia is 8 ft³/lbm and 9.5 ft³/lbm when leaving. Assuming no kinetic or potential energy changes, determine the temperature of the ammonia leaving if the initial temperature is 200°F.

Section 6.2

6.6. A rigid container is filled with a perfect gas having a gas constant of 35 ft-lbf/lbm°R. The volume is 3 ft³, the gas pressure is 20 psia, and the temperature is 98°F. If the gas is heated to 400°F determine the final pressure and the work.

6.7. For the container and constants described in problem 6.6, assuming the specific heat at constant volume is 0.12 Btu/lbm°R, determine the reversible work and irreversible work if the temperature of the gas is increased from 98°F to 400°F by a paddle stirring the gas instead of by heat transfer.

6.8. Methyl alcohol having a density of 0.80 g/cm³ is pumped with no kinetic or potential energy changes from a tank at 10^6 dyne/cm² to a pressurized container at 2×10^6 dyne/cm². Determine the work required for a reversible process.

6.9. Kerosene having a density of 51.0 lbm/ft³ is pumped from a tank at 14.7 psia to a reservoir 30 feet above the tank where the pressure is 20 psia. If 8 lbm/s are transferred, determine the power required by the pump if we assume complete reversibility.

6.10. A round rod 50 cm long and 5 cm in diameter is subjected to a load of 350,000 dynes in an axial direction, as shown in figure 6–9. If the rod is made of aluminum having a modulus of elasticity of 12×10^{12} dyne/cm², determine the work or energy absorbed by the rod.

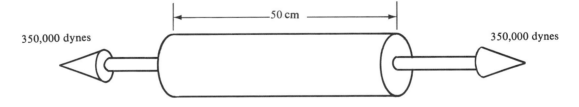

350,000 dynes

50 cm

350,000 dynes

Figure 6–9   **Problem 6.10.**

6.11. A block of wood is subjected to a uniform load of 800 lbf, as shown in figure 6–10. If the modulus of elasticity of the wood is 6×10^5 lbf/in², determine the energy absorbed by the wood for this loading.

Section 6.3

6.12. Air contained in a frictionless piston-cylinder is subjected to an isothermal process where the pressure increases from 15 psia to 50 psia. Determine the work per unit mass of air for this process if the air temperature is 90°F.

6.13. One lbm of air is compressed in a piston-cylinder from 15 psia to 100 psia. If the air temperature remains constant at 85°F, determine the work required.

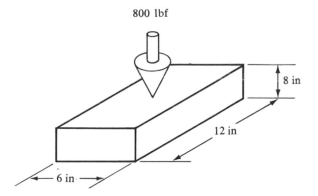

800 lbf

8 in

12 in

6 in

Figure 6–10   Problem 6.11.

6.14. Determine the heat transfer during an isothermal expansion of 2 lbm of air from 15 ft³ to 30 ft³ if the temperature is 110°F and the process is carried out in a closed container.

6.15. Two ft³ of air are expanded in a piston-cylinder from 600 psia and 3400°F to a final volume of 6 ft³. Assuming the process is isothermal, determine the final pressure, the work done, the heat transferred, and the internal energy change. Also sketch the p-V diagram of this process.

6.16. Ammonia gas is compressed in a piston-cylinder at a constant temperature of 130°C from a pressure of 1 bar ($= 10^6$ dyne/cm²) to 3.5 bars. Determine the work required, the heat transferred, and the change in internal energy. Also plot the p-V diagram of this process.

6.17. The following data are obtained from the pump shown in figure 6–11:

$T_1 = T_2 = 300°F$

$p_1 = 20$ psia

$p_2 = 36$ psia

$\dot{m} = 1$ lbm/s of air

Determine the power required by the pump, the heat transfer rate, and sketch the p-v diagram.

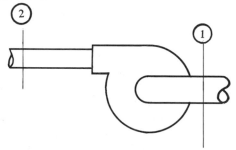

2

1

Figure 6–11   Problem 6.17.

6.18. A fan draws 2 lbm/s of air through a wind tunnel shown in figure 6–12. The following data are given:

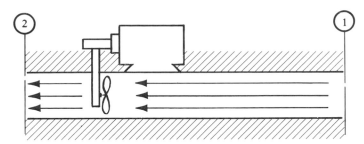

Figure 6–12   Problem 6.18.

$$T_1 = T_2 = 85°F$$
$$p_1 = 14.7 \text{ psia}$$
$$p_2 = 15.2 \text{ psia}$$
$$\overline{V}_1 = V_2$$

Determine the fan motor power requirement and the heat transfer through the wind tunnel wall.

Section 6.4

6.19. During a reversible adiabatic process of 1 lbm of an ideal gas in a piston-cylinder, the pressure decreases from 183 psig to 2 psig. The atmosphere pressure is 13.8 psia, the gas constant is 37.7 ft-lbf/lbm°R, and the initial temperature of the gas is 863°F. Determine the initial and final gas density, the heat transferred, the work produced, and the change in internal energy. Assume the specific heat at constant pressure is 0.22 Btu/lbm°R.

6.20. In a piston-cylinder device, 0.03 lbm of air is compressed reversibly and adiabatically. The air is initially at 78°F and 16 psia; and after being compressed, the temperature is 878°R. Determine the final pressure, heat transferred, increase in internal energy, and work required.

6.21. Prove the identity

$$c_v = \frac{-R}{1 - k}$$

6.22. Prove the identity

$$c_p = \frac{kR}{k - 1}$$

6.23. In a gas turbine 3000 lbm/min of air are expanded in a reversible adiabatic manner. If we neglect kinetic and potential energy changes of the air flow, and if we find the entering air to have a temperature of 2100°R and a pressure of 230 psia, determine the power developed by the turbine. Assume the exhaust air pressure is 15 psia.

6.24. A rotary compressor handles 400 grams/s of air, increasing the pressure, reversibly and adiabatically, from 1 bar to 22 bars. If the air temperature initially is 10°C and $k = 1.4$, determine the power required by the compressor. Also plot the *p-v* diagram and indicate the work per unit mass on this diagram.

6.25. One lbm/s of oxygen gas is expanded through a nozzle reversibly and adiabatically. The dimensions of the nozzle are shown in figure 6–13 and the temperature and pressure of the air are 200°F, and 89 psia entering the nozzle. Calculate

   (a)   Final temperature.
   (b)   Final pressure.
   (c)   Heat transfer rate.
   (d)   Final density of air.

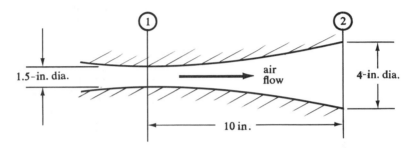

Figure 6–13   **Problem 6.25.**

6.26. Five lbm/s of helium at 540°R are expanded through an adiabatic nozzle and turbine so that the pressure is dropped to one-half. Assuming the kinetic energy decreases 30 Btu/s during this process, determine the power produced if the process is irreversible and the following process equation holds:

$$pv^{1.4} = C$$

Also determine the change in internal energy per unit time.

Section 6.5

6.27. Two lbm of CO_2 are expanded reversibly from 90°F and 32 psia to 16 psia. If the expansion is polytropic with $n = 1.27$, determine the work, heat transferred, and energy change. Assume a closed system with no kinetic or potential energy changes.

6.28. One lbm of air at 82°F and 14.7 psia is compressed reversibly and polytropically ($n = 1.25$) to 6.5 atm in a piston-cylinder. Determine the work required.

6.29. A gas at 160 psia and 300°F expands to 28 psia in a closed container. If the final temperature is 125°F, determine the value of n for this expansion, assuming it to be reversible and polytropic.

6.30. Ammonia is compressed polytropically in a piston-cylinder from 15 psia and 3 ft³ to 150 psia and 0.4 ft³. Determine the value for n and the heat transferred.

6.31. A compressor handles 3500 lbm/min of air in a steady flow polytropic process ($n = 1.36$). The entering air is at 11.5 psia, 85°F, and has negligible velocity, while the compressed air is at 115 psia and has a velocity of 30 ft/s. Determine the power required of the compressor, the heat transfer rate, and the rate of enthalpy change of the air.

6.32. Three lbm/s of air are reversibly and polytropically ($n = 1.48$) expanded in a gas turbine from 2100°K to 900°K. If the exhaust pressure is 15 psia, determine the work produced per lbm of air, the power generated, and the rate of heat transfer.

6.33. Seventy g/s of helium flows through a gas turbine, expanding from 15×10^6 dyne/cm^2 to 1×10^6 dyne/cm^2 in a reversible polytropic manner. If the inlet air temperature is 2400°K and $n = 1.5$, determine the power generated, the rate of heat transfer, and the rate of enthalpy change. Assume that kinetic and potential energy changes are negligible.

*6.34. Derive an expression for Wk_{cs} in terms of n, pressure, and volume for a reversible polytropic process.

*6.35. Derive equation (6–34) for a reversible adiabatic process of an open system, neglecting kinetic and potential energy changes.

*6.36. Derive an expression for Wk_{os} (in terms of n, pressure, and volume for a reversible polytropic process) if kinetic and potential energy changes are neglected.

7

Entropy, Heat, and Temperature

One of the most important properties, *entropy*, is presented in this chapter. It is introduced as the property that relates temperature and heat — not like the caloric equations of state, which related temperature and energy — but like volume, which related pressure and work. Analogies are presented to give the student a broad concept for this very important property. Emphasis is placed, however, on presenting tools for evaluating entropy and its changes during processes.

The *third law of thermodynamics* is shown to be a limiting condition for the value of entropy, a law which probably has more qualitative value than quantitative.

**7.1
Entropy**

In chapter 3 we arrived at equations for calculating heat transfer rates. These equations were concerned primarily with conditions at the boundary of a system, that is, the system's outer surface temperature and the heat transfer conductivity of that boundary. In this section let us see if a work-heat analogy is profitable. For the reversible work of a closed system we have written

$$Wk_{cs} = \int pdV$$

which looks like an intensive, surface property (pressure) mechanically motivating a change in an extensive, bulk property (volume). In a like manner, temperature induces heat transfer and changes some extensive system property, i.e.

$$\text{reversible } Q = Q_{rev} = \int TdS \qquad (7\text{--}1)$$

where we quite arbitrarily call entropy S. It is defined from a rearrangement of the equation,

$$dQ_{rev} = TdS$$

and then

$$dS = \frac{dQ_{rev}}{T}$$

or

$$\Delta S = \int \frac{dQ_{rev}}{T} \qquad (7\text{--}2)$$

Entropy is an extensive property, dependent on the mass of the system, and is described by the units of energy per unit temperature. It can, however, be converted into an intensive property — specific entropy — by the operation

$$\frac{S}{m} = s = \text{specific entropy} \qquad (7\text{--}3)$$

where s has units of energy per unit temperature per unit mass. Notice from equation (7–2) that as heat is transferred into the system, entropy will increase and as heat is transferred out of the system, it will decrease.

Entropy is a measure of unavailable energy in a system; the greater the entropy of a system, the less available is that system's energy for doing work or transferring heat. In mechanical systems we have seen that as a system increases in volume, it can perform work, but thereby also has a reduced capability to perform other work. It seeks an equilibrium in pressure with its surroundings and when it expands in volume so that the pressures are equal, the system cannot perform additional work. Thus, a system at high pressure (with respect to its surroundings) has a lower entropy than does the same system upon equilibrating its pressure with the surroundings. Also, if we have a system at a high temperature, it will tend to reach a temperature equal to the surroundings. Here the volume may not change, but entropy will. The high temperature system can perform work or heat processes and has a low entropy; but after reducing its temperature, it does not have the same capability — it has greater entropy.

In more general terms, entropy is a measure of disorder or randomness. Suppose we have ten red marbles and ten green marbles in separate boxes. All the red are in one box and all the green in another—we thus have an ordered system of two boxes. We place the boxes on a high shelf so that we can withdraw the marbles, but cannot look in the boxes. Thus, if we wish to get a red marble, we reach up and remove a marble from one of the boxes. If it is green, then we are assured of a red marble by selecting a second from the other box. Here we have a system that is ordered and has a low entropy. But we now mix the red and green marbles in one box and return the box to the shelf. The system is well mixed, disordered, and has a high entropy. If we want to get a red marble, we are never assured that we will pick one. Perhaps, if we are lucky on the first try, we will get a red marble, but we could also get ten green ones before we would get one red one. In a less abstract manner, suppose we have a system composed of pure hydrogen gas (H_2) and pure oxygen gas (O_2), each in a separate container. The system has a low entropy, it is well ordered or well arranged, and can easily be mixed to produce water. Also, we could get work or electrical energy from it if we wished. However, if we mix the hydrogen and oxygen and produce water (H_2O), we immediately have a higher entropy. The system is less ordered or arranged (we no longer know exactly where the oxygen atoms are) and it is only water. We cannot get work or electrical energy from it anymore. We can thus say that a low entropy implies a broader capability for energy systems, and if we perform reversible processes only, the world will retain this low entropy. The world, as we are reminded by environmentalists, is increasing its entropy, and the choices of power supplies are indeed not as many today as some persons would have us believe.

The property entropy gives us the tools (theoretically, at least) to calculate heat transfer by measuring only the system's properties. Previously, in the heat transfer relationships of conduction, convection, and radiation, we needed to know the properties of the system *and* the surroundings in order to calculate heat transfer. By plotting entropy versus temperature on a diagram, called a *T-S diagram*, we can then represent heat as the area under a curve, as shown in figure 7–1, which is reminiscent of work in an area in a *p-V* diagram. Of course, the heat determined is the *reversible* heat; but under any circumstance, the entropy change will *be at least as great as that determined from* equation (7–2). We can have an

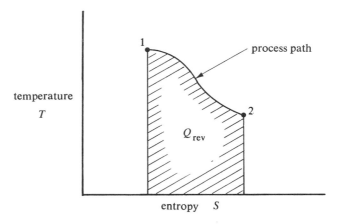

Figure 7–1   Characteristics of a process on a *T-S* plane.

irreversible entropy change, ΔS_{irr}, and for any process

$$\Delta S_{actual} = \Delta S_{irr} + \int \frac{dQ_{rev}}{T} \tag{7-4}$$

For a reversible process, ΔS_{irr} is zero and the irreversible entropy change for a given process can only, at the present state of science, be determined empirically; that is, measure it or find the entropy values from tables. The appendix tables contain listings of specific entropy, values which, in general, were calculated using statistical thermodynamic concepts and computer techniques which we will not discuss.

In the following sections, we will introduce ways of calculating entropy changes, with particular emphasis on perfect gases.

**7.2
Isothermal
Entropy
Change**

For those reversible processes where temperature is fixed (isothermal), we can readily calculate changes in entropy by

$$\Delta S = \int \frac{dQ_{rev}}{T} = \frac{1}{T} \int dQ_{rev} = \frac{Q_{rev}}{T} \tag{7-5}$$

The heat transfer is then

$$Q_{rev} = T\Delta S \tag{7-6}$$

as long as we treat reversible cases. While most processes are not isothermal, equations (7–5) and (7–6) can be used in some of these other processes. If the heat transfer is carried on with constant temperature surroundings, for instance, the properties T and S of the surroundings will suffice to calculate Q_{rev}, even if the system is exhibiting a temperature change. We see this in the following problem.

Example 7.1

In figure 7–2, a system has a loss of 300 Btu of energy in the form of reversible heat transfer to the surroundings. If the temperature of the surroundings is 40°F and constant, determine the entropy change of the system.

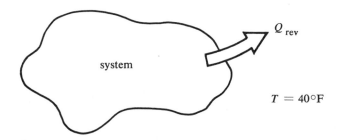

Figure 7–2

Example 7.1

Solution

The entropy change of the surroundings can easily be determined. It is $\Delta S_{surr} = Q/T$ and T is 40°F + 460°F or 500°R. The entropy change is therefore

$$\Delta S_{surr} = 300 \text{ Btu}/500°R = 0.6 \text{ Btu}/°R$$

For a reversible process it can be shown that the sum of all entropy changes resulting from the process is zero. That is,

$$\Sigma \Delta S = 0 = \Delta S_{surr} + \Delta S_{system} \tag{7–7}$$

and then

$$\Delta S_{system} = -\Delta S_{surr} = -0.6 \text{ Btu}/°R \qquad \textit{Answer}$$

The entropy has, therefore, decreased in the system and increased in the surroundings. Equation (7–7) was utilized to arrive at this result, but notice that this result is true for reversible processes only. For irreversible processes, the entropy must increase or remain constant, i.e.

$$\Sigma \Delta S \geq 0 \tag{7–8}$$

or

$$\Delta S_{surr} + \Delta S_{sys} \geq 0 \tag{7–9}$$

and

$$\Delta S_{sys} \geq -\Delta S_{surr}$$

Then for an irreversible process for example 7.1, equation (7–9) demands that

$$\Delta S_{\text{sys}} \geq -0.6 \text{ Btu/}°\text{R}$$

We have previously inferred that the relation $\int dQ/T$ is equal to or less than the actual ΔS. We must, therefore, never assume that entropy is conserved; it is conserved for reversible processes only. It is continually being increased, and rather effectively, by our technological mills in irreversible processes.

For a perfect gas, we can obtain another relation for calculating entropy change during an isothermal process. We first note that

$$Q = Wk_{cs} = mC \ln\left(\frac{V_2}{V_1}\right)$$

for the isothermal process. The constant C is equal to p_2V_2 or any other value of pV during the process. Then from equation (7–5), we obtain

$$\Delta S = \left(\frac{mp_2V_2}{T_2}\right) \ln\left(\frac{V_2}{V_1}\right) = mR \ln\left(\frac{V_2}{V_1}\right) \qquad (7\text{–}10)$$

or for specific entropy changes

$$\Delta s = \Delta\frac{S}{m} = R \ln\left(\frac{V_2}{V_1}\right) \qquad (7\text{–}11)$$

Example 7.2

During a reversible isothermal compression of 0.5 lbm of air, the pressure ratio is 10. Determine the entropy changes and the specific entropy changes for this process.

Solution

Since the pressure ratio is 10, we know that

$$\frac{p_2}{p_1} = 10 = \frac{V_1}{V_2}$$

Then the entropy change is calculated from equation (7–10)

$$\Delta S = (0.5 \text{ lbm})(53.3 \text{ ft-lbf/lbm}°\text{R}) \ln\left(\frac{1}{10}\right)$$

$$= 61.36 \text{ ft-lbf/}°\text{R} = 0.0787 \text{ Btu/}°\text{R} \qquad \textit{Answer}$$

The specific entropy change is obtained from equation (7–11)

$$\Delta s = \Delta\frac{S}{m}$$

$$= \frac{(0.0787 \text{ Btu/}°\text{R})}{(0.5 \text{ lbm})}$$

$$= 0.1577 \text{ Btu/lbm}°\text{R} \qquad \textit{Answer}$$

7.3 Adiabatic Entropy Change

For the adiabatic process, no heat is transferred and consequently there is no change in entropy due to reversible heating. The only entropy change that can occur must be due to irreversible internal or external effects. Then

$$\Delta S = \Delta S_{irrev} \qquad \text{(7–12)}$$

and entropy changes can only be determined by knowing the initial state and final state. From this information, the entropy values can then be extracted from thermodynamic tables. For reversible adiabatic processes, of course, equation (7–12) reduces to

$$\Delta S = 0 \qquad \text{(7–13)}$$

and since this implies a constant value for entropy, the reversible adiabatic process is frequently called the *isentropic process*. The reversible adiabatic process, however, is a more restricted process than the isentropic one. We can allow an irreversible process to progress, yet still retain a constant entropy by proper heat transfers; consequently we will generally use the term *reversible adiabatic* instead of *isentropic*.

Example 7.3

Steam at 200 psia and 1200°F is expanded in a reversible adiabatic manner through a steam turbine until the pressure is 10.0 psia. Determine the final steam temperature.

Solution

The entropy change for this process is zero so that

$$s_2 = s_1$$

and from table B.2, we find that steam at 200 psia and 1200°F has a specific entropy s of 1.9109 Btu/lbm°R. Therefore, the final specific entropy has this same value, and at 10.0 psia the temperature must be between 300°F and 400°F as indicated in table B.2. We will interpolate to determine the precise value:

$$\Delta s = 1.8593 \text{ at } 300°F$$
$$= 1.9173 \text{ at } 400°F$$

then

$$\Delta S = 0.0580 \text{ with } \Delta T = 100°F$$

and

$$s_2 = 1.9109$$

so,

$$\frac{1.9109 - 1.8593}{0.0580} = \frac{T_2 - 300°F}{100°F}$$

$$\frac{0.0516}{0.0580} \times 100 + 300 = T_2$$

yielding

$$T_2 = 388.9°F \qquad\qquad Answer$$

**7.4
Polytropic
Entropy
Change**

With the introduction of the entropy function, the first law can now be written in the following manner:

$$dU = TdS - pdV \qquad\qquad (7\text{--}14)$$

or,

$$du = Tds - pdv \qquad\qquad (7\text{--}15)$$

In addition, if we replace pdv by the sum $dpv - vdp$ we have, for equation (7–15) that

$$du = Tds - dpv + vdp$$

or,

$$du + dpv = Tds + vdp$$

But, $du + dpv$ equals dh so that

$$dh = Tds + vdp \qquad\qquad (7\text{--}16)$$

What we are seeking is a general relation for the property entropy involving the more common properties of pressure, volume, and temperature. From equation (7–15) we find

$$ds = \frac{du}{T} + \frac{pdv}{T}$$

or, integrating both sides

$$\Delta s = \int \frac{du}{T} + \int \frac{pdv}{T} \qquad\qquad (7\text{--}17)$$

From equation (7–16) we obtain, similarly, that

$$\Delta s = \int \frac{dh}{T} - \int \frac{vdp}{T} \qquad\qquad (7\text{--}18)$$

These last two equations are general; they can be used for open or closed systems and for any material. If we assume a perfect gas with constant specific heats, then equation (7–17) can be reduced to

$$\Delta s = c_v \int_{T_1}^{T_2} \frac{dT}{T} + R \int_{v_1}^{v_2} \frac{dv}{v} \qquad\qquad (7\text{--}19)$$

or

$$\Delta s = c_v \ln \left(\frac{T_2}{T_1}\right) + R \ln \left(\frac{v_2}{v_1}\right)$$

and equation (7–18) can become

$$\Delta s = c_p \ln \left(\frac{T_2}{T_1}\right) - R \ln \left(\frac{p_2}{p_1}\right) \qquad (7\text{–}20)$$

These two equations then are correct for determining the change in enthalpy of any polytropic process involving a perfect gas.

The isobaric (constant pressure) process implies $p_2 = p_1$ and then in equation (7–20)

$$R \ln \left(\frac{p_2}{p_1}\right) = R \ln 1 = 0$$

so that,

$$\Delta s = c_p \ln \left(\frac{T_2}{T_1}\right) \qquad (7\text{–}21)$$

For the isometric process, v_2 is equal to v_1, reducing equation (7–19) to

$$\Delta s = c_v \ln \left(\frac{T_2}{T_1}\right) \qquad (7\text{–}22)$$

A process involving constant temperature perfect gases implies that $\ln T_2/T_1$ is zero so that

$$\Delta s = R \ln \left(\frac{v_2}{v_1}\right) \qquad (7\text{–}23)$$

from equation (7–19) or

$$\Delta s = -R \ln \left(\frac{p_2}{p_1}\right) = R \ln \left(\frac{p_1}{p_2}\right) \qquad (7\text{–}24)$$

from equation (7–20). For the isothermal process we found $p_1/p_2 = v_2/v_1$ so equations (7–23) and (7–24) are identical. We could also use the equation $\Delta S = Q/T$ as we did in section 7.2. If we note that $v_2 = V_2/m$ and $v_1 = V_1/m$ so that

$$\frac{v_2}{v_1} = \frac{mV_2}{mV_1} = \frac{V_2}{V_1}$$

then equation (7–11) is identical to equation (7–23), as it should be.

Example 7.4   At constant pressure 0.8 lbm of carbon dioxide is expanded from 10 ft³ to 30 ft³. Determine the entropy change and the specific entropy change.

Solution

The process is an isobaric one and the entropy can therefore be determined from equation (7–23) in the form

$$\Delta S = m\Delta s = mR \ln \frac{v_2}{v_1} \qquad (7\text{–}25)$$

The specific volumes are

$$v_2 = \frac{V_2}{m} = \frac{30 \text{ ft}^3}{0.8 \text{ lbm}}$$

$$v_1 = \frac{V_1}{m} = \frac{10 \text{ ft}^3}{0.8 \text{ lbm}}$$

and the gas constant is 35.12 ft-lb/lbm°R from table B.4.
Then

$$\Delta S = (0.8 \text{ lbm})(35.12 \text{ ft-lb/lbm°R}) \ln \left(\frac{30 \text{ ft}^3/0.8 \text{ lbm}}{10 \text{ ft}^3/0.8 \text{ lbm}}\right)$$

$$= 28.096 \text{ ft-lbf/°R} \ln 3 = 30.95 \text{ ft-lbf/°R}$$

$$= +0.0498 \text{ Btu/°R} \qquad\qquad\qquad Answer$$

Also,

$$\Delta s = \frac{\Delta S}{m} = \frac{0.0498}{0.8} = 0.06225 \text{ Btu/lbm°R} \qquad Answer$$

Example 7.5

Air expands polytropically through a nozzle such that the exponent n is to be 1.45. The exhaust pressure of the air is 15 psia and the temperature is 200°F. If the inlet pressure is 60 psig, determine the change in specific entropy of the air as it passes through the nozzle.

Solution

For any polytropic process we can use equation (7–20),

$$\Delta s = c_p \ln \frac{T_2}{T_1} - R \ln \frac{p_2}{p_1}$$

and from table B.4 find that

$$c_p = 0.24 \text{ Btu/lbm°R}$$

and

$$R = 53.3 \text{ ft-lbf/lbm°R} = 0.0686 \text{ Btu/lbm°R}$$

Also, the pressure ratio p_2/p_1 is 15 psia/(60 + 15) psia or $\frac{1}{5}$, assuming an atmospheric pressure of 15 psia. Then,

$$\frac{T_2}{T_1} = \left(\frac{p_2}{p_1}\right)^{n-1/n}$$

from equation (6–47). This gives us

$$\frac{T_2}{T_1} = \left(\frac{1}{5}\right)^{0.45/1.45} = \left(\frac{1}{5}\right)^{0.3} = \frac{1}{1.647}$$

Consequently the specific entropy change can easily be determined:

$$\Delta s = (0.24 \text{ Btu/lbm}°\text{R}) \ln \frac{1}{1.647} - (0.0686 \text{ Btu/lbm}°\text{R}) \ln \frac{1}{5}$$

$$= -0.120 \text{ Btu/lbm}°\text{R} + 0.111 \text{ Btu/lbm}°\text{R}$$

$$= -0.009 \text{ Btu/lbm}°\text{R} \qquad\qquad \textit{Answer}$$

The entropy is decreasing in this process, due to a drop in the air temperature.

7.5
The Third Law of Thermodynamics

In calculating entropy changes due to processes, temperature directly affects the entropy function. A decrease in temperature will induce a decrease in entropy and an increase in temperature will induce increased entropy. We might ask, how low can entropy be? The answer is given by the third law of thermodynamics.

> **Third Law of Thermodynamics:** *Entropy tends to a minimum constant value as temperature tends to absolute zero. For a pure element this minimum value is zero, but for all other substances it is not less than zero, but possibly more.*

This third statement or law is a result of experimentation in the temperature regime near absolute zero and has not been violated—therefore it is considered a "law." From a practical viewpoint, it tells us that it is impossible to reach an absolute zero temperature by other than a reversible process since near the zero point (as illustrated in figure 7–3), the change in entropy is zero and the only irreversible means of lowering entropy further is to have a surrounding which is cooler yet than absolute zero. This is impossible, so the final approach to absolute zero for cooling any material must be reversible and adiabatic (isentropic). Another manner of observing this is to take the definition, equation (7–2), and set S equal to zero.

$$\Delta s = \int \frac{dQ}{T} = 0$$

In figure 7–3, for substance A near absolute zero, point (1), the entropy s_1 is equal to the entropy when the temperature will be zero, s_0. Then

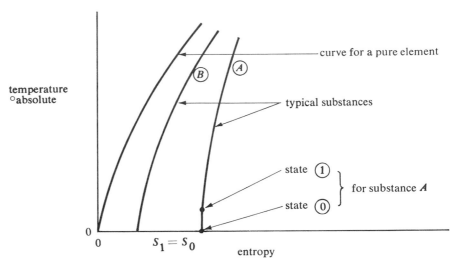

Figure 7-3   Behavior of entropy near absolute zero of temperature.

$$\Delta s = s_0 - s_1 = 0$$

and for equation (7–2) to be satisfied dQ must be zero (adiabatic), else the term dQ/T will approach infinity, which is impossible. The process from 1 to 0 must therefore be adiabatic and reversible.

The point in this discussion is that the state of absolute zero temperature is a highly desirable one for making energy available, but it is impossible to achieve. Materials close to zero temperature are, however, attractive as sources of heat or work.

**7.6
Summary**

The property entropy, S, has been introduced, along with equations for calculating changes in it. These equations are listed in tables 6–2 and 6–3 for ready references. In these tables are listed specific entropy changes s but the relation

$$\Delta S = m\Delta s$$

can readily allow conversion from specific to total entropy if desired.

Entropy is defined mathematically in terms of equation (7–2)

$$dQ_{\text{rev}} = TdS$$

as a tool for evaluating heat transfer. Entropy is a measure of the order of a system or the unavailability of thermal energy.

If we wish to know the entropy change for a given process, we may calculate such change from various equations developed in this chapter (depending on what type of process is occurring), or we may determine the change from thermodynamic tables listing entropy values. Care must be exercised, however, when using equations to calculate entropy changes. For instance, unless a perfect gas (or a substance reasonably like a perfect gas) represents the system, we should not use those equations developed from the perfect gas assumption. Generally, greater accuracy can be achieved by using entropy values from thermodynamic tables, even for those gases behaving like perfect gases.

The third law of thermodynamics was stated and represents a restriction of all physical systems to the temperature regime which excludes absolute zero.

Practice Problems

Problems designated with an asterisk * preceding the number should only be attempted by those having a background which includes the knowledge of integral calculus.

Sections 7.1 and 7.2

7.1. During a reversible, isothermal process, 700 Btu of heat transfer are directed into a system at 80°F. Determine the entropy change.

7.2. Seventy-two grams of air in a piston-cylinder are heated by 135 calories while going through a reversible, isothermal process at 70°C. Determine the specific and total entropy changes for the air.

7.3. Helium gas is increased in pressure from 15 psia to 35 psia. Determine the specific entropy changes of the helium if the compression is done reversibly under constant temperature conditions.

7.4. Steam is decompressed from 220 psia to 200 psia at 800°F. Determine the specific entropy change.

7.5. The entropy function of a certain gas is plotted in figure 7–4. Determine the energy required in the form of heat transfer to heat the material to 300°K from absolute zero temperature.

Section 7.3

7.6. Superheated steam is reversibly and adiabatically expanded from 250 psia and 800° to 300°F. Determine the final pressure.

7.7. During an adiabatic process, argon gas exhibits an increase of specific entropy of 0.038 Btu/lbm°R. What is the irreversible entropy increase of 20 lbm of argon?

7.8. Determine the final temperature of 7,000 grams of steam at 75 psia and 1200°F contained in an insulated chest. The chest expands isentropically until the steam is at 10 psia.

Section 7.4

7.9. A perfect gas contained in a rigid container in 90°F surroundings exhibits a pressure drop from 28 psia to 20 psia. Assuming the gas constant is 63 ft-lbf/lbm°R and the specific heat, c_p, is 0.30 Btu/lbm°R, calculate the temperature change and the specific entropy change for the gas.

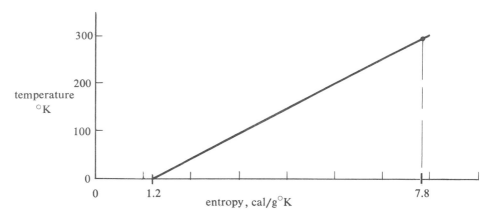

Figure 7–4   Problem 7.5.

7.10. During an isometric process of 1 lbm of nitric oxide (NO), the specific entropy decreases by 0.073 Btu/lbm°R. If the nitric oxide was initially at 12°F, determine the final temperature.

7.11. In a rigid stainless steel tank, 540 grams of dodecane, ($C_{12}H_{26}$) are heated from 130°C to 250°C. Determine the entropy change in calories per degree Kelvin.

7.12. Acetylene gas is contained in a rigid steel tank while its pressure increases from 10 psig to 15 psig. Determine the change in specific entropy of the gas during this process.

7.13. Determine the specific entropy change of superheated steam during an isometric process where the initial state is 1500°F and 150 psia and the final is 700°F.

7.14. During a constant pressure process, 1.5 kg of sulfur dioxide (SO_2) are heated to 70°C from an initial temperature of 10°C. Determine the total and specific entropy changes.

7.15. A drop in entropy of 10 cal/g-mole °K is calculated for propylene gas as it is cooled to 35°C during an isobaric process. Determine the initial temperature.

7.16. In a flexible container, 1.8 lbm of air are expanded at constant pressure from an initial volume of 20 ft³ to 40 ft³ finally. If the initial temperature is 195°F, determine the final pressure and the change in total and specific entropy.

7.17. Mercury vapor is heated from a saturated liquid to saturated vapor. Determine the specific entropy change if the pressure is
   (a) 1.0 psia.
   (b) 6 psia.
   (c) 160 psia.

7.18. Freon, F-12, is cooled from 50° to 20° at
   (a) 30 psia.
   (b) 1 psia.
   (c) 5 psia.
   Determine the specific entropy decreases in these cases.

7.19. Freon, F-22, is heated from saturated vapor at 31.16 psia to superheated vapor at
(a) 10°F at 30 psia.
(b) 20°F at 40 psia.
(c) 0°F at 5.0 psia.
Determine the specific entropy changes for these cases.

7.20. Steam is heated from saturated liquid at 15 psia to superheated steam at 200 psia and 600°F. Determine the specific entropy change.

7.21. Steam is expanded from 300 psia and 800°F to a saturated vapor in a reversible adiabatic manner. Determine the final pressure of the steam.

7.22. During a polytropic process involving air, the specific entropy is found to have decreased by 0.008 Btu/lbm°R. The pressure ratio (final to initial pressure) was known to be 12.1 to 1 and the initial temperature was 180°F. Obtain the polytropic exponent, n, and the final temperature.

7.23. A gas having the same thermodynamic properties as air is contained in the combustion chamber of an internal combustion engine. The gas is at 200°F and 200 psia from which it expands polytropically to 15 psia. If the polytropic exponent is 1.51, determine the specific entropy change of the gas during this process.

7.24. A polytropic process with an exponent, n, of 1.43 is found to describe an expansion through a gas turbine where the entrance pressure is 180 psia and the exhaust is 18 psia. Assuming the gas flowing through the turbine is air, determine its entropy change per pound-mass.

*7.25. A perfect gas with constant heat is found to have an entropy function given by the equation
$$S = (0.006\ T^2 + 0.031\ T)\ \text{Btu/°R}$$
Determine the heat transferred during a reversible process of this gas when the temperature increases from 100°F to 300°F.

*7.26. The equation
$$S = (15.0\ T^3 + 325)\ \frac{\text{calories}}{\text{°K}}$$
describes the entropy function of a particular gas. Determine the heat transfer of a reversible process involving this gas if the temperature decreases from 425°K to 25°C.

8

The Heat Engine and the Second Law of Thermodynamics

In this chapter, cyclic devices which have heat exchanges with the surroundings, called *heat engines*, are described. The *Carnot engine* is cited and discussed as an example of a heat engine.

The general definition of *thermodynamic efficiency* is given, and a development is made of the *Clausius inequality* which provides a criterion for a comparison of actual to ideal reversible power cycles. The *second law of thermodynamics* is stated as a physical limitation on real and ideal cyclic devices. *Cyclic heat* and *work* are then discussed, leading into the differentiation of *closed* and *open cycles* — synonymous with the closed and open systems.

Two rather extended example problems are presented at the close of this chapter; one treats the *Carnot cycle* which produces power, and the other treats a *reversed Carnot cycle* which is the basis for a type of refrigerator or heat pump.

8.1
Cyclic Devices
— System
Diagrams

A thermodynamic system which progresses through a set of processes and periodically returns to its initial or beginning state, we have defined as a *cyclic device*. One complete set of processes which allows the first return of a system to its beginning state we call a *cycle*. There are many examples of cycling devices, some of which are as follows: electrical power plants which use coal or nuclear fuel; electric motors; internal combustion engines using gasoline; jet and steam engines; refrigerators; and air conditioners. Items which are *not* cyclic devices include the solid or liquid rocket engines; electrical batteries; fuel cells; and most biological systems, for example, trees.

A particular type of cyclic device which transfers heat to its surroundings we call a *heat engine*. A heat engine and its important thermodynamic characteristics are visualized concisely in a *system diagram* in figure 8–1, where two types of cyclic devices are presented, the open system and the closed system. While these diagrams contribute no information regarding the details of how the heat engine mechanically functions, they do provide a complete and clear picture of the thermodynamic

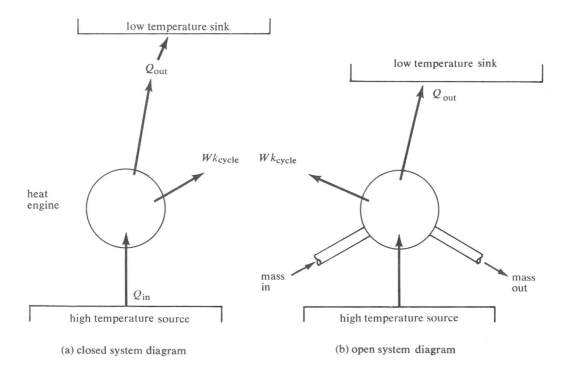

(a) closed system diagram   (b) open system diagram

Figure 8–1   The system diagram.

operation of the devices. Thus, a system diagram can be a worthwhile exercise in the analysis or synthesis of a cycling heat engine. Keep in mind that the term *heat engine* does not necessarily require mechanical components such as gears, shafts, or piston-cylinders, but it is any device which involves heat and work.

Notice in figure 8–1 that heat is transferred from some high temperature source (which may physically be within the system boundary or may be more than one source) to the heat engine. The heat engine then converts the transferred energy to mechanical work and dumps the excess into a low temperature sink. This is the general operation of all heat engines, and where these phenomena occur in a particular device can be found from an initial study of a system diagram. Note, also, that in the cycling device no energy is accumulating or being drained from the engine itself — all energy is taken from some foreign source and dumped in another place.

If we consider the heat engine to be reversible, then we can visualize the energy transfers of heat and work to take either direction. In figure 8–2 is shown such a reversible heat engine where the discontinuous arrows represent the set of transfers which would occur reversed from the set

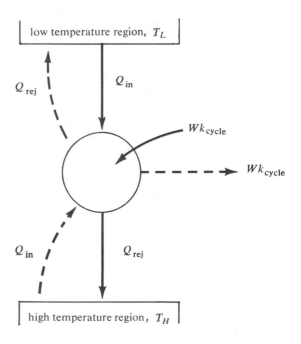

Figure 8–2   Reversible heat engine.

represented by solid arrows. It can be seen that the arrangement of energy transfers diagrammed with the solid arrows is just a device which will provide refrigerating capability; that is, it pumps energy from a lower temperature into a higher temperature region by expending work. This type of heat engine is generally referred to as a *heat pump*.

8.2 The Carnot Engine

In 1824, Sadi Carnot published a treatise on thermodynamics in which he invented a cycle composed of four special processes. The heat engine which would run on this cycle (but which no one has yet built) has since been called the *Carnot engine* and the cycle, the *Carnot cycle*. The four processes, in order, which comprise this cycle are as follows:

(1–2)   Reversible isothermal compression at temperature T_L.
(2–3)   Reversible adiabatic compression from the low temperature T_L to a higher temperature T_H.
(3–4)   Reversible isothermal expansion at temperature T_H.
(4–1)   Reversible adiabatic expansion from temperature T_H to T_L.

Notice that the above descriptions of the processes indicate that the cycle is reversible since all the individual processes are. The *p-V* and *T-s* diagrams are displayed in figures 8–3 and 8–4 respectively, and we see from figure 8–3 that work is put into the engine during processes (1–2) and (2–3) while processes (3–4) and (4–1) produce work. If we add up the area under each of these curves we have

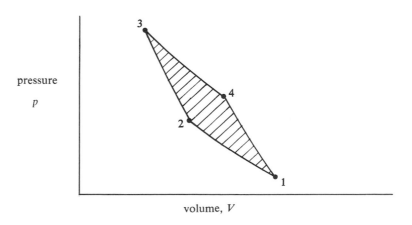

Figure 8–3   *p-V* diagram for Carnot cycle.

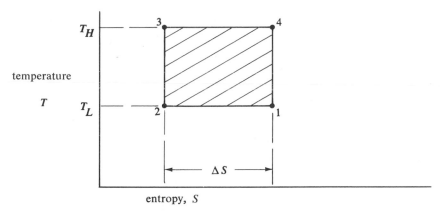

Figure 8–4   *T-s* diagram for Carnot cycle.

$$\int_{p_1}^{p_2} pdV + \int_{p_2}^{p_3} pdV + \int_{p_3}^{p_4} pdV + \int_{p_4}^{p_1} pdV$$

and this sum we denote by

$$\oint pdV = \oint dWk$$

We will use the sign $\oint$ to represent a cyclic sum or sum of the integrals during one complete cycle. We do not have the details here to arithmetically add up the area under curves, but the shaded area in figure 8–3 should be recognized as just this sum we have discussed, i.e.

$$\oint dWk = \text{net work of a Carnot cycle} = Wk_{\text{rev cycle}}$$

In a similar manner, in figure 8–4 the shaded area enclosed by the curves, a simple rectangle, represents the sum of the four $\int Tds$ terms which together represent the net heat transferred into the engine. Then

$$\oint Tds = \oint dQ_{\text{rev}} = \text{net heat of a cycle} = Q_{\text{rev cycle}}$$

Interestingly, only two processes contribute to the sum of the heat transfer integrals — process (2–3) and (4–1). These areas are easily seen to be $T_H(\Delta s)$ for the area under (2–3), and $T_L(-\Delta S)$ for the area under (4–1). The sum of the heat transfers for a Carnot cycle, is thus

$$\oint Tds = \oint dQ_{\text{rev}} = T_H\Delta S - T_L\Delta S$$

or

$$Q_{\text{Carnot}} = (T_H - T_L)\Delta S \qquad \textbf{(8–1)}$$

This result is not true for all reversible cycles and particularly not for irreversible cycles; use it for Carnot cycles only, and correctness will be assured.

Since the Carnot engine is a cycle, we may say that $\oint dU$ is zero. That is, the sum of internal energy changes for one cycle is zero, as it should be for any property. We have

$$\oint dU = \int_1^2 dU + \int_2^3 dU + \int_3^4 dU + \int_4^1 dU$$

or

$$\oint dU = U_2 - U_1 + U_3 - U_2 + U_4 - U_3 + U_1 - U_4 = 0$$

and we see that this statement is true; in fact it is true for *all cycles*, reversible or not. The first law then can be written

$$\oint dU = \oint dQ - \oint dWk \qquad (8\text{--}2)$$

but

$$\oint dU = 0$$

so

$$\oint dQ = \oint dWk \qquad (8\text{--}3)$$

or

$$Q_{\text{cycle}} = Wk_{\text{cycle}} \qquad (8\text{--}4)$$

For the Carnot cycle, then

$$(T_H - T_L)\Delta S = Wk_{\text{Carnot}}$$

from the result in equation (8–1).

It was stated that no one has yet built a Carnot engine; there are at least two reasons:

(1) Reversible cycles can be approached but never fully realized.

(2) While reversible adiabatic processes can be closely approximated in technology, the reversible isothermal process is difficult to put into practice and still have sufficient amounts of heat transfer in a reasonable time period.

The Carnot cycle/engine concepts are theoretical tools which have added immensely to the thermodynamic method; we will refer to them frequently as a standard of comparison.

**8.3
Thermo-
dynamic
Efficiency**

In this section we will consider the methods of determining the effectiveness of heat engines; that is, we will define the efficiencies of cyclic devices. As we will see, one efficiency is not an accurate description of an engine. Rather, we will have many efficiencies with precise meanings for each. For the heat engine analysis, the term we will use is *thermodynamic efficiency,* η_T. It is defined as

$$\eta_T = \left(\frac{Wk_{\text{cycle}}}{Q_{\text{add}}}\right)(100) \qquad (8\text{--}5)$$

and it gives us the ratio of the net work *out* to the heat *in*. For some of the other information we might want, such as the reversibility of the cycle, the unnecessary degradation of available energy, or the rate at which work can be furnished, we must seek other efficiency terms.

We can obtain some idea of the thermodynamic irreversibility from the heat engine efficiency η defined as

$$\eta = \left(\frac{Wk_{\text{actual}}}{Wk_{\text{cycle}}}\right)(100) \qquad (8\text{--}6)$$

where Wk_{actual} is the actual work gained from the heat engine and the Wk_{cycle} is the ideal or reversible work.

If we consider systems which have no thermal effects, the mechanical efficiency η_{mech} is defined as

$$\eta_{\text{mech}} = \left(\frac{Wk_{\text{out}}}{Wk_{\text{in}}}\right)(100) \qquad (8\text{--}7)$$

and gives us some idea of the energy dissipated during mechanical conversions of energy. This is probably the most familiar statement of efficiency — output divided by input — but the reciting of all these dimensionless efficiency numbers is not the heart of a true technical analysis.

Example 8.1

A Carnot engine operates in an atmosphere at 343°K, with an addition of heat from combusting gases burning at 1473°K. Determine the thermodynamic efficiency of the engine.

Solution

We are concerned here with a reversible heat engine which operates between high and low temperature regions. We can visualize this Carnot engine better from a system diagram as shown in figure 8–5. In this figure we see that

$$Q_{\text{added}} - Q_{\text{rejected}} = Wk_{\text{cycle}}$$

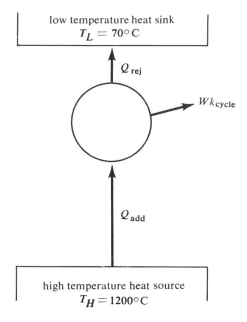

Figure 8–5   Carnot system diagram for example 8.1.

which is a restatement of equation (8–4). The term Q_{added} is equal to $T_H \Delta S$, where T_H is the high temperature region at 1473°K. Similarly, $Q_{rejected}$ is $T_L \Delta S$ where T_L is 343°K.
Then

$$(T_H - T_L)\Delta S = Wk_{cycle}$$

and from the equation,

$$\eta_T = \frac{Wk_{cycle}}{Q_{added}} \times 100 \qquad (8–5)$$

we have, for the general Carnot engine

$$\eta_T = \frac{(T_H - T_L)\Delta S}{T_H \Delta S} \times 100$$

or

$$\eta_T = \left(1 - \frac{T_L}{T_H}\right) \times 100 \qquad (8–8)$$

Specifically for this problem, we obtain

$$\eta_T = \left(1 - \frac{343}{1473}\right) \times 100 = 76.7\% \qquad \textit{Answer}$$

Notice that for the Carnot heat engine, the thermodynamic efficiency is only a function of the two temperature regions, and no other cycle can have a higher efficiency.

8.4 Clausius' Inequality

We have considered the property *entropy* and noted that for irreversible processes the entropy change is limited by the inequality

$$\Delta S > \int \frac{dQ}{T} \qquad (8\text{--}9)$$

For a heat engine which has irreversibilities in it, we say the sum of ΔS is

$$\Sigma \Delta S = \oint dS$$

and

$$\oint dS > \oint \frac{dQ}{T} \qquad (8\text{--}10)$$

But for any cycle the sum of the property changes for one cycle must be zero so that

$$\oint dS = 0 > \oint \frac{dQ}{T} \qquad (8\text{--}11)$$

This statement is called *Clausius' inequality*. For reversible cycles the equality

$$\oint \frac{dQ}{T} = 0 \qquad (8\text{--}12)$$

holds; but for real, irreversible cycles, equation (8–11) must hold. This inequality has been the beginning of much conjecture and advanced discussion in thermodynamic analysis. For us, the use of Clausius' inequality is in its inference that the entropy increase of a heat engine cannot be determined solely from the heat transfer data. For instance, for the Carnot reversible engine we found that

$$\frac{Q_{add}}{T_H} - \frac{Q_{rej}}{T_L} = 0$$

but for an irreversible heat engine this sum does not add to zero. From Clausius' inequality, we see that

$$\frac{Q_{add}}{T_H} - \frac{Q_{rej}}{T_L} < 0$$

or

$$\frac{Q_{add}}{T_H} < \frac{Q_{rej}}{T_L}$$

where the left side represents entropy *reduction* of some energy source while the right represents *increases* to a surrounding energy sink. Obviously the entropy has a net increase to the surroundings for irreversible processes.

8.5 The Second Law of Thermodynamics

We have considered the first law of thermodynamics and from it came an elementary restriction on any device we wish to analyze or design — energy cannot be created. It would be a tremendous feat to have an engine that would operate as diagrammed in figure 8-6, but this device is creating energy from nothing and the first law tells us this is impossible. Engines such as that shown in figure 8-6 are called *perpetual motion machines of the first kind* (PMM1) and occasionally we hear pronouncements that someone has invented one. They are impossible to achieve, however, because of the natural restrictions stated in the first law, and a simple system diagram can usually pinpoint the fallacy in any such supposed device.

There is, however, a more subtle restriction we must contend with above and beyond the first law, contained in the second law.

Second Law of Thermodynamics: *No heat engine can produce a net work output by exchanging heat with a single fixed temperature region.*

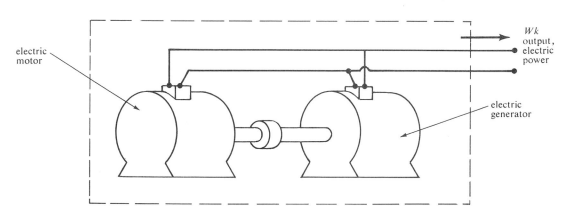

electric motor

Wk output, electric power

electric generator

Figure 8-6   Example of a perpetual motion machine.

The effect of this categoric statement is that a restriction is placed on all heat engines, a restriction which requires that both Q_{add} and Q_{rej} be nonzero. You may have considered that to achieve 100% efficiency for the Carnot engine all that was required was for the heat rejected to be zero, or that the heat sink to be at 0°R. That is

$$\eta_T = \left(1 - \frac{T_L}{T_H}\right) \times 100 = 100\% \tag{8–8}$$

when $T_L = 0$. The second law now tells us it is impossible to reach 100% efficiency for a heat engine since the heat rejected must be greater than zero. For the bulk of technological applications, we must be further restricted to a low temperature sink which is at the atmospheric state. The term T_L in the efficiency equation (8–8) has a value of between 0°C and 50°C for many common cycles.

The second law will not provide us with any new computational or bookkeeping techniques as the first law did, but its underlying usefulness cannot be overestimated. Suppose, for instance, that we have a reversible heat engine which is capable of producing net work from only one temperature region. This machine, called a *perpetual motion machine of the second kind* (PMM2), would be converting 100% of the heat into work. In addition, since we need be concerned with only one temperature region, our heat engine can run at any lower temperature and the efficiency of the engine (100%) is independent of everything. This type of device does not in any way violate the first law, and yet it has never been achieved. It has a certain advantage over a PMM1 in that it can provide work and refrigeration by extracting heat from any temperature region. The PMM1 merely provided work from nothing. We thus are led to the conclusion that for a heat engine to operate within the laws of nature, it must be somehow dependent upon the surroundings. The second law provides the answer.

There are other statements of the second law which are equivalent to the one given above. One of these alternatives is attributed to Clausius, who stated that it is impossible for a heat pump to have as its only effect the transfer of heat from a low to a high temperature region. (Thus, a refrigerator cannot operate without work supplied to it.) The equivalence of the two second law statements can easily be shown. Suppose, as indicated in figure 8–7(a), that we have a reversible heat pump requiring no work input and a reversible heat engine. The arrangement of (a) is then the same as (b), and in this second system diagram there is a transfer of heat with only one temperature region. This is an obvious violation of the initial statement of the second law so that we see Clausius' statement is its equivalent.

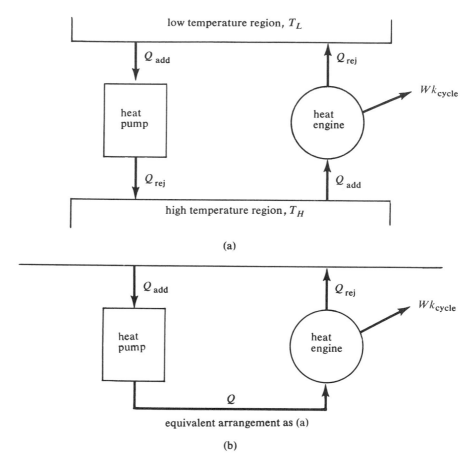

(a)

equivalent arrangement as (a)

(b)

Figure 8–7   Hypothetical heat pump and engine combination.

8.6
Heat and Work
of a Cycle

During the execution of a set of processes on a system or heat engine, which together constitute a cycle, none of the properties change during the complete cycle; but for each individual process the properties will definitely change. Using the equation of state, caloric equations of state, and process equation, we may obtain values of heat and/or work for each process, and by arithmetic addition the net work and heat of a cycle are determined. These net values of a cycle have been denoted by the cyclic integral sign, $\oint$, and from the first law it was shown

$$\oint dWk = \oint dQ \tag{8–13}$$

for any cycle. This is equivalent to the statement

$$Wk_{\text{cycle}} = Q_{\text{net}} \tag{8–14}$$

where in general,

$$Q_{net} = Q_{added} - Q_{rejected} \qquad \text{(8–15)}$$

Most references to the work or heat of cycle mean the net values, as these are the only quantities which affect the surroundings. Caution should be exercised, however, in interpreting this terminology, as there are possibilities for referring to individual work or heat quantities for the separate processes making up the cycle of a heat engine or pump.

8.7
Closed Cycle

System identification and classification as to open or closed has been the approach in analyzing a complete thermodynamic problem. In this chapter, we introduced the system as a device which is rotating or in some manner going through a cycle — and called it a *heat engine* or *pump*. Just as we classified systems, we can classify heat engines as open or closed, that is, open or closed cycles. The closed cycle heat engine/pump is a device which allows no mass transfer across the boundaries; the device retains the same mass during the cycle. The system diagram for a

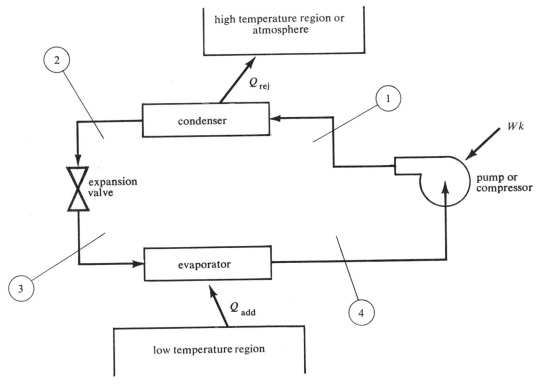

Figure 8–8   Typical refrigerating cycle.

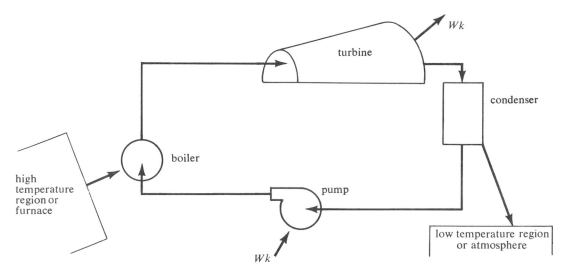

Figure 8–9   Steam turbine cycle.

closed cycle is shown in figure 8–1(a). Examples of closed cycles are the heat pump system of the refrigerator (as shown in figure 8–8), the steam turbine - electric generating system (shown in figure 8–9), and a piston-cylinder device (shown in figure 8–10).

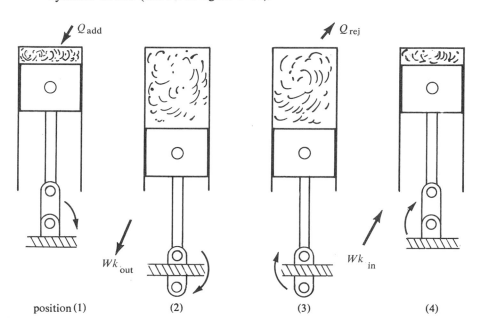

Figure 8–10   Piston-cylinder device executing one cycle of a closed cycle.

Notice in the first examples that while the complete devices were closed, there were four components in which each was individually represented as an open system. In figure 8–8, the pump, the condenser, the expansive valve, and the evaporator each contribute a process to the closed cycle of the refrigerator; but each of the processes is carried out in open systems. A similar situation exists in the steam turbine cycle of figure 8–9, but in figure 8–10 we have a closed system contributing all the processes to a complete closed cycle.

It is then apparent that a closed cycle and closed system are not one and the same. Open systems can be a part of a closed cycle and obviously, so can closed systems.

8.8
Open Cycle

We consider an open cycle heat engine/pump to be one which involves mass transfer across the device boundary. Since we are considering cyclic devices, there is no change in the amount of mass over a cycle, but the mass may be flowing through continuously or flushed and rinsed periodically during specific cycle processes. The system diagram of a typical, generalized open cycle is shown, in figure 8–1(b). Examples of open cycles are given in figures 8–11 and 8–12. Notice that in figure 8–11 during process (3), the mass is being exchanged; this exchange represents the only difference in theory from that shown in figure 8–10.

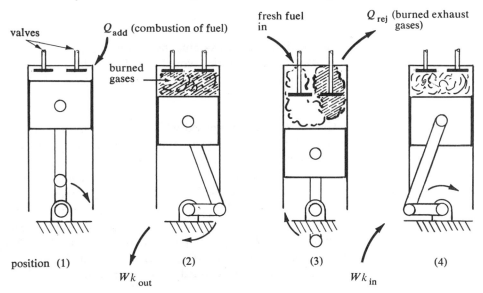

Figure 8–11   Open cycle of a piston-cylinder device exchanging fresh fuel and air mixture for spent exhaust gases.

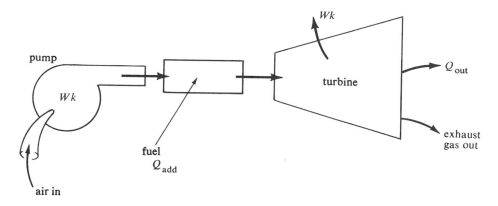

Figure 8–12   Typical gas turbine cycle.

The gas turbine shown in figure 8–12 is obviously open — all the components are open with air or gases flowing through each of them under steady flow conditions.

For the open cycle then, we may have some processes which are individually closed, or we may have all processes open.

8.9
Carnot Power
Cycle

In this section, we will examine the details of a Carnot engine and attempt to devise a machine which would function under the requirements of the Carnot cycle.

Example 8.2

A Carnot engine is proposed which has the physical characteristics of a piston-cylinder while the gas contained within the cylinder is air having a mass of amount 0.002 lbm. The piston has a diameter of 3 inches and travels a distance of 4 inches, reciprocating from within $\frac{1}{4}$ inch of the cylinder and to $4\frac{1}{4}$ inches. The arrangement is shown in figure 8–13. Heat proceeds to be added when the piston is at the extreme inward position (1) and proceeds to be rejected when the piston is in the extreme outward position (3). If the high temperature region is at 1200°F and the low temperature region is at 80°F, determine the engine's thermal efficiency, net work per cycle, heat added, and heat rejected.

Solution

The system is identified as the air in the cylinder and the system diagram is sketched in figure 8–14. The *p-V* diagram is given in figure 8–15 and

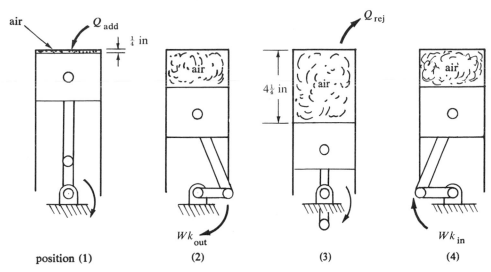

Figure 8–13   Carnot engine using the piston-cylinder mechanism.

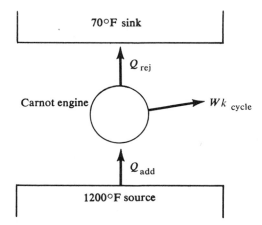

Figure 8–14   System diagram for reciprocating piston-cylinder Carnot engine.

the physical positions denoted in figure 8–13 correspond to the numbers in this diagram. Then in figure 8–16 is shown the *T-S* diagram for the given Carnot engine.

We now have two separate methods of obtaining our desired answers:

(1)   Determine the net heat transfer $\oint dQ$ by obtaining the entropy change ΔS from our process equations. Then we get Q_{add} from the product $T_1 \Delta S$ and the Q_{rej} from $T_4 \Delta S$. Using the sum

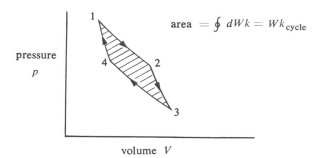

Figure 8–15   Pressure-volume diagram for engine of example 8.2.

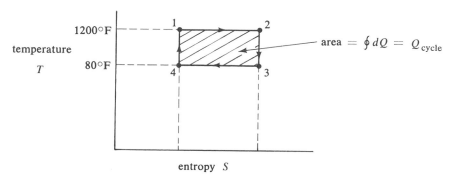

Figure 8–16   Temperature-entropy diagram for engine of example 8.2.

$$\oint dQ = Q_{\text{add}} - Q_{\text{rej}} = \oint dWk = Wk_{\text{cycle}}$$

will yield our answers.

(2)   Reversing the procedure of (1) by first determining the work
of a cycle and subsequently finding the heat transfer terms.
While method (2) is a little lengthier than (1), we will use it to
help clarify the computational techniques we have accumulated.

The thermal efficiency we obtain from

$$\eta_T = \left(1 - \frac{T_L}{T_H}\right) \times 100$$

which gives us,

$$\eta_T = \left(1 - \frac{540}{1660}\right) \times 100 = 67.5\% \qquad \textit{Answer}$$

The work of the cycle we identify by the sum

$$\oint dWk = Wk_{\text{cycle}} = Wk_{(1-2)} + Wk_{(2-3)} + Wk_{(3-4)} + Wk_{(4-1)}$$

The first term of this sum we find from equation (6–17)

$$Wk_{(1-2)} = p_1V_1 \ln \frac{V_2}{V_1} = p_2V_2 \ln \frac{V_2}{V_1}$$

since process (1–2) is isothermal. Similarly

$$Wk_{(3-4)} = p_4V_4 \ln \frac{V_4}{V_3} = p_3V_3 \ln \frac{V_4}{V_3}$$

for the reason that process (3–4) is isothermal. Processes (2–3) and (4–1) are both reversible adiabatic (or isentropic as well) and the work is gotten from equation (6–33)

$$Wk_{(2-3)} = \frac{mR}{1-k}(T_3 - T_2)$$

and

$$Wk_{(4-1)} = \frac{mR}{1-k}(T_1 - T_4)$$

respectively. But we see that T_3 equals T_4, and T_1 equals T_2 so that $Wk_{(2-3)} = -Wk_{(1-4)}$, and we need only worry about the two isothermal work terms. The volume V_1 we can calculate at position (1) of the piston

$$V_1 = \pi \times \left(\frac{1}{4} \text{ in}\right)(1.5 \text{ in})^2 = 1.767 \text{ in}^3$$

The volume V_3 is also directly obtained at position (3)

$$V_3 = \pi \left(4\frac{1}{4} \text{ in}\right)(1.5 \text{ in})^2 = 30.0 \text{ in}^3$$

Then we can obtain V_2 and V_4 from the reversible adiabatic process relations, equation (6–41)

$$\frac{V_2}{V_3} = \left(\frac{T_3}{T_2}\right)^{1/k-1}$$

and

$$\frac{V_1}{V_4} = \left(\frac{T_4}{T_1}\right)^{1/k-1} = \left(\frac{T_3}{T_2}\right)^{1/k-1}$$

From this we get

$$\frac{V_2}{V_3} = \left(\frac{540}{1660}\right)^{1/0.4} = 0.060$$

or

$$V_2 = 30 \times 0.060 \text{ in}^3 = 1.80 \text{ in}^3$$

and

$$V_4 = \frac{1.767}{0.060} \text{ in}^3 = 29.5 \text{ in}^3$$

The work then is obtained from

$$Wk_{\text{cycle}} = p_1 V_1 \ln \frac{V_2}{V_1} + p_3 V_3 \ln \frac{V_4}{V_3}$$

and since $pV = mRT$, we have

$$Wk_{\text{cycle}} = mR\left[T_1 \ln \frac{V_2}{V_1} + T_3 \ln \frac{V_4}{V_3} \right]$$

Substituting values in, we obtain

$$Wk_{\text{cycle}} = (0.002 \text{ lbm})(53.3 \text{ ft-lbf/lbm°R})\left[1660°\text{R} \ln \frac{1.80}{1.767} + 540 \ln \frac{29.5}{30.0} \right]$$

$$= (0.1066 \text{ ft-lbf/°R})[1660°\text{R} \ln 1.019 - 540°\text{R} \ln 1.019]$$

$$= 2.25 \text{ ft-lbf} \qquad\qquad\qquad\qquad Answer$$

This answer represents the work produced by the Carnot engine per cycle. In heat units it is

$$Wk_{\text{cycle}} = \frac{2.25}{778} \text{ Btu} = 0.00289 \text{ Btu}$$

and the heat added can be quickly determined from the efficiency

$$\eta_T = 67.5\% = \left(\frac{Wk_{\text{cycle}}}{Q_{\text{add}}} \right)(100)$$

so that

$$Q_{\text{add}} = 0.00289/0.675 = 0.00428 \text{ Btu} \qquad\qquad Answer$$

The heat rejected can be calculated from the relation

$$\oint dQ = \oint dWk = 0.00289 = Q_{\text{add}} - Q_{\text{rej}}$$

so that

$$Q_{\text{rej}} = 0.00428 - 0.00289 = 0.00139 \text{ Btu} \qquad\qquad Answer$$

This problem, while somewhat lengthy, should illustrate the drawing together of the various ideas and equations so necessary in solving thermodynamics problems in engineering.

8.10 Reversed Carnot Cycle

Consider the Carnot cycle operating in reverse. This is the heat pump which, from the second law of thermodynamics, is required to have a net work or power put into the device to make it practical. With the following example problem we introduce the full significance of its operation.

Example 8.3   A Carnot heat pump operates between the limits of 10°F and 120°F. Its mechanical aspects are diagrammed in figure 8–17, and it operates with ammonia as the working fluid. At point (1) in the closed cycle, the ammonia is completely saturated liquid at 120°F while at point (4) the ammonia is saturated vapor at 120°F. The ammonia expands adiabatically from (1) to (2), furnishing some work to help drive the compressor. The operating temperature at point (2) is 10°F. Heat is added to the working fluid in the evaporator and rejected in the condenser. The compressor supplies the energy to the ammonia, which increases in pressure and temperature, making the heat rejection to the high temperature (120°F or less) region possible. Determine Q_{add}, Q_{rej}, Wk_{cycle}, and heat pump efficiency.

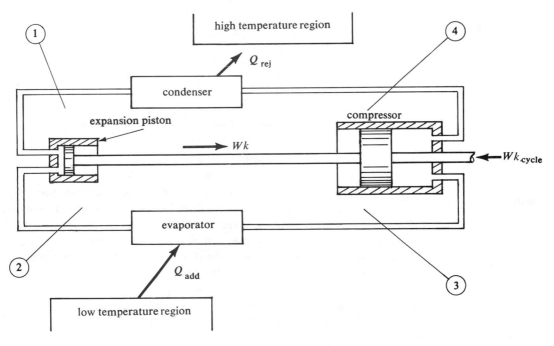

Figure 8–17   An arrangement for a Carnot heat pump.

Solution   The T-S and p-v diagrams are shown in figures 8–18 and 8–19 respectively. Notice that a line, called the *saturation curve*, has been included on the T-S diagram and the area under this bell-shaped curve is designated as "liquid and vapor mixture." The significance here is that the liquid and

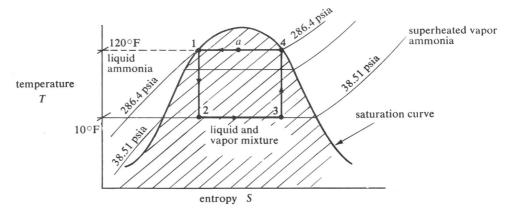

Figure 8–18   *T-S* diagram for Carnot heat pump using ammonia (not drawn to scale).

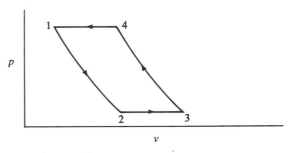

Figure 8–19   *p-v* diagram for Carnot heat pump.

vapor mixture represents the state a material (in this case ammonia) must pass through to change from a liquid phase to a gaseous or vapor phase. This area under the saturation curve we call the *saturated region;* a *saturated liquid* is a point on the left side of the saturation curve, and a *saturated vapor* is a point on the right side. These points are tabulated for various temperatures or pressures in the table B.11 and to find states in between, we need to know the percent vapor or the ratio of liquid to vapor. For instance, a point midway between (1) and (4), point (a), would be 50% quality. We denote quality by the Greek letter chi χ and define it as follows:

$$\chi_a = \left(\frac{\text{mass of vapor at } a}{\text{mass of vapor and liquid at } a}\right) \times 100 \qquad \textbf{(8–16)}$$

which can be reduced to the equation

$$\chi_a = \left(\frac{s_a - s_1}{s_4 - s_1}\right) \times 100 \qquad \textbf{(8–17)}$$

Points (1) and (4) in figure 8–18 correspond to the saturated liquid and vapors respectively, the entropies of which are denoted in the appendix tables as s_f for saturated liquid, and for saturated vapor, s_g. In some tables the difference $s_g - s_f$ is denoted s_{fg} so that equation (8–17) can then be written

$$\chi_a = \left(\frac{s_a - s_f}{s_g - s_f}\right) \times 100 = \left(\frac{s_a - s_f}{s_{fg}}\right) \times 100 \qquad (8\text{–}18)$$

or, equivalently

$$s_a = s_{fg}\frac{\chi_a}{100} + s_f \qquad (8\text{–}18)$$

and the other properties at state a are as follows:

$$h_a = h_{fg}\left(\frac{\chi_a}{100}\right) + h_f \qquad (8\text{–}19)$$

$$u_a = u_{fg}\left(\frac{\chi_a}{100}\right) + u_f \qquad (8\text{–}20)$$

and

$$v_a = v_{fg}\left(\frac{\chi_a}{100}\right) + v_f \qquad (8\text{–}21)$$

The quantity h_{fg} is equal to the difference of the entropy of saturated vapor, h_g, and saturated liquid, h_f, or

$$h_{fg} = h_g - h_f$$

and is called the *latent heat, heat of vaporization,* or *heat of condensation.* A better term would be *energy of vaporization* since it is energy — not heat — which is required to convert a liquid to a vapor, but the initial term has stuck from earlier days.

In figure 8–18 are drawn two typical constant pressure lines (for 286.4 psia and 38.51 psia) which correspond to isothermal lines in the phase change region but which diagonally cross isothermal lines elsewhere. The fact that pressure and temperature both remain constant during a phase change accounts for the slight difference in the p-v diagram in figure 8–19 from the normal Carnot cycle of figure 8–3.

We will consider in further detail the mechanisms of phase change in chapter 12 when we consider steam and water transformations, but now let us return to our reversed Carnot engine. The first law, applied to the condenser, is

$$\Delta h = q_{\text{rej}} - wk$$

since the condenser is an open system and the ammonia flows through without kinetic or potential energy changes. Also, wk is zero in the condenser so we obtain

$$\Delta h = h_1 - h_4 = q_{rej}$$

but, from table B.11 we have

$$h_1 = h_f \text{ (at } 120°F) = 179.0 \text{ Btu/lbm}$$

and

$$h_4 = h_g = 634.0 \text{ Btu/lbm}$$

We get then

$$q_{rej} = 179.0 - 634.0 = -455.0 \text{ Btu/lbm} \qquad \textit{Answer}$$

We are using the tabulated values because the equations developed previously are primarily of value for perfect gases and this obviously is not a suitable assumption for this problem. The heat added q_{add} can be gotten from a first law application to the evaporator;

$$\Delta h = h_3 - h_2 = q_{add} - wk$$

Again, wk is zero in an evaporator and we need only determine the enthalpy values from the tables. To find these enthalpy values we need to know the quality. At state (2) this is from equation (8–17)

$$x_2 = \frac{s_2 - s_f}{s_g - s_f} \times 100$$

where, at 10°F

$$s_f = 0.1208 \text{ Btu/lbm°R}$$

and

$$s_g = 1.3157 \text{ Btu/lbm°R}$$

Then the entropy at (2) is the same as it is at (1) so that in reversible adiabatic or isentropic process (1)

$$s_1 = s_2 = 0.03576 \qquad \text{(from table B.11)}$$

and the quality then can be calculated

$$x_2 = \frac{0.3576 - 0.1208}{1.3157 - 0.1208} \times 100 = \frac{0.2368}{1.1949} \times 100 = 19.8\%$$

The enthalpy at point (2) is then obtained from equation (8–19)

$$h_2 = x_2 h_{fg} + h_f$$

and the table B.11 data gives us, at 10°F

$$h_g - h_f = h_{fg} = 561.1 \text{ Btu/lbm}$$

$$h_f = 53.8 \text{ Btu/lbm}$$

from which we calculate h_2

$$h_2 = \frac{19.8}{100} \times 561.1 \text{ Btu/lbm} + 53.8 \text{ Btu/lbm}$$

$$= 164.9 \text{ Btu/lbm}$$

In a like manner, we calculate the enthalpy at point (3) from

$$h_3 = x_3 h_{fg} + h_f$$

where the h_{fg} and h_f values are the same as at point (2) since the temperature is the same. The quality, however, is obtained from

$$x_3 = \left(\frac{s_3 - s_f}{s_g - s_f}\right) \times 100$$

where

$$s_f = 0.1208 \text{ Btu/lbm}°\text{R} \qquad (\text{at } 10°\text{F})$$
$$s_g = 1.3157 \qquad (\text{at } 10°\text{F})$$

and

$$s_3 = s_4 = 1.1427 \qquad (\text{at } 120°\text{F})$$

This last equality $s_3 = s_4$ results from the reversible adiabatic compression process. The quality then is

$$x_3 = \frac{1.1427 - 0.1208}{1.3157 - 0.1208} \times 100$$

$$= 85.5\%$$

and the enthalpy at point (3) is calculated

$$h_3 = \frac{85.5}{100} \times 561.1 + 53.8 = 533.5 \text{ Btu/lbm}$$

We then obtain the heat added from the first law equation:

$$q_{\text{add}} = h_3 - h_2$$

or

$$q_{\text{add}} = 533.5 - 164.9 = 368.6 \text{ Btu/lbm} \qquad \textit{Answer}$$

By using the first law for the full cycle

$$\oint dwk = \oint dq$$

we obtain

$$wk_{\text{cycle}} = q_{\text{add}} + q_{\text{rej}}$$

$$= 368.6 - 455.0 = -86.4 \text{ Btu/lbm} \qquad \textit{Answer}$$

The efficiency or measure of effectiveness of heat pumps is given by the coefficient of performance (COP), defined as

$$COP = \frac{Q_{rej}}{Wk_{cycle}} \qquad (8\text{--}22)$$

which is the ratio of the heat produced by the cycle to the amount of work required to drive the device. For refrigerators or other cooling devices, the coefficient of refrigeration (COR) is a more appropriate measure of the cycle effectiveness. This is defined as

$$COR = \frac{-Q_{add}}{Wk_{cycle}} \qquad (8\text{--}23)$$

where the term Q_{add} can be associated with the "cooling" or amount of heat withdrawn from a region by the refrigerator. The quantities COP and COR are positive and greater than 1 for all heat pump and refrigerating cycles.

For our problem the COP is easily determined

$$COP = \frac{-455}{-86.4} = 5.27$$

and the coefficient of refrigeration, is

$$COR = \frac{-368.6}{-86.4} = +4.27$$

We will return to other details concerning the heat pump in chapter 14.

Practice Problems

Section 8.2

8.1. Determine the work generated by a Carnot engine operation, between 2000°F and 100°F, if the entropy change is 10 Btu/°R.

8.2. Determine the net heat of the Carnot engine of problem 8.1 and the heat rejected.

8.3. A Carnot engine operates at 200 rpm with an entropy change of 0.015 Btu/°R. Determine the high temperature required to generate 100 kW of power.

8.4. Determine the heat added in the engine of problem 8.3.

Section 8.3

8.5. A Carnot heat engine operates between 1000°F and 200°F. If the entropy change is 7 Btu/lbm°R, determine the cycle efficiency.

8.6. At 1000°C, 8000 Btu/s of heat are added to a Carnot engine. If the surroundings are at a temperature of 20°C, determine the cycle efficiency and the output power.

8.7. What maximum efficiency may a heat engine have if it can exchange heat by radiation with the sun and outer space? Assume the temperature of the sun is 10,000°C and outer space is −60°F.

8.8. A Carnot heat engine is designed to produce 70 horsepower at 3000 rpm. If the energy source is at 1500°F and sink at 90°F, determine the cycle efficiency and the ratio of heat transfer to the cycle Q_{rej}/Q_{add}.

8.9. Determine the rate of heat rejection of the engine in problem 8.8.

8.10. An inventor claims an efficiency of 90% for an engine he has built. He claims that the highest temperature is 130°F. Is his claim possible? Discuss why or why not.

Sections 8.4 and 8.5 **8.11.** An engine is proposed which extracts energy from sea water by heat transfer and then returns the cooled water to the ocean. No other heat transfer occurs with the engine. Does this violate any of our principles or laws of thermodynamics?

8.12. A device is proposed which is essentially a heat pump driven by a heat engine. A fluid is taken from a low temperature area and by means of the heat pump, is deposited in a high temperature sink. This sink is subsequently used to furnish energy to drive the heat engine which in turn rejects heat and fluid to the low temperature region. Does this device violate the second law? Does this device violate any principle of thermodynamics?

Sections 8.8, 8.9, **8.13.** A Carnot engine operating with a perfect gas between 800°C and 25°C
and 8.10 has an operating pressure range between 0.2 bar and 60 bars. Determine the work of the cycle, heat added, heat rejected, and efficiency, if the gas has physical properties equivalent to those of air.

8.14. A Carnot heat engine operating at 3000 rpm with a perfect gas having the properties listed below has a heat source at 1500°F and a sink at 85°F. Determine
(a) Cycle efficiency.
(b) Work of the cycle.
(c) Heat added and heat rejected.
The gas properties known are

$$R = 48.3 \text{ ft-lbf/lbm°R} \qquad k = 1.396$$
$$c_p = 0.22 \text{ Btu/lbm°R} \qquad p_1 = 800 \text{ psia}$$
$$c_v = 0.157 \text{ Btu/lbm°R} \qquad p_2 = 70 \text{ psia}$$

8.15. A Carnot heat pump uses nitrogen gas (which behaves as a perfect gas), and operates between 376°F and −15°F. The gas is at 15 psia when entering the compression and 270 psia when leaving. Determine the following for this cyclic device:
(a) Work of the compression per unit mass.
(b) COP.
(c) COR.

8.16. A Carnot engine composed of a piston-cylinder uses 0.006 lbm of air per cycle. The minimum volume of the enclosed cylinder is 3 in³ and the maximum is 70 in³, while the temperature range is 1500°F and 150°F.

Determine the following if the engine is running at 400 rpm.

(a) Power output.

(b) Rate of heat addition.

(c) Cycle efficiency.

8.17. A Carnot heat pump, diagrammed in figure 8–17, operates between 0°F and 80°F with ammonia as a working fluid. Assume the ammonia is a saturated vapor at point (4) while it is saturated liquid at point (1). Determine

(a) Heat added per lbm of ammonia.

(b) Heat rejected per lbm.

(c) Net work per lbm required.

(d) COP and COR.

8.18. For the heat pump of problem 8.17, assume 6 lbm/s of ammonia flows through the system. Determine

(a) Power required.

(b) Rate of heat added.

(c) Rate of heat rejected.

8.19. A Carnot heat pump uses Freon-12 as a working fluid. The cycle operates between −10°F and 70°F, and pumps 60 lbm/min of refrigerant. The physical aspects are similar to those of the device in figure 8–17, but the compressor does not utilize the work developed by the expansion of the fluid from point (1) to (2). If the freon is saturated liquid at (1) and saturated vapor at (4), determine

(a) Power required.

(b) Rate of heat addition from low temperature source.

(c) Rate of heat rejection to sink.

(d) COP and COR.

9

Availability and Useful Work

Some concepts are presented here which have general value in determining the limitations imposed on thermodynamic systems as they execute processes. From the first law of thermodynamics, a general equation is developed for useful work. We see that useful work is precisely what the name implies — work that we can directly use. From this concept we progress to a definition of *availability* and a discussion of its practical meaning. Emphasis is placed on evaluating the change in availability, which is equal to *useful work*, but theory is not ignored entirely. Example problems have been included which should aid in clarifying the *energy-entropy-work relationships*, including the ways in which energy progresses from an available to an unavailable condition. Available and unavailable energy are introduced as special cases of our availability function, and the free energy functions are given as alternate statements resulting from availability.

**9.1
Useful Work**

The simple, homogeneous system can increase in volume and thereby provide work to be used external to the system. For a specific system, such as the piston-cylinder depicted in figure 9–1, we found the work was

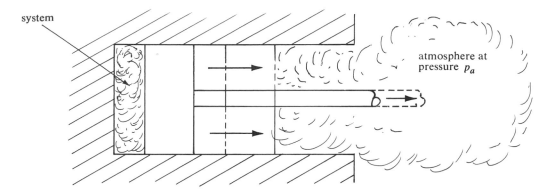

Figure 9-1   The piston-cylinder increasing the system volume.

described by the relationship

$$Wk_{cs} = \int p\,dV \qquad (9\text{--}1)$$

if the process were reversible. This equation then takes on various forms (see tables 6–2 and 6–3), depending on the type of process. The piston-cylinder is a detailed configuration of a general system, as shown in figure 9–2, which is capable of changing volume. However, almost all of the systems we can visualize will ultimately be used in an atmosphere (the Earth, Mars, or somewhere else, other than in a vacuum), and as such the system must displace some of the atmosphere, namely the amount by which the system itself changes volume. This requires work, which detracts from the $\int p\,dV$ term, and consequently we say that useful work, Wk_{use}, is obtained from the equation

$$Wk_{use} = \int_{V_1}^{V_2} p\,dV - p_a(V_2 - V_1) \qquad (9\text{--}2)$$

for a reversible process. The term p_a is the atmospheric or surrounding pressure.

We can also recall the first law for the closed system

$$E_2 - E_1 = Q - \int p\,dV \qquad (9\text{--}3)$$

and substitute the result from equation (9–2), and get

$$E_2 - E_1 = Q - Wk_{use} - p_a(V_2 - V_1) \qquad (9\text{--}4)$$

Since we are here considering ideal or reversible processes and assuming that heat transfers are conducted with the atmosphere at isothermal

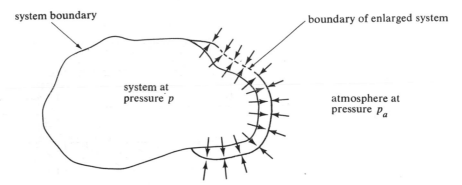

system boundary

boundary of enlarged system

system at
pressure p

atmosphere at
pressure p_a

Figure 9–2   System increasing in volume.

processes, we can then write

$$Q = T_a(S_2 - S_1) \tag{9–5}$$

when T_a is the atmospheric temperature. If we substitute this into
equation (9–4) and rearrange slightly we have

$$Wk_{\text{use}} = T_a(S_2 - S_1) - p_a(V_2 - V_1) - (E_2 - E_1) \tag{9–6}$$

We can easily include irreversible or real processes by using an inequality sign

$$Wk_{\text{use}} \leq T_a(S_2 - S_1) - p_a(V_2 - V_1) - (E_2 - E_1) \tag{9–7}$$

Of course, we identify the "less than" with an irreversible process and the
"equality" with an ideal or a reversible one.

Various information can be drawn from either equation (9–6) or (9–7).
For instance, from equation (9–6) we see that a decrease in a system's
energy does not, of itself, assure a supply of useful work. The initial
energy E_1 might very well be greater than the final energy E_2 but the
terms $T_a(S_2 - S_1)$ may be negative and/or $p_a(V_2 - V_1)$ may be large;
this could cancel any energy decreases.

Also, it is quite possible to get work out of a system during a single
process and not have a decrease in energy — or it is possible to have a heat
transfer, as the following example shows.

Example 9.1   A piston-cylinder encloses a perfect vacuum, shown in figure 9–3. The
pin holds the piston in the extended position. If the piston diameter is 6
inches, determine the useful work obtained from this device when the
pin is removed.

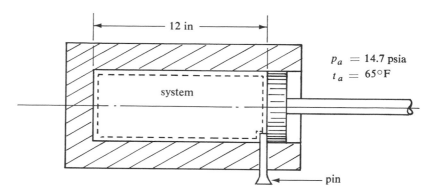

$p_a = 14.7$ psia
$t_a = 65°$F

system

pin

Figure 9–3   Vacuum system of example 9.1.

Solution   We identify the system as the enclosed vacuum and observe that the system has no entropy or energy values before, during, or after the process. The useful work we obtain from

$$Wk_{\text{use}} = T_a(S_2 - S_1) - p_a(V_2 - V_1) - (U_2 - U_1) \qquad \textbf{(9–6)}$$

and since

$$U_2 = 0 \qquad U_1 = 0 \qquad S_2 = 0 \qquad S_1 = 0$$

We have

$$Wk_{\text{use}} = p_a(V_2 - V_1)$$

The initial volume V_1 is calculated

$$V_1 = \pi \times (9 \text{ in}^2)(12 \text{ in}) = 339.1 \text{ in}^3$$

The final volume is zero since the atmosphere will push the piston into the cylinder completely. Then

$$Wk_{\text{use}} = 14.7 \text{ lbf/in}^2 \times 339.1 \text{ in}^3 = 4985 \text{ in-lbf}$$

or

$$Wk_{\text{use}} = 0.53 \text{ Btu} \qquad\qquad \textit{Answer}$$

and we see that a significant amount of work can be derived from an absolute vacuum.

9.2 Availability

We have seen that the useful work obtained from a closed system is given by equation (9–6)

$$Wk_{\text{use}} = T_a(S_2 - S_1) - p_a(V_2 - V_1) - (E_2 - E_1)$$

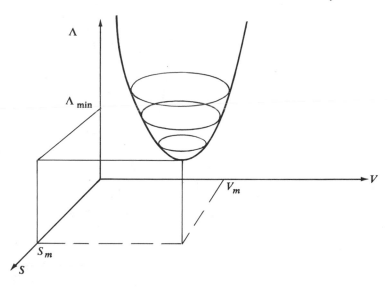

Figure 9–4   Graph of Λ-function in Λ-V-S space.

or, per unit mass

$$wk_{\text{use}} = T_a(s_2 - s_1) - p_a(v_2 - v_1) - (e_2 - e_1) \qquad (9\text{–}8)$$

where wk_{use} is Wk_{use}/m. If we now group some properties together and define a new property Λ as

$$\Lambda = E + p_a V - T_a S \qquad (9\text{–}9)$$

then, from equation (9–6), the useful work can also be given by the following equation:

$$Wk_{\text{use}} = \Lambda_1 - \Lambda_2 = -\Delta\Lambda \qquad (9\text{–}10)$$

For a given atmosphere, holding T_a and p_a constant, we can plot the property Λ. In figure 9–4 is plotted the surface of the Λ-function along the entropy-volume axis. Notice that at one unique point, Λ is a minimum — when entropy has a value S_m and volume, a value V_m. This point, Λ_{min}, is a "lowest point" and intuitively we see that a system will gravitate toward the minimum point if it is at some other state. In figure 9–5 is shown the Λ-function for entropy having a constant value S_m. It is indicated that during a process where no work is put into the given system (natural process) and where the process is merely allowed to proceed in its course, the value of Λ invariably takes a lower value, or the system approaches the state given by Λ_{min}.

We will define the total availability, Φ, as

$$\Phi = \Lambda - \Lambda_{\text{min}} \qquad (9\text{–}11)$$

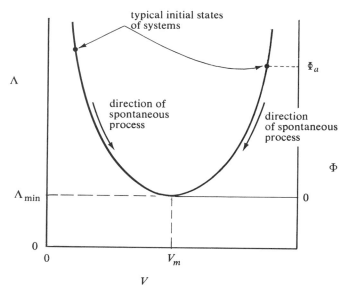

Figure 9–5   Graph of Λ or Φ and volume showing direction of a spontaneous or natural process of a system.

and remark that Φ has a value of zero when it is at the state specified by the property values S_m, V_m, and E_m. Availability is plotted in the graph of figure 9–5 to show this relationship and it can be seen that since Λ_{min} represents a constant value

$$\Phi_2 - \Phi_1 = \Lambda_2 - \Lambda_1$$

or, generally

$$\Delta\Phi = \Delta\Lambda = (E_2 - E_1) + p_a(V_2 - V_1) - T_a(S_2 - S_1) \quad \textbf{(9–12)}$$

and using equation (9–10) we obtain

$$Wk_{use} = -\Delta\Phi \quad \textbf{(9–13)}$$

Again, we can include all processes, reversible and irreversible, by writing

$$Wk_{use} \leq -\Delta\Phi \quad \textbf{(9–14)}$$

The availability then gives us an upper limit or greatest expected output of a particular system. Let us look at a problem solved by the use of the availability concept.

Example 9.2   Determine the vertical distance through which ice can raise itself in an atmosphere at 14.7 psia and 70°F. The energy required to change ice to water at 32°F is 80 cal/g and the specific heat of water can be taken as 1 Btu/lbm°R between 32°F and 70°F.

Solution

First, we must realize that the answer we will get is one that is a maximum value; for real irreversible processes, the vertical distance will be less than the calculated answer. The system is, obviously, the ice (or water) and may also be visualized as cubes of ice — 1-lbm cubes as shown in figure 9–6. Also in this figure we see how the ice is going to lift itself up; it acts as a low temperature reservoir (but ultimately reaches equilibrium with the atmosphere) for a reversible heat engine and the heat engine then can drive an appropriate device to hoist the ice, melting ice, or water. This heat engine can, of course, run only until the ice has melted and attained a temperature of 70°F. We can easily identify the work to be used for each pound-mass of ice, assuming a gravitational acceleration of 32.2 ft/s² as 1 lbf × y, which is just the increase in potential energy of the water. The change in availability of the ice (or water) is obtained from the relationship

$$\Delta\Phi = \Delta\Lambda = \Lambda_a - \Lambda_1$$

and then

$$\Delta\Phi = E_a + p_aV_a - T_aS_a - E_1 - p_aV_1 + T_aS_1$$

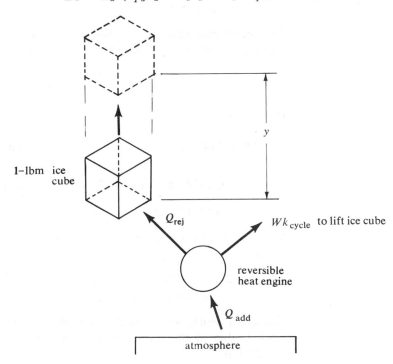

1–lbm ice cube

Q_{rej}

Wk_{cycle} to lift ice cube

reversible heat engine

Q_{add}

atmosphere

Figure 9–6   The availability of an ice cube.

The energy of the water is in the form of internal energy and potential energy (since it is being hoisted y feet up). We then write

$$\Delta\Phi = U_a - U_1 + W(y) + p_a(V_a - V_1) - T_a(S_a - S_1)$$

Since no appreciable volume change is exhibited by the water in going from ice to 70°F water, we can ignore the $p_a(V_a - V_1)$ term.

As we discussed in problem 8.3, most materials require a proportionately large amount of energy to change phase and this change occurs at constant temperature. This generally is true for the liquid-solid phase change as well as for the vapor-liquid. The energy required to convert a solid to a liquid is frequently called the *latent heat of liquefaction* or just *latent heat*. We know it to be 80 cal/g for water at 32°F, so our energy change can be found from the following:

$$\begin{aligned}
U_a - U_1 &= \text{(latent heat)} + mc(T_a - T_1) \\
&= 80 \text{ cal/g} + (1 \text{ lbm})(1 \text{ Btu/lbm°R})(70°F - 32°F) \\
&= 80 \text{ cal/g} \times \frac{454 \text{ g/lbm}}{252 \text{ cal/Btu}} + 38 \text{ Btu} = 144 \text{ Btu} + 38 \text{ Btu} \\
&= 182 \text{ Btu}
\end{aligned}$$

The energy change for each pound-mass of water is then 182 Btu. For the entropy change we use the definition of entropy

$$\Delta S = \frac{Q_{\text{rev}}}{T} = \frac{\text{latent heat}}{T_1} + \int_{T_1=32°F}^{T_2=70°F} \frac{mcdT}{T}$$

from which we obtain

$$\begin{aligned}
\Delta S &= 144 \text{ Btu}/492°R + (1 \text{ lbm})(1 \text{ Btu/lbm°R}) \int_{T_1}^{T_2} \frac{dT}{T} \\
&= 0.2927 \text{ Btu/°R} + (1 \text{ Btu/°R})\left(\ln \frac{530}{492}\right) = 0.2927 + 0.074 \\
&= 0.3667 \text{ Btu/°R}
\end{aligned}$$

We then have for a change in availability of each pound-mass of the water

$$\begin{aligned}
\Delta\Phi &= 182 \text{ Btu} + (1 \text{ lbf})(y) - (530°R)(0.3667 \text{ Btu/°R}) \\
&= -12.4 \text{ Btu} + 1 \text{ lbf}(y)
\end{aligned}$$

The useful work, which we derive directly from the ice, is zero (all our work is due to the reversible heat engine), so we get from equation (9–12) that

$$Wk_{\text{use}} = 0 = -\Delta\Phi$$

and

$$\Delta\Phi = -12.4 \text{ Btu} + (1 \text{ lbf})(y)$$

From this we obtain

$$y = 12.4 \text{ Btu}/1 \text{ lbf} = 12.4 \times 778 = 9{,}609 \text{ ft} \qquad \textit{Answer}$$

Notice in this problem that the actual lifting of the water was contingent on a number of mechanical apparatus: a reversible heat engine; a transmission to physically hoist the water; and some type of reversible heat transfer pipe from the heat engine to both atmosphere and water.

We will now concern ourselves with some remarks of the degradation of energy to an unavailable state.

**9.3
Energy
Degradation**

While we have seen that it is not energy but availability which is desirable, we will use the term *energy degradation* to imply a degrading of availability or a moving down the curve in figure 9–5. We lose availability in a real process and in a reversible process it merely remains the same. Loss of availability is subtle and difficult to detect, but predicting irreversibilities, degradation of energy, or increases in entropy in a general process is still more difficult. One of the important areas of research in thermodynamics is the development of more general and accurate equations or theorems which can allow these predictions of irreversibility — it is a crucial area since most of our present tools of thermodynamics can only be used accurately for reversible and static systems which are indeed not the real world.

Example 9.3

We seek a comparison of the maximum useful work obtainable from two systems which begin at identical states, proceed through different processes, and reach states which are not too different.

System (1), shown in figure 9–7, is an adiabatic chamber enclosing air in the left half. The right half is a vacuum separated from the left by a removable wall. The air pressure is 50 psia, the temperature 100°F, and the volume 2 ft³. The volume of the vacuum is also 2 ft³ so that when the removable wall is taken away, the air can occupy the full volume of 4 ft³. When this happens the air pressure will drop to 25 psia, but the temperature will remain at 100°F. Why?

System (2) in figure 9–7 is an adiabatic piston-cylinder device which initially encloses 2 ft³ of air at 50 psia and 100°F. The system then

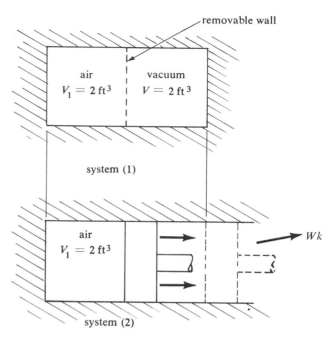

Figure 9–7   Two systems which are going through adiabatic processes with identical volumes but which will reach different states and have different amounts of work produced.

progresses through a reversible adiabatic process until the air fills a volume of 4 ft³.

Both systems are at final states that are slightly different; from system (2) we have extracted some useful work while system (1) has its initial energy value. Heat has not been transferred in either system but we can ask, "Have we lost any availability in system (1) by not utilizing the expansion of the air from 2 to 4 ft³?" The answer, as we will subsequently see is "yes."

First let us look at system (1). We assume the air to be a perfect gas so we have the following results:

$$\Delta s = s_2 - s_1 = R \ln \frac{V_2}{V_1} \tag{9–15}$$

$$= 53.3 \text{ ft-lbf/lbm°R } \ln \frac{4}{2} \times 1 \text{ Btu/778 ft-lbf}$$

$$= 0.0475 \text{ Btu/lbm°R}$$

$$wk = 0 = q \tag{9–16}$$

$$\Delta u = u_2 - u_1 = 0 \tag{9-17}$$

$$\Delta T = T_2 - T_1 = 0 \tag{9-18}$$

$$p_2 = p_1\left(\frac{V_1}{V_2}\right) = 50 \text{ lbf/in}^2\left(\frac{2}{4}\right) = 25 \text{ psia}$$

For system (2), with air behaving as a perfect gas again, we have these results:

$$\Delta S = 0$$

$$q = 0$$

$$wk = \frac{R}{1-k}(T_2 - T_1) \tag{9-19}$$

$$T_2 = T_1\left(\frac{V_1}{V_2}\right)^{k-1} = (560°R)\left(\frac{2}{4}\right)^{0.4} = 424°R$$

The work we then obtain

$$wk = \frac{53.3 \text{ ft-lbf/lbm°R}}{1.0 - 1.4}(424°R - 560°R)$$

$$= 18{,}122 \text{ ft-lbf/lbm} = 23.3 \text{ Btu/lbm}$$

Also we get

$$p_2 = p_1\left(\frac{V_1}{V_2}\right)^{1.4} = 50 \text{ lbf/}m^2\left(\frac{2}{4}\right)^{1.4} = 18.95 \text{ psia}$$

and

$$\Delta u = u_2 - u_1 = -wk = -23.3 \text{ Btu/lbm}$$

Obviously, the availability was the same for both systems at their initial states. To determine the availability of the systems at their final states, we need the change in availability. This change we find from equation (9-12)

$$\Delta\Phi = \Phi_2 - \Phi_1 = U_2 - U_1 + p_a(V_2 - V_1) - T_a(S_2 - S_1)$$

For system (1) the change is obtained

$$\Phi_2 - \Phi_1 = 0 + p_a(4 \text{ ft}^3 - 2 \text{ ft}^3) - T_a m(0.0475 \text{ Btu/lbm°R})$$

We now assume the surroundings are at a temperature T_a of 70°F and pressure p_a of 14.7 psia. The system mass m we get from

$$m = \frac{p_1 V_1}{R T_1} = \frac{50 \text{ lbf/in}^2 \times 144 \text{ in}^2/\text{ft}^2 \times 2 \text{ ft}^3}{53.3 \text{ ft-lbf/lbm°R} \times 560°R}$$

$$= 0.483 \text{ lbm}$$

and then

$$\Phi_2 - \Phi_1 = 14.7 \text{ lbf/in}^2 \times 144 \text{ in}^2/\text{ft}^2 \times 2 \text{ ft}^3 - 530°R \times 0.483 \text{ lbm}$$
$$\times 0.0475 \text{ Btu/lbm°R}$$
$$= 4230 \text{ ft-lbf} - 12.1 \text{ Btu}$$
$$= 5.4 \text{ Btu} - 12.1 \text{ Btu} = -6.7 \text{ Btu}$$

For system (2), the change in availability is calculated

$$\Phi_2 - \Phi_1 = m(u_2 - u_1) + p_a(V_2 - V_1) - T_a\Delta S$$
$$= (0.483 \text{ lbm})(-23.3 \text{ Btu/lbm}) + \frac{14.7 \times 144 \times 2}{778} \text{ Btu}$$
$$= -11.2 \text{ Btu} + 5.4 \text{ Btu} = -5.8 \text{ Btu}$$

Notice that since system (2) involved a reversible process, the change in availability is exactly the amount of useful work gotten from the piston-cylinder. System (1), by going through an irreversible expansion, decreased its availability more than system (2), even though its energy has remained constant. The apparent inequality which gives system (2) a higher final availability than system (1) is due to the fact that the final temperature of system (2) is 424°R — well below the temperature of the surroundings and therefore providing a capability for heat transfer. This aspect goes back to the concepts of example 9.2.

One of the practical conclusions to be drawn from this problem is that a gas in a cylinder or any other source of availability must be used in a work process during the actual expansion or else the availability is lost forever. It is exactly like water flowing over a drop; if the potential energy of the water is not immediately converted into some other energy, such as electric power, then the water will flow down and dissipate (quite irreversibly) the kinetic energy and degrade the availability of the energy of that portion of water. Thus, engineers and technologists must be opportunistic in generating power for society by extracting energy from processes at the appropriate times and by better utilizing the naturally occurring processes of nature (such as tides, tornadoes, hurricanes, and winds) for energy sources.

One other task of those who furnish power or availability to society is the storage of such products. In mechanics, a flywheel rotates at high speed, storing kinetic energy for later use. Electric energy is stored in batteries or capacitors, while thermal or internal energy is stored in material which is retained in insulated or adiabatic chambers. When the energy is put into storage, we lose some availability in the mere process of storing. The term *unavailable energy* has traditionally been associated with the amount of energy which is lost during a process and which can never again be put to useful work. *Available energy*, on the other hand, has been associated with that energy which can be reconverted to useful work;

it is efficiently stored for future use. These two terms are tied up in the availability concept, but an example problem might provide some enlightenment.

Example 9.4

A heat exchanger is a device which transfers thermal energy from one material to another. In figure 9–8 is shown such a heat exchanger which allows hot exhaust gases to transfer some of their energy to air flowing in the opposite direction. Outside the heat transfer q, between the two streams, the system can be assumed to be adiabatic. Assume 61,200 lbm/hour of hot gases enter at 900°F and leave at 400°F, all at a pressure of 14.7 psia. There are 60,000 lbm/hour of air flowing into the exchange at 250°F and at 14.7 psia. The specific heats at constant pressure are 0.26 Btu/lbm°R for the gases and 0.24 for air; both are assumed to be perfect gases having R values of 70 ft-lbf/lbm°R and 53.3 ft-lbf/lbm°R respectively. If the atmospheric conditions are 80°F and 14.7 psia, determine the loss of availability per hour of the system due to this process.

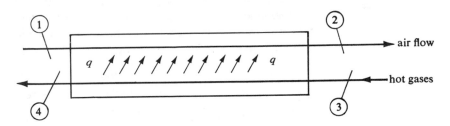

Figure 9–8   Heat exchanger.

Solution

We recognize that here we have an open system and that our concepts of useful work and availability were adapted to closed systems. It can be shown, however, that for open, steady flow systems, the rate of change in availability, $d\Phi/d\tau$, is obtained from;

$$\dot\Phi = \dot m\left((h_2 - h_1) + \frac{V_2^2 - V_1^2}{2g_c} + \frac{g}{g_c}(z_2 - z_1) - T_a(s_2 - s_1)\right) \quad (9\text{–}20)$$

The total rate of change in availability, for this system is given by

$$\dot\Phi = \dot\Phi_{\text{air}} + \dot\Phi_{\text{gas}} \quad (9\text{–}21)$$

In figure 9–9 is shown the T-s diagram for the air and gas simultaneously. The total areas under each curve are equal and represent the heat transferred between the gas and air, q. That is, the total area under curve (3–4) is equal to the total area under (1–2). Notice also that the areas under the T_a abscissa line are labeled as "unavailable energy." This is a

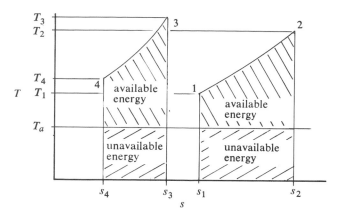

Figure 9-9   *T*-s diagram for heat exchanger.

good way of visualizing equation (9–20) or (9–12), but in this graphical representation, the $p_a(V_2 - V_1)$ term is neglected. We see also that the "available energy" areas of figure 9–9 are equal to the availability per unit mass for this steady flow problem. The generation of unavailable energy then is given by the expression

$$\dot{m}(T_a \Delta s) \qquad\qquad (9\text{–}22)$$

For the gases we determine the following properties:

$$h_4 - h_3 = c_p(T_4 - T_3) = 0.26 \text{ Btu/lbm}^\circ\text{R}(400 - 900)^\circ\text{R}$$
$$= -130 \text{ Btu/lbm}$$

$$s_4 - s_3 = c_p \ln \frac{T_4}{T_3} = 0.26 \text{ Btu/lbm}^\circ\text{R} \ln \frac{860}{1360} = -0.119 \text{ Btu/lbm}^\circ\text{R}$$

From these results and using equation (9–20), we calculate the rate of change of availability for the gas assuming no kinetic or potential energy changes

$$\dot{\Phi}_{\text{gas}} = \dot{m}_{\text{gas}}[(h_4 - h_3) - T_a(s_4 - s_3)]$$
$$= 61200 \text{ lbm/hr } [- 130 \text{ Btu/lbm} + 540^\circ\text{R}(0.119 \text{ Btu/lbm}^\circ\text{R})]$$
$$= -4{,}020{,}000 \text{ Btu/hr}$$

The available energy per unit mass of the gas in figure 9–9 is equal to $h_4 - h_3 - T_a(s_4 - s_3)$ or -65.7 Btu/lbm, and the unavailable energy is given by the term $540^\circ\text{R}(-0.119 \text{ Btu/lbm}^\circ\text{R})$ which is equal to -64.3 Btu/lbm.

For the air we have

$$h_2 - h_1 = c_p(T_2 - T_1) = 0.24 \text{ Btu/lbm}^\circ\text{R}(T_2 - 250^\circ\text{F})$$

$$s_2 - s_1 = c_p \ln\left(\frac{T_2}{T_1}\right) = 0.24 \text{ Btu/lbm°R } \ln\left(\frac{T_2}{(250 + 460)°R}\right)$$

To determine the final air temperature, we notice from the first law of thermodynamics applied to the heat exchanger that the rate of increase in enthalpy of the air is exactly the same magnitude as the decrease of the gas. Therefore

$$\dot{m}_{air}(h_2 - h_1) = \dot{m}_{gas}(h_3 - h_4)$$

$$(60000 \text{ lbm/hr})(h_2 - h_1) = (61200 \text{ lbm/}h)(130 \text{ Btu/lbm})$$

and using

$$h_2 - h_1 = c_p(T_2 - T_1)$$

we obtain

$$T_2 = T_1 + \frac{(61,200)(130)}{(60,000)(0.24)}$$

assuming $c_p = 0.24$ Btu/lbm°R. Since T_1 equals 250°F, we have

$$T_2 = 250 + 552.5 = 802.5°F$$

Then

$$s_2 - s_1 = 0.24 \ln\frac{802.5 + 460}{710} = 0.138 \text{ Btu/lbm°R}$$

We now calculate the enthalpy change:

$$h_2 - h_1 = \left(\frac{61,200}{60,000}\right)(130) = 132.6 \text{ Btu/lbm air}$$

The rate of change of availability for the air is, using equation (9–20) with kinetic and potential energies neglected

$$\Phi = \dot{m}_{air}[(h_2 - h_1) - T_a(s_2 - s_1)]$$
$$= 60,000 \text{ lbm/hr } [132.6 \text{ Btu/lbm} - 540°R(0.138 \text{ Btu/lbm°R})]$$
$$= 3,484,800 \text{ Btu/hr}$$

and the total rate of change of availability for our system is, from equation (9–21)

$$\Phi = 3,484,800 - 4,020,000 = -535,200 \text{ Btu/hr}   \textit{Answer}$$

The available energy of the air per lbm of gas is given by

$$\frac{\dot{m}_{air}}{\dot{m}_{gas}}[(h_2 - h_1) - T_a(s_2 - s_1)]$$

or

$$\left(\frac{60{,}000}{61{,}200}\right)[(132.6) - 540(0.138)] = 56.9 \text{ Btu/lbm gas}$$

The unavailable energy of the air per lbm of gas is

$$\frac{\dot{m}_{\text{air}}}{\dot{m}_{\text{gas}}}[T_a(s_2 - s_1)] = \frac{60{,}000}{61{,}200}[540(0.138)] = 73.1 \text{ Btu/lbm gas}$$

The total available energy change per lbm of gas is obtained from the difference of the two designated areas under the curves in figure 9–9 or the sum of the available energies of the gas plus the air. Thus

Available energy change per lbm $= -65.7 + 56.9 = -8.8$ Btu/lbm

Likewise, the unavailable energy change per lbm is equal to the sum of the unavailable energies of the gas plus air

$$-64.3 + 73.1 \quad \text{or} \quad +8.8 \text{ Btu/lbm}$$

The decrease in available energy here corresponds to the decrease in availability or increase in unavailable energy. The total generation of unavailable energy is given by

$$\dot{m}_{\text{air}}T_a\Delta s_{\text{air}} + \dot{m}_{\text{gas}}T_a\Delta s_{\text{gas}} = 60{,}000 \times 540(0.138) - 61200 \times 540(0.119)$$
$$= 538{,}488 \text{ Btu/hr}$$

9.4
Free Energy

For a system which is in thermal and pressure equilibrium with its surroundings, that is, $T = T_a$ and $p = p_a$, we can write the change in availability in the form

$$\Delta\Phi = (E_2 - E_1) + p(V_2 - V_1) - T(S_2 - S_1) \qquad (9\text{--}23)$$

If we have no kinetic and potential energy changes we have

$$E_2 - E_1 = U_2 - U_1$$

and then

$$\Delta\Phi = (U_2 - U_1) + p(V_2 - V_1) - T(S_2 - S_1) \qquad (9\text{--}24)$$

The property identified with the quantity, $U + pV - TS$, we call the *total Gibbs free energy* G' and write

$$G' = U + pV - TS \qquad (9\text{--}25)$$

The total Gibbs free energy has units of energy and the Gibbs free energy g′ defined as

$$g' = \frac{G'}{m} = u + pv - Ts \qquad (9\text{-}26)$$

has units of energy per unit mass. These properties are quite useful in analyzing chemical reactions, in predicting heats of combustion tabulated in table B.7, and, of course, in predicting the useful work obtainable from a system whose pressure and temperature are equal to constant atmospheric conditions. If this equilibrium holds for a system then

$$\Delta\Phi = \Delta G' \qquad (9\text{-}27)$$

and we can directly say

$$Wk_{use} = -\Delta G' \qquad (9\text{-}28)$$

For systems which are at constant volume we predict the maximum useful work from

$$Wk_{use} = (E_2 - E_1) - T_a(S_2 - S_1) \qquad (9\text{-}29)$$

and, again, if the system is in thermal equilibrium with the atmosphere at constant temperature $(T = T_a)$

$$Wk_{use} = (E_2 - E_1) - T(S_2 - S_1) \qquad (9\text{-}30)$$

We associate with the quantity $U - TS$ a property of the system called the *total Helmholtz free energy* H′ written

$$H' = U - TS \qquad (9\text{-}31)$$

and, obviously, for the above conditions of the system with no kinetic or potential energy changes

$$Wk_{use} = -\Delta H' \qquad (9\text{-}32)$$

We can consider also the Helmholtz free energy defined by

$$h' = \frac{H'}{m} = u - Ts \qquad (9\text{-}33)$$

and thereby get an intensive property of the system.

The Gibbs and Helmholtz free energy properties are frequently used in thermodynamic literature, and this introduction to their derivation should be helpful to the engineer and technologist seeking the full use of thermodynamic tools.

Practice Problems

Problems designated with an asterisk * preceding the number should only be attempted by those having a background which includes the knowledge of integral calculus.

Assume the atmosphere conditions are 60°F and 14.7 psia for problems 9.1–9.13.

Sections 9.1 and 9.2

9.1. Determine the Λ-function of 2 lbm of H_2O at 500°F and (a) $p = 40$ psia, (b) saturated liquid, (c) saturated vapor.

9.2. Determine the Λ-function of 60 grams of mercury vapor at 90 psia and (a) saturated vapor, (b) saturated liquid. (assume $v_f = 0.0013$ ft³/lbm)

9.3. Determine the total availability of a vacuum of 2 ft³.

9.4. Determine the availability of 3 lbm of dichlorodifluoromethane, F-12, at −20°F and (a) saturated vapor, (b) saturated liquid.

*9.5. Calculate the maximum useful work obtainable for (a) 1 lbm H_2O at 700°F and 200 psia, (b) 1 lbm H_2O at 1000°F and 200 psia, (c) 1 lbm H_2O at 700°F and 300 psia.

9.6. Determine the maximum useful work obtainable from 1 lbm of the following perfect gases at 1000°F and 200 psia:
(a) Air.
(b) Nitrogen, N_2.
(c) Hydrogen, H^2.
(d) Sulfur dioxide, SO_2.

9.7. Determine the availability for the following perfect gases at 500°C and 760 mm Hg:
(a) Oxygen, O_2.
(b) Methane, CH_4.
(c) Carbon dioxide, CO_2.
(d) Carbon monoxide, CO.

9.8. Determine the vertical distance through which 3 lbm of ice can lift itself if $T_a = 40°F$ and $p_a = 14.7$ psia.

Section 9.3

9.9. Calculate the generation or total rate of change of availability for 500 lbm/min of air flowing through a reversible adiabatic turbine from 1200°F and 200 psia, exhausting to 15 psia. (Hint: Assume that there are no kinetic or potential energy changes and the power generated in the turbine is all useful.)

9.10. Calculate the total rate of change of availability for 20 lb/s of air compressed in a reversible adiabatic manner from 100°F and 15 psia to 180 psia. (Hint: Assume no kinetic or potential energy changes. Obviously, the work put into the compressor is useful.)

9.11. A piston-cylinder contains 0.5 lbm of oxygen gas at 30 psia and 100°F. The device then compresses the gas in a reversible manner to a pressure of 300 psia. Assume $n = 1.45$ and the gas behaves as a perfect gas with constant specific heats. Determine the increase in availability of the oxygen and the total change in availability of the universe due to this compression process.

9.12. There is 0.2 kg of air contained in a frictionless piston-cylinder. The air is at a pressure of 20×10^6 dynes/cm² and a temperature of 600°C. Through a reversible adiabatic process the air pressure is reduced to

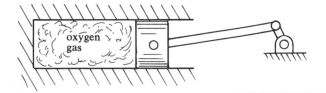

Figure 9–10

Problem 9.11.

1×10^6 dyne/cm². Determine the change in availability of the air and the total change in the availability of the universe.

*9.13. In the Φ-S-V space determine expressions for the tangent or slope of the Φ-surface in the plane of
  (a) Constant entropy.
  (b) Constant volume.
  (c) Constant availability.

Section 9.4

9.14. Determine the Gibbs and Helmholtz free energies for 1 lbm H_2O at 600°F and
  (a) $p = 40$ psia.
  (b) $p = 100$ psia.
  (c) saturated vapor.
  (d) saturated liquid.

9.15. Determine the Gibbs free energy for ammonia, NH_3, at 120°F and 20 psia.

9.16. Determine the Helmholtz free energy for ammonia at 200°F and 60 psia.

10

The Internal Combustion Engine and the Otto and Diesel Cycles

The *internal combustion* (IC) *engine*, characterized by the reciprocating piston-cylinder gasoline engine, is probably the most common power-producing device in our society. It is utilized in all phases of transportation, for auxiliary electrical power generation and innumerable small, portable power tools. This engine is a good example of what we have defined as a *heat engine*, and although the concept of an external heat source may not be clearly compatible with the practicality of an internal combustion (to the system) of fuel, we will see how this ambiguity can be eliminated.

We will define the *ideal Otto cycle* and see how it fits into the analysis of the spark ignited internal combustion engine. Initially we will lean heavily on the assumption that air is the working fluid; but a use of the gas tables in solving Otto cycle problems will help to explain the departures of the air-fuel mixtures and exhaust gases from this assumption.

The analogy between the actual engine and Otto cycle will be sharpened by using the *polytropic process equations*. The actual Otto cycle will then be discussed with some of the more traditional modifications designed to improve the engine power, efficiency, fuel economy, or other engine performance parameters.

We will introduce the *Diesel cycle* and its adaption in the Diesel engine. The *air-standard analysis* will be treated thoroughly and a comparison between the Diesel and Otto cycles will better focus in the reasons for each and their traditional areas of application.

Finally we will discuss some of the design problems inherent in the reciprocating IC engine and the approaches for solving them. The engines described in this chapter have probably absorbed more engineering talent than any other comparable device of man. An overview of the shortcomings and the advantages of these engines should be useful to the student.

10.1 The Ideal Otto Cycle

The ideal Otto cycle is defined as the following set of reversible processes:

(1–2)   Adiabatic compression.
(2–3)   Constant volume heat addition.
(3–4)   Adiabatic expansion.
(4–1)   Constant volume heat rejection.

These processes are indicated in figure 10–1 where *p-V* and *T-S* diagrams are used to describe the Otto cycle. Notice that since these processes are

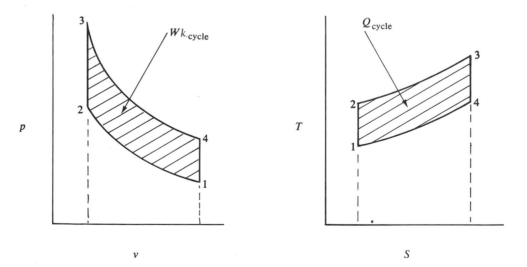

Figure 10–1   Property diagrams for ideal Otto cycle.

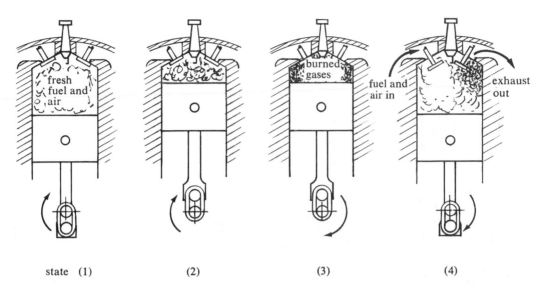

state (1)   (2)   (3)   (4)

Figure 10–2   Otto cycle application in piston-cylinder device.

all reversible, we can easily identify the enclosed area on the *p-V* diagram as the net work of the cycle and the enclosed area on the *T-S* diagram as the net heat added. To understand how these four processes manage to be descriptive of a real machine operation, let us see how the piston-cylinder can be used with an Otto cycle. In figure 10–2 is shown the sequence of motions of the piston corresponding to those processes. If the piston is reciprocating in a continuous manner, the processes (2–3) and (3–4) must be performed in a zero time period since there is no motion. This is, of course, a deviation the ideal cycle has from the actual case, but we will later see that this is not a large error.

Process (1–2) is a compression of a charge of air and unburned fuel. Upon reaching state (2), the spark plug fires, initiating a chemical reaction between the fuel and air. This is the internal combustion which categorizes the cycle and which releases energy from the fuel (or adds heat to the piston-cylinder), producing a high temperature-pressure gas which drives the piston through an expansion process to state (4). Heat is then quickly removed by opening the exhaust and intake valves, discharging the burned exhaust gases, and just as quickly replacing this volume with a fresh charge of air and unburned fuel to proceed through the cycle again. This fast shuttling of gases allows us to have power produced on each cycle, called a *two-stroke cycle* or *two-stroke engine*. It is given this name because two strokes complete the cycle (up-down), but the quick gas shuttle requires imaginative design in valves. The configu-

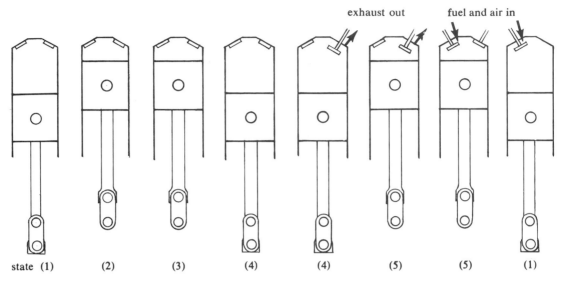

exhaust out   fuel and air in

state (1)   (2)   (3)   (4)   (4)   (5)   (5)   (1)

Figure 10–3   Four-stroke Otto cycle.

ration of figure 10–2 probably would not give a good exchange of fresh charge for exhaust and much effort is expended in attempting to design a better two-stroke engine which does not waste fuel in discharging the exhaust gases. The prize is one power stroke per revolution of the engine. More often though the engine is changed from the Otto cycle by proceeding through one more revolution to rinse the exhaust gases and add fresh charges more effectively. This variation we call the *four-stroke cycle*, characterized by the opening of the exhaust valve only when the piston-cylinder achieves state (4). This valve remains open through a process (1–5), shown in figures 10–3 and 10–4, during which all the exhaust is pushed out by the piston. At state (5), the exhaust valve closes and the intake valve opens. The fresh fuel-air mixture is then taken in by the retreating piston until state (1) is reached, the intake valve closes, and the normal Otto cycle can proceed. The four-stroke cycle is more commonly used because it allows for better control of the gases than the two-stroke cycle, but there is only one power stroke every other revolution, thus the term "four-stroke" since we have an in-out-in-out motion of the piston for each cycle. Theoretically, the four-stroke engine needs to rotate twice as fast as the two-stroke engine to achieve the same power. This does not generally hold true in practice, however, due to other complications which tend to degrade the attractiveness of the two-stroke cycle.

The complications of timing, that is, the opening and closing of valves at opportune moments, is a serious deficiency in the whole design for the

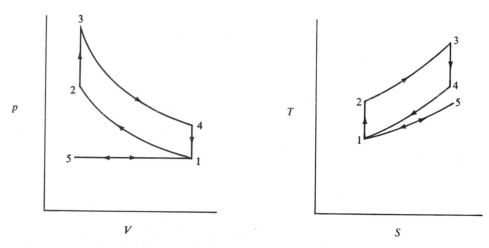

Figure 10-4   Property diagrams for the four-stroke Otto cycle.

normal Otto cycle application. In table 10–1, somewhat of a recapitula-
tion, it is indicated when exhaust and intake valves are open or closed.
This is done for both the two- and four-stroke cycles and should aid in
visualizing the motion of the spark-ignition, internal combustion, piston-
cylinder engine.

Table 10-1

Valve Positions of Otto Cycles

Process	4-stroke cycle		2-stroke cycle	
	Intake Valve	Exhaust Valve	Intake Valve	Exhaust Valve
1–2	closed	closed	closed	closed
2–3	closed	closed	closed	closed
3–4	closed	closed	closed	closed
4–1	closed	open	open	open
1–5	closed	open		
5–1	open	closed		

In figure 10–5 is shown a typical cutaway view of an internal combustion
engine which operates on a cycle approximately like the Otto cycle. In
this figure are shown the major components of the engine, and in figure
10–6 is shown a section of an actual IC engine with water cooling. This
view shows the external characteristics of the typical water-cooled IC

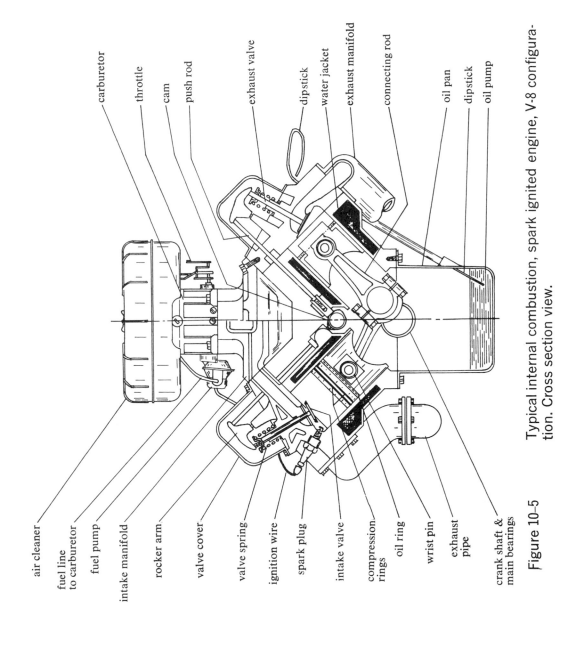

air cleaner

fuel line
to carburetor

fuel pump

intake manifold

rocker arm

valve cover

valve spring

ignition wire

spark plug

intake valve

compression
rings

oil ring

wrist pin

exhaust
pipe

crank shaft &
main bearings

carburetor

throttle

cam

push rod

exhaust valve

dipstick

water jacket

exhaust manifold

connecting rod

oil pan

dipstick

oil pump

Figure 10–5   Typical internal combustion, spark ignited engine, V-8 configuration. Cross section view.

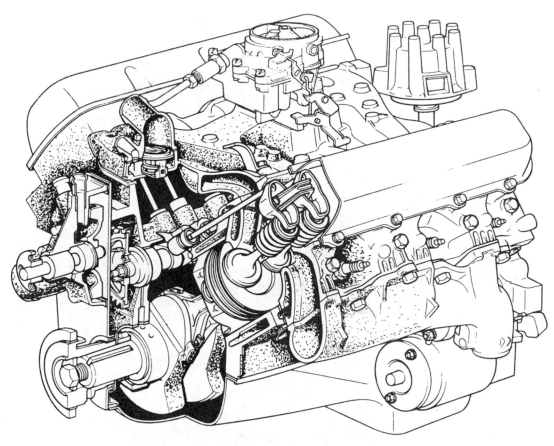

Figure 10–6   The internal combustion engine. Reprinted by permission of
General Motors Corporation, Oldsmobile Division.

engine with a portion of it cut away to expose the major workings of the
internal parts of the engine. Compare this figure with figure 10–5 and the
component parts will be better visualized.

While the Otto cycle is characterized by heat transfer during constant
volume, the real engine is continually experiencing heat transfers. Because of this, water is generally directed through cavities in the engine
block to keep the cylinder and piston from reaching high temperatures,
but the air-cooled engine in figure 10–7 is also effective. In this type of
engine, water is not used, but air is forced around the cylinder block to
provide convective heat transfer, thus keeping the engine cool.

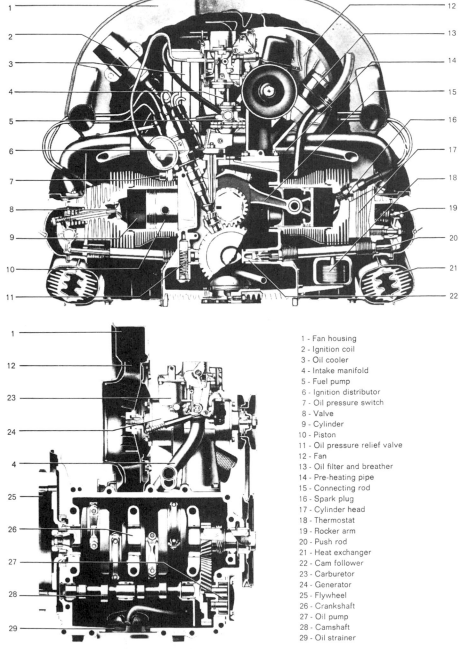

1 - Fan housing
2 - Ignition coil
3 - Oil cooler
4 - Intake manifold
5 - Fuel pump
6 - Ignition distributor
7 - Oil pressure switch
8 - Valve
9 - Cylinder
10 - Piston
11 - Oil pressure relief valve
12 - Fan
13 - Oil filter and breather
14 - Pre-heating pipe
15 - Connecting rod
16 - Spark plug
17 - Cylinder head
18 - Thermostat
19 - Rocker arm
20 - Push rod
21 - Heat exchanger
22 - Cam follower
23 - Carburetor
24 - Generator
25 - Flywheel
26 - Crankshaft
27 - Oil pump
28 - Camshaft
29 - Oil strainer

Figure 10–7   Air-cooled internal combustion engine. Reprinted from Volkswagen Service Manual (VW 1300-1500) (Bielfeld and Berlin, Germany); with permission of Volkswagen of America, Inc., and Delius Klasing and Co.

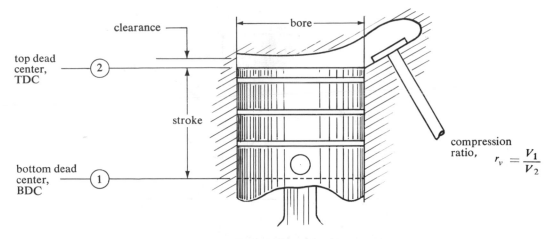

Figure 10–8   Common IC engine parameters.

10.2
The Air-Standard Otto Cycle

Since air is the major constituent of the gases entering the cylinder in the Otto cycle, we shall assume that it is all air. This is the requirement of an air-standard Otto cycle and the analysis will be rather straightforward and familiar. We will use the process equations from chapter 6 in many of the calculating procedures and the student would benefit from referring to chapter 6 to see why certain equations are used. In the following two example problems will be shown the thermodynamics of the Otto cycle.

Example 10.1

An internal combustion, spark-ignited, four-cylinder engine (ICSI), has a bore (piston diameter) of 3 inches, a stroke (distance piston travels in one stroke) of 3 inches, and a clearance of 0.4 inch. (See figure 10–8.) The air in the cylinder at the beginning of compression, state (1), is at 95°F and 14.7 psia. The fuel and air react to release 1200 Btu/lb of air during the internal combustion. Assuming an air-standard Otto cycle, determine the properties p, T, and V at each corner of the cycle, the heat added per cycle, the work produced per cycle, and the heat rejected per cycle. Also determine the power generated if the engine runs at 1200 rpm and we assume a four-stroke cycle.

Solution

The system is the air (and fuel, which we ignore) contained in the cylinders; and for each of the four Otto cycle processes, the system is closed. With T_3 designated as the highest cycle temperature and T_1 as the lowest, the system diagram shown in figure 10–9 is found to operate between

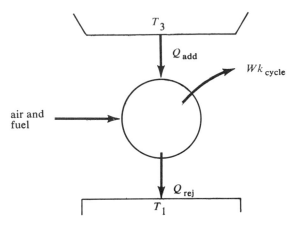

Figure 10-9   System diagram for Otto cycle.

these two temperatures and is an open cycle. Notice that the air and exhaust gas flows are not steady flow during one cycle; but when observed for an extended period of time, their flows are steady.

To determine the properties at each of the four-cycle corners, let us first sketch the p-V and T-s diagrams. (See figure 10–10.) We see that $V_1 = V_4$ and $V_2 = V_3$. Let us then calculate the volumes from the dimensional data:

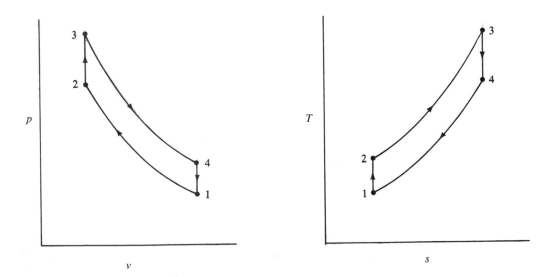

Figure 10-10   Typical Otto engine property diagrams.

$$V_1 = \pi \frac{(3)^2}{4} \times 3.4 = 24.0 \text{ in}^3 = V_4 \qquad \textit{Answer}$$

$$V_2 = \pi \frac{(3)^2}{4} \times 0.4 = 2.83 \text{ in}^3 = V_3 \qquad \textit{Answer}$$

For the mass where m = mass of air in one cylinder per cycle

$$m = \frac{p_1 V_1}{R T_1} = \frac{14.7 \times 24.0}{53.3 \times (95 + 460) \times 12} = 0.000994 \text{ lbm}$$

For the properties at point (2)

$$\frac{p_2}{p_1} = \left(\frac{V_1}{V_2}\right)^k = r_v^k$$

where we define r_v, the compression ratio, by

$$r_v = \frac{V_1}{V_2} \qquad\qquad \textbf{(10–1)}$$

Then

$$p_2 = (14.7 \text{ psia}) \left(\frac{24.0}{2.83}\right)^{1.4} = 293 \text{ psia} \qquad \textit{Answer}$$

and the compression ratio is obtained

$$r_v = \frac{24.0}{2.83} = 8.5$$

The temperature at state (2) we may calculate from either a process equation relating states (2) and (1) or from the perfect gas equation:

$$T_2 = \frac{p_2 V_2}{mR} = \frac{293 \times 2.83}{0.000994 \times 53.3 \times 12} = 1309°R \qquad \textit{Answer}$$

Process (2–3) is a reversible constant volume process involving a heat transfer and no work since the system is closed. Then we can write

$$u_3 - u_2 = q_{\text{add}} = 1200 \text{ Btu/lbm air}$$

and since air is a perfect gas with constant specific heats (we assume) we can write

$$u_3 - u_2 = c_v(T_3 - T_2) = 1200 \text{ Btu/lbm air}$$

Since $c_v = 0.171$ Btu/lbm°R for air we get

$$T_3 - T_2 = \frac{1200}{0.171}°R = 7018°R = T_3 - 1310$$

and

$$T_3 = 7018 + 1310 = 8328°R \qquad \textit{Answer}$$

Readily then, using the process relations between (2) and (3)

$$\frac{p_3}{p_2} = \frac{T_3}{T_2}$$

and

$$p_3 = p_2 \left(\frac{T_3}{T_2}\right) = (293 \text{ psia}) \left(\frac{8328°R}{1310°R}\right) = 1863 \text{ psia}$$

The properties at state (4) are found with process equations relating states (4) to (3)

$$\frac{p_4}{p_3} = \left(\frac{V_3}{V_4}\right)^k = \left(\frac{1}{r_v}\right)^k = \frac{1}{8.5^{1.4}} = \frac{1}{20}$$

so that

$$p_4 = \frac{1863 \text{ psia}}{20} = 93.2 \text{ psia}$$

and states (4) and (1)

$$\frac{p_4}{p_1} = \frac{T_4}{T_1}$$

$$T_4 = (555°R) \left(\frac{93.2}{14.7}\right) = 3519°R$$

We can easily calculate the entropy change per unit air mass from equation (7–22)

$$\Delta s = s_3 - s_2 = c_v \ln \frac{T_3}{T_2} = 0.171 \text{ Btu/lbm°R} \left(\ln \frac{8328}{1310}\right)$$

$$= 0.316 \text{ Btu/lbm°R}$$

Then

$$-\Delta s = s_4 - s_1 = -s_2 + s_3 = -0.316 \text{ Btu/lbm°R}$$

The net work per cylinder per cycle can be found by determining the enclosed area of the *p-V* diagram (figure 10–10 or 10–11)

$$Wk_{\text{cycle}} = Wk_{(3-4)} + Wk_{(1-2)}$$

$$= \frac{mR}{1-k} (T_4 - T_3 + T_2 - T_1) \tag{10–2}$$

$$= \frac{0.000994 \times 53.3}{-0.4 \times 778} (3519 - 8328 + 1309 - 555)$$

$$= 0.69 \text{ Btu/cycle}$$

The engine will then deliver, from four cylinders

$$Wk = 4 \times 0.69 = 2.76 \text{ Btu/cycle} \qquad \textit{Answer}$$

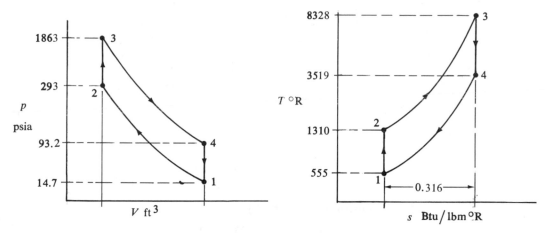

Figure 10–11   Otto engine diagrams with values from calculations.

The power produced Wk will be the work produced per cycle times the number of cycles per unit time. This last term is 1200 rpm, but only with every other revolution do we get work so the term becomes 600 rpm. Then

$$Wk = 2.76 \text{ Btu/cycle} \times 600 \text{ cycle/min} = 1656 \text{ Btu/min}$$
$$= 39.1 \text{ hp} \qquad \qquad \qquad \textit{Answer}$$

The heat rejected per cycle can be determined from the following analysis:

$$Q_{add} + Q_{rej} = Q_{cycle} = Wk_{cycle} \qquad \qquad \textbf{(10–3)}$$
$$Q_{rej} = 2.76 \text{ Btu/cycle} - (4)(1200 \text{ Btu/lbm})(0.000994)$$
$$= 2.76 - 4.77 = -2.01 \text{ Btu/cycle} \qquad \textit{Answer}$$

In this analysis of a simplified Otto cycle engine, we used many formulas which have been previously introduced; and here their utility in a more descriptive problem was brought out. Let us now look at the same physical configuration of problem 10.1, but now introduce the polytropic process to account for heat transfers during compression and expansion.

Example 10.2   For the engine of problem 10.1, assume that the compression process (1–2) is polytropic with $n_{21} = 1.30$; and assume that the expansion (3–4) is also polytropic with $n_{43} = 1.45$. If all else is the same, determine the work of the cycle, the power produced at 1200 rpm, and the heat rejected per cycle.

Solution   The system under analysis is the same as in example 10.1, that is, the enclosed volume in the cylinder. The engine, a reversible heat engine, can be described by a system diagram just like the one in figure 10–8 and the air in the chamber at the beginning of compression, state (1), is at 14.7 psia and 95°F. The compression ratio is still 8.5 and the volumes remain the same

$$V_1 = V_4 = 24.0 \text{ in}^3$$
$$V_2 = V_3 = 2.83 \text{ in}^3$$

We calculate the pressure at state (2)

$$\frac{p_2}{p_1} = \left(\frac{V_1}{V_2}\right)^n$$

$$p_2 = (14.7 \text{ psia}) \left(\frac{24.0}{2.83}\right)^{1.30} = 237 \text{ psia}$$

and the temperature at state (2)

$$T_2 = \frac{p_2 V_2}{mR} = \frac{237 \times 2.83}{0.000994 \times 53.3 \times 12} = 1055°\text{R}$$

We can easily obtain the temperature at state (3) now:

$$c_v(T_3 - T_2) = 1200 \text{ Btu/lbm air}$$

$$T_3 = \frac{1200}{0.171}°\text{R} + T_2 = 7018 + 1055$$

$$= 8073°\text{R}$$

The temperature at state (4) can then be calculated from the process equation (6–47)

$$\frac{T_4}{T_3} = \left(\frac{V_3}{V_4}\right)^{n-1}$$

from which

$$T_4 = T_3 \left(\frac{V_2}{V_1}\right)^{n-1} = 8073°\text{R} \left(\frac{2.83}{240}\right)^{1.45-1}$$

$$= (8073°\text{R})\left(\frac{1}{8.5}\right)^{0.45} = 3082°\text{R}$$

The work per cylinder per cycle is quickly obtained then

$$Wk_{\text{cycle}} = \frac{mR}{1 - n_{43}}(T_4 - T_3) + \frac{mR}{1 - n_{21}}(T_2 - T_1) \qquad \textbf{(10–4)}$$

$$= \frac{0.000994 \times 53.3}{778}\left[\frac{3082 - 8073}{1.00 - 1.45} + \frac{1055 - 555}{1.00 - 1.30}\right]$$

$$= 0.642 \text{ Btu/cycle} \qquad\qquad\qquad \textit{Answer}$$

The total power for four cylinders is calculated

$$\dot{W}k = 4 \times Wk_{\text{cycle}} \times 1200 \text{ cycles/min} \times \frac{1}{2}$$

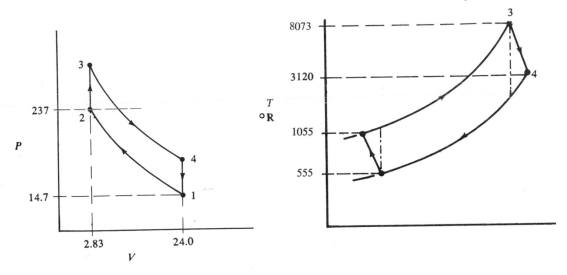

Figure 10–12   Polytropic Otto cycle property diagrams for example 10.2.

$$= 4 \times \frac{0.642 \times 1200}{2} \text{ Btu/min} = 36.3 \text{ hp} \qquad \textit{Answer}$$

The heat rejected per cycle, Q_{rej}, we get from equation (10–3)

$$Q_{rej} = Q_{add} - Wk_{cycle}$$

which gives us

$$Q_{rej} = 0.000994 \text{ lbm/cycle} \times 1200 \text{ Btu/lbm} \times 4 - 0.642 \text{ Btu/cycle} \times 4$$
$$= -2.20 \text{ Btu/cycle} \qquad \textit{Answer}$$

A comparison of these results with the answers from problem 10.1 indicates a decrease in work or power produced by the engine. This is a typical result of using polytropic processes to represent better the actual engine — the ideal Otto cycle is just not quite as accurate as the polytropic cycle, provided the exponents n are selected correctly.

In figure 10–12 are shown the property diagrams resulting from problem 10.2. Notice that entropy changes occur in all four of the cycle processes and if we desired, the magnitudes of these changes could be easily calculated from equations listed in tables 6–2 and 6–3. We will not go through the calculations here, but the direction of entropy changes in processes (1–2) and (3–4) should be observed (an increase in both). The shape of the curves will be characteristic of irreversible effects in processes and we will later use this.

**10.3
Otto Cycle
Efficiency**

The efficiency of the Otto cycle, as indeed for any cycle, is defined as

$$\eta_T = \frac{(Wk_{\text{cycle}})}{Q_{\text{add}}} (100) \qquad (10\text{--}5)$$

For the ideal Otto cycle and assuming a perfect gas working medium, we can reduce this to more specific equations. First we have that

$$Q_{\text{add}} = Q_{(2\text{--}3)} = U_3 - U_2 = mc_v(T_3 - T_2)$$

and in problem 10.1 we saw that

$$Wk_{\text{cycle}} = Wk_{(1\text{--}2)} + Wk_{(3\text{--}4)} = \frac{mR}{(1-k)} (T_2 - T_1 + T_4 - T_3)$$

Since $R = c_p - c_v$ and $k = c_p/c_v$ we can then write

$$Wk_{\text{cycle}} = \frac{m(c_p - c_v)}{(1 - c_p/c_v)} (T_2 - T_1 + T_4 - T_3)$$

or

$$Wk_{\text{cycle}} = \frac{mc_v(c_p - c_v)}{(c_p - c_v)} (T_1 - T_2 + T_3 - T_4)$$

$$= mc_v(T_1 - T_2) + mc_v(T_3 - T_2)$$

We then obtain the efficiency by substituting these results into equation (10–5)

$$\eta_T = \frac{mc_v(T_1 - T_4) + mc_v(T_3 - T_2)}{mc_v(T_3 - T_2)} \times 100$$

or

$$\eta_T = \left[1 - \frac{T_4 - T_1}{T_3 - T_2}\right] \times 100 \qquad (10\text{--}6)$$

Now using some more algebraic manipulations we obtain the following form

$$\eta = \left[1 - \frac{T_1(T_4/T_1 - 1)}{T_2(T_3/T_2 - 1)}\right] \times 100$$

and

$$\frac{T_4}{T_3} = \left(\frac{V_3}{V_4}\right)^{k-1} = \left(\frac{V_2}{V_1}\right)^{k-1} = \frac{T_1}{T_2}$$

so that

$$\frac{T_4}{T_1} = \frac{T_3}{T_2}$$

and we have

$$\eta_T = \left(1 - \frac{T_1}{T_2}\right) \times 100 \qquad (10\text{--}7)$$

Equation (10–7) is frequently reduced further by noting that

$$\left(\frac{V_2}{V_1}\right)^{k-1} = \frac{1}{(r_v)^{k-1}}$$

which gives us

$$\eta_T = \left(1 - \frac{1}{(r_v)^{k-1}}\right) \times 100 \qquad (10\text{–}8)$$

This result, as mentioned, is good only for ideal Otto cycles using perfect gases with constant specific heats. However, it is used for all sorts of engine analyses, sometimes when compression ratio means absolutely nothing for the device being considered. We must remember to use equation (10–8) with discretion; it may be approximately true for an irreversible (real) internal combustion engine but not completely — there are invariably some extenuating complications.

Example 10.3

Determine the efficiency of the engine of problem 10.1.

Solution

The problem we consider here involves an ideal Otto cycle with a perfect gas. The efficiency can be calculated using either equation (10–5) or (10–8). Using equation (10–5), we have

$$\eta_T = \frac{Wk_{\text{cycle}}}{Q_{\text{add}}} \times 100 = \frac{0.69 \text{ Btu/cycle}}{1200 \text{ Btu/lbm} \times 0.000994 \text{ lbm/cycle}} \times 100$$

$$= 57.8\% \qquad Answer$$

Since the compression ratio was 8.5, we easily obtain the efficiency from equation (10–8) as well:

$$\eta_T = 1 - \left(\frac{1}{r_v}\right)^{1.4-1.0} \times 100 = 1 - \left(\frac{1}{8.5}\right)^{0.4} \times 100 = 57.4\% \qquad Answer$$

We see then that the efficiency indeed can be found either way with an agreement within 1% of the answer. It might also be interesting to compare our result here with the standard of comparison — the Carnot cycle. Operating between the high temperature of the engine T_3 and the low temperature T_1 the Carnot engine efficiency is calculated using the values from example 10.1:

$$\eta_{\text{Carnot}} = \left(1 - \frac{T_1}{T_3}\right) \times 100 = \left(1 - \frac{555}{8328}\right) \times 100$$

$$= 93.3\%$$

This comparison is somewhat biased, however, for if we compare T-s diagrams of both Otto and Carnot cycles, we must conclude that for differential changes in entropy, both cycles have the same efficiency. This

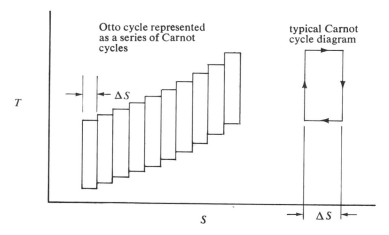

Figure 10–13   Comparison of Carnot and Otto cycles using *T-s* diagram.

we see in figure 10–13 and from this we see that using this criterion, the Carnot efficiency of the differential cycles is

$$\eta_{\text{Carnot}} = \left(1 - \frac{T_1}{T_2}\right) \times 100 = \left(1 - \frac{555}{1309}\right) \times 100$$

$$= 57.4\%$$

which is the same as for the Otto engine.

One other parameter frequently mentioned with efficiency to describe an Otto engine performance is the mean effective pressure (mep) which is here defined:

$$\text{mep} = \frac{Wk_{\text{cycle/cylinder}}}{\text{stroke} \times \pi \times (\text{piston diameter})^2/4} \tag{10–9}$$

The divisor here is referred to as the *cylinder displacement*. The engine displacement, or just *displacement*, is the cylinder displacement times the number of cylinders.

The mep is a design parameter. It is a theoretical average pressure of the engine gases which produces the same work as the actual engine and its variation in gas pressure.

Example 10.4   Determine the engine displacement and mep for the engines of examples 10.1 and 10.2.

Solution   The engine displacement is the same for both engines. We obtain

$$\text{Engine displacement} = (3 \text{ inches})(\pi)\left(\frac{9 \text{ inches}^2}{4}\right) \times 4 \text{ cylinder}$$

$$= 84.8 \text{ cubic inches} \qquad \textit{Answer}$$

The mep for the engine of problem 10.1 we then calculate

$$\text{mep} = \frac{0.69 \text{ Btu/cycle}}{84.8 \text{ in}^3}(4) = 0.0326 \text{ Btu/in}^3$$

Pressure is normally described in psia units so if we convert to ft-lbf units, using 778 ft-lbf/Btu, we obtain

$$\text{mep} \times 0.0326 = 778 \text{ ft-lbf/in}^3 \times 12 \text{ in/ft} = 304 \text{ psi} \quad \textit{Answer}$$

For problem 10.2, the mep is now calculated

$$\text{mep} = \frac{0.642}{84.8} \times 4 \times 778 \times 12 = 283 \text{ psi} \qquad \textit{Answer}$$

10.4
The Gas Tables

In adapting the theoretical processes of the Otto cycle to the actual IC engine workings, we used polytropic processes as opposed to reversible adiabatic. Another method of better predicting the actual processes is the utilizing of gas tables. In the appendix is an abbreviated air table (table B.6) which will serve us in showing how tabulated data can be used in Otto cycle analyses.* In table B.6 are listed temperature, enthalpy, internal energy, and three other properties: the relative pressure p_r; the relative specific volume v_r; and the ϕ-function. All these properties are functions of temperature only, and in particular we define

$$\phi = \int_0^T \frac{C_p dT}{T} \tag{10–10}$$

Notice that this is not the same phi as was defined in chapter 9 (Φ), but when gas tables are used, the definition (10–10) should be understood. The relative pressure p_r we define as

$$p_r = e^{\phi/R} \quad \text{or} \quad \phi = R \ln p_r \tag{10–11}$$

from which the relative volume is written:

$$v_r = \frac{RT}{p_r} \tag{10–12}$$

The general entropy equation resulting from equation (7–18) is

---

*A more complete air table as well as properties of various gases can be found in the book, *Gas Tables*, Keenen and Kaye, John Wiley & Sons Inc. (New York, 1948).

$$\Delta s = \int_{T_1}^{T_2} \frac{c_p dT}{T} - R \int_{P_1}^{P_2} \frac{dp}{p}$$

where we have assumed $dh = c_p dT$ and $V/T = R/p$. Then we can write this as

$$\Delta s = \phi_2 - \phi_1 - R \ln \frac{p_2}{p_1} \tag{10–13}$$

from the definition (10–10). For the reversible adiabatic process, equation (10–13) becomes (since $\Delta s = 0$)

$$\phi_2 - \phi_1 = R \ln \left(\frac{p_2}{p_1}\right)$$

and since we defined relative pressure p_r by the equation (10–11)

$$\phi = R \ln p_r$$

we have

$$R \ln \left(\frac{p_2}{p_1}\right) = R(\ln p_{r_2} - \ln p_{r_1})$$

$$= R \ln \frac{p_{r_2}}{p_{r_1}}$$

Consequently,

$$\frac{p_2}{p_1} = \frac{p_{r_2}}{p_{r_1}} \tag{10–14}$$

Remember, this result is only true for the reversible adiabatic case and should not be used under other circumstances. Now let us look at some applications of the air tables.

Example 10.5 There are 2 lbm of air heated from 240°F and 15 psia to 440°F in a rigid container. Determine the entropy change and the enthalpy change.

Solution The system, 2 lbm of air, reaches a final pressure p_2 calculated from

$$\frac{p_2}{p_1} = \frac{T_2}{T_1}$$

since we have an isometric process of a perfect gas. Then

$$p_2 = (15 \text{ psia}) \frac{900}{700} = 19.3 \text{ psia}$$

and from table B.6

$$\phi_1 = 0.66321 \text{ @ } T_1 = 240°F = 700°R$$

$$\phi_2 = 0.72438 \ @ \ T_2 = 440°F = 900°R$$

The entropy change is calculated from equation (10–13)

$$\Delta s = 0.72438 - 0.66321 - \frac{53.3}{778} \ \text{Btu/lbm°R} \ \ln \frac{19.3}{15}$$

$$= 0.72438 - 0.66321 - 0.01727 = 0.04390 \ \text{Btu/lbm°R}$$

and for 2 lbm of air

$$\Delta S = m\Delta s = 2 \ \text{lbm} \times 0.04390 = 0.08780 \ \text{Btu/°R} \qquad \textit{Answer}$$

The enthalpy change is quickly determined from values taken from table B.6

$$\Delta H = H_2 - H_1 = m(h_2 - h_1)$$
$$= 2 \ \text{lbm}(216.26 - 167.56) = (2 \ \text{lbm})(48.7 \ \text{Btu/lbm})$$
$$= 97.4 \ \text{Btu} \qquad \textit{Answer}$$

Example 10.6

There are 0.03 lbm of air compressed reversibly and adiabatically from 100°F and 14.8 psia to 120 psia. Determine the final temperature, change in entropy, and change in enthalpy for the process.

Solution

Since the system is undergoing a reversible adiabatic process, from equation (10–14) we have

$$\frac{p_{r_1}}{p_{r_2}} = \frac{p_1}{p_2} = \frac{14.8}{120}$$

but from table B.6 we find that $p_{r1} = 1.5742$ and then

$$p_{r_2} = (1.5742) \left(\frac{120}{14.8} \right) = 12.8$$

This is approximately the relative pressure corresponding to a gas temperature of 1000°R, and we therefore say

$$T_2 = 1010°R \qquad \textit{Answer}$$

The entropy change is obviously zero, but the enthalpy change we determine using tabular values from table B.6

$$H_2 - H_1 = m(h_2 - h_1) = 0.03 \ \text{lbm} (243.47 - 133.86)$$
$$= (0.03)(109.61) \ \text{Btu} = 3.29 \ \text{Btu} \qquad \textit{Answer}$$

The gas tables (or air tables) can be used to reduce calculations if used with care. They are not meant to replace fully the formulas of chapter 6 but can be used with confidence as additional thermodynamics tools.

**10.5
The Actual
Otto Cycle**

Using our previously developed tools, we will analyze the operation of an Otto engine which could conceivably be used to power an automobile, truck, tractor, or other mechanical device; that is, we will consider the actual or "practical" Otto engine. Before doing this, however, let us introduce additional terminology and cycle parameters, namely, *indicated horsepower* (ihp), *brake horsepower* (bhp), *fuel heating value* (HV), and *brake specific fuel consumption* (bsfc).

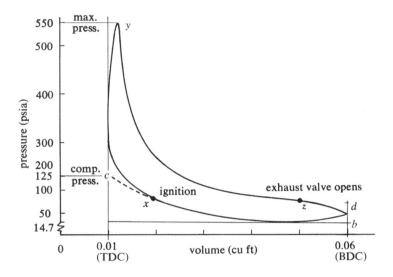

Figure 10–14

Typical *p-V* diagram for spark ignition engines at wide open throttle. Reproduced from E. I. Obert, *Internal Combustion Engines*, 2nd ed. (Scranton, 1959); with permission of International Textbook Co.

The indicated horsepower is determined from a *p-V* diagram of the test engine. This diagram can be obtained by using an engine indicator, which is a mechanism capable of measuring and recording the pressure in a cylinder while concurrently recording the piston position. The engine indicator is a common test device with a history of use dating to before 1900. At any rate, a detailed description of the indicator may be found in various reference literature; here let us look only at the result of the device. In figure 10–14 is shown the typical pressure-volume diagram, of an Otto engine, specifically of an engine with the throttle wide open. The curve can easily be seen to differ somewhat from the typical Otto cycle as shown in figure 10–1. The comparison is made easier by noting figure 10–15 where the polytropic Otto cycle is superimposed on an actual Otto cycle. For further analysis of this example we will use the polytropic

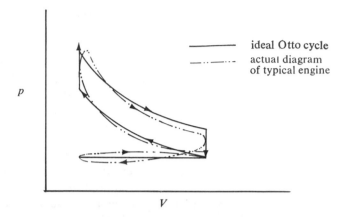

ideal Otto cycle
actual diagram
of typical engine

Figure 10–15   p-V diagram of ideal Otto cycle with actual cycle superimposed.

Otto cycle curves to calculate the area of the enclosed *p-V* diagram. This area, however, could be quite authentically calcuated using various geometric and mechanical devices (for example, the planimeter), but in any case, the accuracy of the recorded data is probably no less than 2% errors. The concern for accuracy here is, of course, due to the fact that the enclosed area of the *p-V* diagram is work of the cycle and the rate of this work, power, is referred to as *indicated horsepower*, ihp. It is not the power actually produced by the engine but rather the power which could be produced if we had a reversible engine.

The actual power produced by an engine we call *brake horsepower*, bhp, and the mechanically efficiency η_{mech} is then obtained for the actual Otto engine from

$$\eta_{\text{mech}} = \frac{\text{bhp}}{\text{ihp}} \times 100 \qquad (10\text{–}15)$$

The brake horsepower results strictly from test data. We have no thermodynamic tools at this time to predict the value of bhp, even though we can make some crude guesses from a knowledge of ihp. There are various test devices to measure bhp, all of which utilize some brake or resistance to measure engine output. Dynamometers, prony brakes, and water brakes are common devices for measuring bhp, but we will not here concern ourselves with the details of their workings. Using this measured value bhp, however, we can determine a parameter called the *brake mean effective pressure*, bmep. We define this as

$$\text{bmep} = \frac{\text{bhp}(4)}{\text{stroke} \times \pi \times (\text{piston dia})^2 \times N \times n} \qquad (10\text{–}16)$$

where *n* is the number of cylinders and *N* is the number of power strokes per unit time. The bmep is generally described in units of pressure, just like mep.

The fuel consumed by an engine is of major concern to both designer and user. In particular, the amount of fuel used per bhp per hour is called the *brake specific fuel consumption*, bsfc, and calculated from

$$\text{bsfc} = \frac{m}{(\text{bhp})\tau} \qquad (10\text{--}17)$$

where m is the mass of fuel consumed by the test engine during time τ, while producing power, bhp. The amount of fuel used in relation to the amount of *clean air* required, is of course meager (about 15 lbm of air per lbm fuel), and we could discuss a brake specific air consumption. However, a term called *volumetric efficiency*, η_v, defined as the ratio of the mass of air taken into the engine cylinder, m_a, to the mass which theoretically could have been taken in at atmospheric conditions, m_t, or

$$\eta_v = \frac{m_a}{m_t} \qquad (10\text{--}18)$$

is a common parameter used to describe engine performance. Notice that volumetric efficiency is not a volume efficiency, but rather a mass efficiency.

The chemical reaction between the air and fuel produces heat, which in turn increases the temperature and pressure of the compressed gases and ultimately results in work *out*. The heat released during the air-fuel reaction is called the *heating value* of the fuel. When this reaction occurs, water is formed (either in a vapor or a liquid), and if we allow the water to condense to a liquid, the heating value is called the *higher heating value* HHV, while the lower heating value LHV results if we retain the water in a vapor state. We can write

$$\text{HHV} - \text{LHV} = \text{Heat of vaporization of water}$$
$$\text{formed during reaction} \qquad (10\text{--}19)$$

and note that the heating values given here are tabulated in table B.7.

Example 10.7   A four-stroke engine having the *p-V* diagram shown in figure 10–14 has four cylinders and is operating at 4800 rpm. Using the polytropic Otto cycle and gas tables, determine the indicated horsepower and mechanical efficiency if bhp is found to be 183 hp. Also determine the bmep, the fuel heating value, and bsfc if 80 lbm per hour of fuel are consumed.

Solution   For the engine considered, the ihp can be determined by the relation

$$\text{ihp} = Wk_{\text{cycle}} \times 4 \text{ cylinders} \times \frac{4800}{2} \text{ cycles/min}$$

and the work is determined from the enclosed area of the curve of figure 10–14. From the diagram we get the following data:

$$p_1 = 14.7 \text{ psia} \qquad V_1 = 0.06 \text{ ft}^3 = V_4$$
$$p_2 = 125 \text{ psia} \qquad V_2 = 0.01 \text{ ft}^3 = V_3$$
$$p_3 = 550 \text{ psia} \qquad p_4 = 50 \text{ psia}$$

We will also assume the air at state (1) has a temperature of 65°F. To determine the temperatures at the remaining states, we need the polytropic exponents of processes (1–2) and (3–4). For process (1–2)

$$\frac{p_1}{p_2} = \left(\frac{V_2}{V_1}\right)^n \qquad \text{or} \qquad \ln \frac{p_1}{p_2} = n \ln \frac{V_2}{V_1}$$

from which

$$\ln \frac{14.7}{125} = n \ln \frac{0.01}{0.06} = -\ln \frac{125}{14.7} = -n \ln 6$$

and

$$-\ln 8.5 = -n \ln 6$$

Therefore

$$n = \frac{\ln 8.5}{\ln 6} = \frac{2.14}{1.79} = 1.196 = n_{21}$$

For process (3–4)

$$\frac{p_3}{p_4} = \left(\frac{V_4}{V_3}\right)^n$$

$$\ln \frac{550}{50} = n \ln 6$$

yielding

$$n = \frac{\ln 11}{\ln 6} = 1.338 = n_{43}$$

From this we calculate the temperatures:

$$\frac{T_2}{T_1} = \left(\frac{V_1}{V_2}\right)^{n_{21}-1} = 6^{0.196} = 1.412$$

so that

$$T_2 = T_1(1.42) = (65 + 460)(1.417) = 746°R$$

and, since process (1–4) is isometric

$$\frac{T_4}{T_1} = \frac{p_4}{p_1} = \frac{50}{14.7} = 3.4$$

giving us

$$T_4 = (525°R)(3.4) = 1785°R$$

Also

$$\frac{T_3}{T_4} = \left(\frac{V_4}{V_3}\right)^{n_{45}-1} = 6^{0.338} = 1.832$$

yielding

$$T_3 = (1785)(1.832) = 3270°R$$

The work can then be easily obtained;

$$Wk_{\text{cycle}} = \frac{mR}{1-n}(T_4 - T_3 + T_2 - T_1)$$

$$= mR\left(\frac{T_4 - T_3}{1 - n_{43}} + \frac{T_2 - T_1}{1 - n_{21}}\right)$$

where

$$m = \frac{p_1 V_1}{RT_1} = \frac{14.7 \times 144 \times 0.06}{53.3 \times 525} = 0.00454 \text{ lbm}$$

Then

$$Wk_{\text{cycle}} = (0.00454 \text{ lbm})(53.3 \text{ ft-lbf/lbm°R})\left(\frac{1785 - 3270}{1 - 1.338} + \frac{746 - 525}{1 - 1.196}\right)$$

$$= 790 \text{ ft-lbf/cycle} = 1.016 \frac{\text{Btu}}{\text{cycle/cylinder}}$$

The ihp then is found from

ihp $= Wk_{\text{cycle}} \times 4 \times 2400 = 1.016 \times 4 \times 2400$
  $= 9,754 \text{ Btu/min} = 9,754 \text{ Btu/min} \times 1.41 \text{ hp-s/Btu} \times 1 \text{ min/60 s}$
  $= 229 \text{ hp}$ *Answer*

Since the bhp is 133 hp, we easily calculate the mechanical efficiency from equation (10–15)

$$\eta_{\text{mech}} = \frac{\text{bhp}}{\text{ihp}} \times 100 = \frac{183}{229} \times 100 = 79.9\%   \textit{Answer}$$

The bmep can be determined from equation (10–16)

$$\text{bmep} = \frac{\text{bhp}}{\text{stroke} \times \pi \times (\text{piston dia})^2/4 \times N \times n}$$

We know

$$\text{Cylinder displacement} = V_1 - V_2 = 0.05 \text{ ft}^3$$

$$\text{cylinder displacement} = \text{stroke} \times \pi \times \frac{(\text{piston dia})^2}{4}$$

and then

$$\text{bmep} = \frac{183 \text{ hp}}{0.05 \text{ ft}^3 \times 2400 \text{ strokes/min} \times 4}$$

Since 33000 ft-lbf/mm = 1 Btu we can readily give bmep the units of pressure:

$$\text{bmep} = \frac{183 \text{ hp} \times 33000 \text{ ft-lbf/mm} - \text{hp}}{0.05 \text{ ft}^3 \times 2400 \text{ strokes/min} \times 4} = 12581 \text{ lbf/ft}^2$$

or

$$\text{bmep} = 12581 \times \frac{1}{144} \text{ psi} = 87.4 \text{ psi} \qquad \qquad \textit{Answer}$$

The brake specific fuel consumption is obtained from equation (10–17)

$$\text{bsfc} = \frac{m}{(\text{bhp})t}$$

yielding

$$\text{bsfc} = \frac{80 \text{ lb/hr}}{183 \text{ hp}} = 0.437 \text{ lbm/bhp-hr} \qquad \qquad \textit{Answer}$$

To obtain the fuel heating value, we notice that the energy increase (manifested by the pressure and the temperature increase) of process (2–3) represents approximately the energy released by the fuel during the combustion. From table B.6 we find, by interpolating

$$U_2 = 127.56 \text{ Btu/lbm @ } T_2 = 746°\text{R}$$
$$U_3 = 646.30 \text{ Btu/lbm @ } T_3 = 3270°\text{R}$$

and

$$U_3 - U_2 = 518.74 \times 0.00454 \text{ Btu/cycle} = 2.36 \text{ Btu/cycle}$$

The heating value of the fuel is the energy released by the fuel per unit mass of fuel. The energy released per minute is given by

$$2.36 \text{ Btu/cycle} \times 2400 \text{ cycle/min}$$

while the fuel consumed is 80/60 lbm/min.

We therefore calculate the heating value HV from

$$\text{HV} = \frac{2.36 \text{ Btu/cycle} \times 2400 \text{ cycle/min}}{80/60 \text{ lbm/min}}$$

$$= 4248 \text{ Btu/lbm} \qquad \qquad \textit{Answer}$$

This value is significantly below the heating values (or heat of combustion) listed in table B.7. The reason would no doubt be attributed to the incomplete combustion occurring in the test engine. This could be an area of improvement in the engine design.

10.6 Modifications

Let us discuss briefly the major design or technological changes or modifications made to Otto engines in attempting to increase efficiency, power, or some other performance parameter.

The designs of Otto engines have been altered to increase thermodynamic efficiency by increasing the piston stroke or decreasing the cylinder clearance volume. Equation (10–8) is invariably cited in justifying the increase of compression ratios in Otto engines, but while this relation is approximately correct for the actual cycles, the combustion of fuels is made less effective in some cases; and frequently a phenomenon known as *preignition* or *knock* is observed in high compression engines. Knock occurs because all fuel-air mixtures have a temperature at which they spontaneously combust or "explode." This temperature is reached in some cylinder chambers before the spark plug can ignite the mixture and consequently a sharp explosion ensues, producing a noise (knock) and a sharp reduction in power. This may be noted in figure 10–14 by visualizing the effects of the burning of fuel before the ignition point x (to the right on curve xb).

To combat preignition, the shape of the combustion chamber is frequently revised, as shown in figure 10–16. Designs are generally submitted to provide more continuous, expanding combustion of the fuel-air mixture with an elimination of potential corners or "hot spots" which could cause knock. The most common method of preventing preignition has been to add "ignition depressants," such as tetraethyl-lead, to fuel. It is also worth mentioning that the fuel required in high compression engines ($r_v \geq 6$, approximately) is a more refined extraction from crude oil than that required for low compression engines or steady state combustion processes. Water injected into the combustion chamber during compression has an effect similar to the additives (tetraethyl-lead) in reducing preignition.

It has been mentioned that the combustion process is that point of the Otto cycle during which energy is extracted from fuel and added to the cycle in the form of heat. We hope to extract the HHV or at least the LHV from the fuel, but this does not happen in the reciprocating IC

Figure 10–16   Possible configurations of pistons and cylinder chambers for improving the chemical-thermal-mechanical energy conversion in the fuel combustion process of the Otto engine.

engine. Combustion is a grossly irreversible process and in the Otto cycle, it is first suppressed from burning, then is ignited by a spark plug and finally extinguished — all this once every power revolution. The best that can be said for this combustion is that much higher temperatures are feasible than in a steady state combustion, because the materials making up the Otto engine will be subjected to the high temperatures for only small portions of the cycle. As a consequence, higher theoretical thermodynamic efficiencies [see equation (8–8)] are possible for the reciprocating IC engine than for engines which operate at steady state conditions.

Increasing the power or work of an engine is achieved by two general methods: (1) increasing the amount of fuel and air in the cylinder chamber; and (2) increasing the engine size, or displacement.

The second method is strictly a geometric problem. Increasing the piston and cylinder diameters is the only actual way to increase engine displacement, unless additional cylinders are added. The first method of increasing power is carried out in numerous ways: (1) fuel injection (2) supercharging, and (3) exhaust manifold tuning. We will discuss each briefly.

Fuel injection   In this technique, fuel is forced into the cylinder by a pump. The devices are generally quite accurate in controlling the precise air-fuel ratio and they normally replace the carburetor. The carburetor, a device which allows air to mix with fuel by natural convection, is normally used to furnish the fuel-air mixture to the engine.

Supercharging   This is a procedure in which the fuel and air mixture is forced into the cylinder under pressure. This is achieved by means of a pump, fan, or compressor, and can be operated in conjunction with fuel injection or with a carburetor. The pressure of the fuel-air mixture entering the cylinder is generally above atmospheric pressure, but if the pressure is less than atmospheric, the technique is called *scavenging*.

Manifold Tuning   This technique uses a shock wave to force air back into the cylinder. When the exhaust gases are freed from the cylinder, they rapidly travel down the exhaust manifold, led by a high pressure shock wave. If the manifold is such a length that the shock wave has not reached the exit, and is still contained in the manifold when the piston begins to draw a fresh charge into the cylinder chamber, the shock wave may reverse and force the exhaust gases back into the cylinder, thus increasing the mass in the cylinder. This technique is not common in commercial engine design. There are, of course, many other methods of modifying the Otto engine to improve its performance. A brief superficial sketch is given; for more comprehensive information you are referred to appendix C.

The great use of the Otto engine is attributed to the fact that it is an internal combustion device delivering a positive, forceful output by means of a piston-cylinder. That is, it is expected to start easily and reach quickly its normal operating thermal condition, and it is expected to deliver high accelerations and decelerations when demanded. These characteristics are exhibited by the piston-cylinder Otto engine, but the configuration which provides those advantages is also the engine's greatest weakness. The combustion is unsteady (as we have said before); and the complicated mechanical devices required to keep the machine running (such as cams, valves, rocker arms, timing belts, distributors, oil and fuel pumps, crank shafts, connecting rods, and various assorted gadgets) are numerous, and as a result, it is noisy. It is a device which has caused a revolution in social patterns, but its continued popularity has stemmed more from vested financial interests than from its technical superiority over other power devices. Perhaps the most recent scientific improvement to come from this type of engine was the Diesel cycle concept introduced in 1892.

10.7
The Diesel Cycle

The ideal Diesel cycle is defined by the following four reversible processes:

(1–2)   Adiabatic compression.

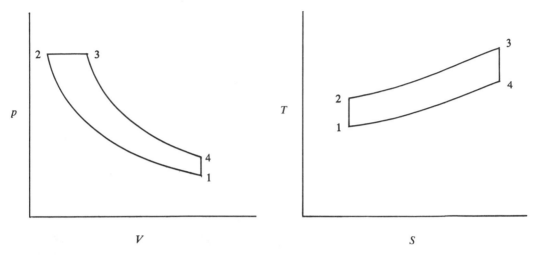

Figure 10–17   Property diagrams for ideal Diesel cycle.

  · (2–1)   Constant pressure heat addition.
    (3–4)   Adiabatic expansion.
    (4–1)   Constant volume heat rejection.

In figure 10–17 are shown the diagrams illustrating the paths of these processes on p-V and T-S coordinates, analogous to the Otto cycle. The enclosed area on the p-V plane is the work of the cycle, and the enclosed area on the T-S plane is the net heat added; the areas must be equal. The mechanism normally used to carry out these processes is the piston-cylinder device in figure 10–18. It has the same basic configuration and operation as that depicted in figure 10–2 except that no spark plug is present in the Diesel cycle and the fuel-air mixture is introduced by a fuel injector rather than by a carburetor and intake valve. Process (1–2) is a compression of a fresh charge of air and fuel (or just air); reaching state (2), the gases combust spontaneously due to the high pressure and the process continues at constant pressure to state (3). Most commonly, only air is introduced and compressed in process (1–2) while fuel is injected at a pressure near that of state (2). Process (2–3) classifies the Diesel engine as an internal combustion, compression-ignited engine (IC-CI engine). Process (3–4) is an expansion of the exhaust gases until the piston reaches BDC (bottom dead center) at state (4). Then the exhaust is rejected into the atmosphere in process (4–1). Simultaneously air and fuel (or just fresh air) is injected into the chamber to be ready for a new cycle. This constitutes the two-stroke Diesel cycle with the exhaust and intake strokes of the piston together taking one revolution of the engine.

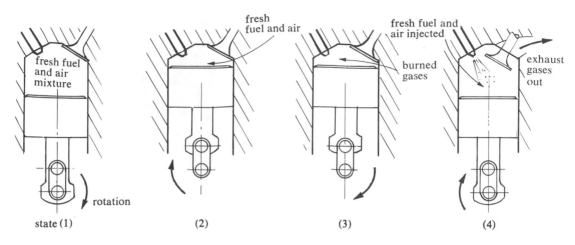

Figure 10–18   Physical operation of the Diesel engine.

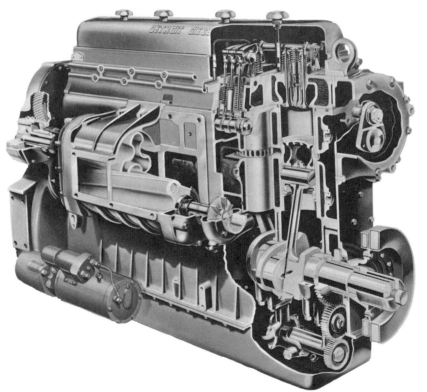

Figure 10–19   Cutaway view of typical Diesel engine. Reproduced with permission of Detroit Diesel Allison, Division of General Motors Corporation.

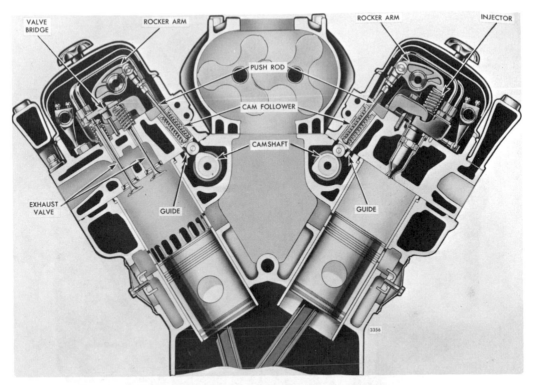

VALVE BRIDGE | ROCKER ARM | ROCKER ARM | INJECTOR

PUSH ROD

CAM FOLLOWER

CAMSHAFT

EXHAUST VALVE | GUIDE | GUIDE

Figure 10–20   Cross section view of typical Diesel engine. Reproduced with permission of Detroit Diesel Allison, Division of General Motors Corporation.

An actual Diesel engine is shown in figure 10–19. This illustration serves to give the reader a better picture of the major components of a Diesel engine. Notice the counter-weighted crankshaft located at the bottom of the engine. This is connected to the piston by means of a connecting rod. Four valves situated at the top of the cylinder are used for exhausting the spent gases. Midway down the cylinder, a row of openings around the periphery of the cylinder can be seen. These are the intake openings where fresh air is introduced into the cylinder when the piston is at the bottom of its stroke. Fuel is injected from above during the compression stroke. The fuel injector used for this can be seen directly between the two exposed valves.

In figure 10–20 are shown the details of the critical parts which convert thermal energy to work through the piston-cylinder device. This view illustrates the manner of operation of the intake of air and fuel and the exhausting of the spent gases. At the top center of the figure are shown

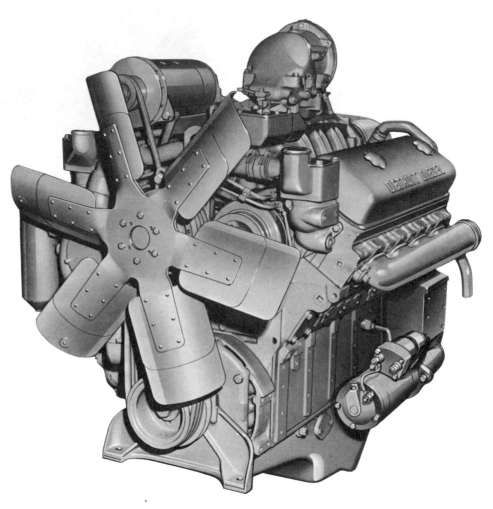

Figure 10–21   Diesel engine. Reproduced with permission of Detroit Diesel
                      Allison, Courtesy of General Motors Corporation.

two 3-lobed rotors which pump air into the center cavity where it is then
introduced into the cylinder through the peripheral openings in the
cylinder walls when the piston is at the bottom of its travel. The exhaust
valves are actuated through a linkage system by the camshaft to allow
the exhaust to escape from the cylinder at an appropriate time during the
cycle. Notice that fuel is introduced into the cylinder through the injector while fresh air is introduced separately through openings in the
cylinder walls. Finally, an external view of a typical Diesel engine is
shown in figure 10–21. This view shows the external components of a

typical Diesel engine. The large six-bladed fan is needed to draw air past a radiator to cool the engine when running. The radiator is not shown in this figure.

10.8 Air-Standard Analysis — Diesel Engine

To simplify calculations, the Diesel engine can be analyzed with the assumption that air is the only working medium. This is the air standard analysis and it ignores the fuel, except as a heat source. This was also done with the Otto engine. The work of a cycle, as was mentioned, is the enclosed area of the p-V plane which gives us the relation

$$Wk_{cycle} = Wk_{12} + W_{23} - Wk_{34} \qquad (10\text{--}20)$$

For an air standard analysis we have

$$Wk_{12} = \frac{1}{1-k}(p_2V_2 - p_1V_1) \qquad (10\text{--}21)$$

or

$$Wk_{12} = \frac{mR}{1-k}(T_2 - T_1) \qquad (10\text{--}22)$$

Also,

$$Wk_{23} = p_2(V_3 - V_2) \qquad (10\text{--}23)$$

and

$$Wk_{34} = \frac{mR}{1-k}(T_4 - T_3) \qquad (10\text{--}24)$$

These results can quickly give us the net work of an ideal cycle, provided we know the properties at the four cycle corners. If we wish to fit the Diesel cycle to actual engines and find that processes (1–2) and (3–4) are not adiabatic, polytropic Diesel processes can easily be substituted. Then the work can be obtained from the equations (10–20), (10–21), (10–22), (10–23), and (10–24), but k will be replaced by the polytropic exponent n. An analysis of the actual Diesel closely parallels the Otto cycle analysis. The terms bhp, ihp, mep, bmep, bsfc, stroke, bore, and p-V diagrams are all determined in a similar manner for the Diesel as for the Otto engine. A typical p-V diagram from an actual engine is shown in figure 10–22 from which we will develop an example problem to show the thermodynamic analysis of a Diesel engine.

Example 10.8

From a two-stroke, six-cylinder Diesel engine operating at 1200 rpm is developed the p-V diagram shown in figure 10–22. If we assume an air

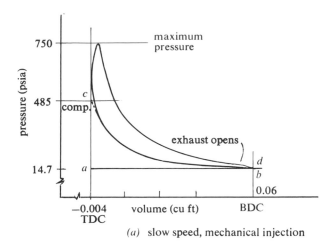

(a) slow speed, mechanical injection

Figure 10–22   Typical *p-V* diagram for compression ignition engines operating at full load. Reproduced from E. F. Obert, *Internal Combustion Engines*, 2nd ed. (Scranton, 1959); with permission of International Textbook Company.

standard analysis and have the following additional information:

$$bhp = 115 \text{ hp @ 1200 rpm}$$
$$= 58 \text{ hp @ 600 rpm}$$

and 75.0 lbm/hr fuel usage at 1200 rpm, determine the Wk_{cycle}, ihp, mechanical efficiency, mep, and bsfc.

Solution   The following property values we extract from the curves of the *p-V* diagram in figure 10–22:

$p_1 = 14.7$ psia $V_1 = 0.06 \text{ ft}^3 = V_4$

$p_2 = 750$ psia $V_2 = 0.004 \text{ ft}^3$

$p_3 = 750$ psia $V_3 = 0.008 \text{ ft}^3$ (estimated)

$p_4 = 20$ psia (estimated) $T_1 = 70°F$ (assumed)

We now recognize that the polytropic exponents must be determined. This we do with the following calculations:

$$\frac{p_1}{p_2} = \left(\frac{V_2}{V_1}\right)^n$$

$$n_{21} = \frac{\ln (p_1/p_2)}{\ln (V_2/V_1)}$$

$$= \frac{\ln 14.7/750}{\ln 0.004/0.06} = \frac{\ln 750/14.7}{\ln 0.06/0.004} = \frac{3.94}{2.71} = 1.45$$

Similarly

$$\frac{p_3}{p_4} = \left(\frac{V_4}{V_3}\right)^n$$

and

$$n_{43} = \frac{\ln p_3/p_4}{\ln V_4/V_3} = \frac{\ln 750/20}{\ln 0.06/0.008} = \frac{3.62}{2.01} = 1.80$$

Notice that the compression ratio r_v, given by

$$r_v = \frac{V_1}{V_2} \tag{10-25}$$

is, for our engine, equal to 15. The cutoff ratio r_c is frequently mentioned with compression ratios when CI engines are discussed. This is defined by

$$r_c = \frac{V_3}{V_2} \tag{10-26}$$

and in this engine, the cutoff ratio is easily obtained:

$$r_c = \frac{0.008}{0.004} = 2$$

The amount of air introduced into the cylinder for each power stroke is obtained from the perfect gas equation

$$p_1 V_1 = mRT_1$$

or

$$m = \frac{p_1 V_1}{RT_1} = \frac{14.7 \text{ lbf/In}^2 \times 144 \text{ in}^2/\text{ft}^2 \times 0.06 \text{ ft}^3}{53.3 \text{ ft-lbf/lbm}°\text{R} \times (460 + 70)°\text{R}}$$

$$= 0.0045 \text{ lbm/cycle}$$

Since the engine is a two-stroke cycle arrangement, the number of cycles equals the number of revolutions. Then

$$m = 0.0045 \text{ lbm/rev}$$

The work per cycle or revolution is obtained after we determine the temperature at states (2), (3), and (4):

$$T_2 = T_1 \left(\frac{V_1}{V_2}\right)^{n_{21}-1} = 530°\text{R} \left(\frac{0.06}{0.004}\right)^{0.45} = 1792°\text{R}$$

$$T_4 = T_1 \left(\frac{p_4}{p_1}\right) = 530°\text{R} \left(\frac{20.0}{14.7}\right) = 721°\text{R}$$

and

$$T_3 = T_2 \left(\frac{V_3}{V_2}\right) = 1792°\text{R} \left(\frac{0.008}{0.004}\right) = 3584°\text{R}$$

Check each process to make sure these equations are applicable.

Now the work is calculated

$$Wk_{cycle} = \frac{mR}{1 - n_{21}}(T_2 - T_1) + p_2(V_3 - V_2) + \frac{mR}{1 - n_{43}}(T_4 - T_3)$$

$$= \left(\frac{0.0045 \text{ lbm} \times 53.3 \text{ ft-lbf/lbm}°R}{1.00 - 1.45}\right)(1792 - 530)°R$$

$$+ (750 \text{ lbf/in}^2 \times 144 \text{ in}^2/\text{ft}^2)(0.008 - 0.004)\text{ft}^3$$

$$+ \frac{0.0045 \text{ lbm} \times 53.3 \text{ ft-lbf/lbm}°R}{1.0 - 1.8}(721 - 3584)°R$$

$$= (-673 + 432 + 861) \text{ ft-lbf/cycle}$$

$$= 620 \text{ ft-lbf/cycle} = 0.797 \text{ Btu/cycle} \qquad Answer$$

The total work for all six cylinders is

$$Wk_{cycle} = 6 \times 0.797 = 4.782 \text{ Btu/cycle} \qquad Answer$$

and the indicated horsepower at 1200 rpm is obtained from

$$ihp = Wk_{cycle} \times N = 4.782 \times 1200 \text{ Btu/min} = 5738 \text{ Btu/min}$$

$$= 5738 \text{ Btu/min} \times \frac{60 \text{ min/hr}}{2545 \text{ Btu/hp-hr}} = 135.3 \text{ hp} \qquad Answer$$

Since the brake horsepower is 115 hp at 1200 rpm, the mechanical efficiency may be obtained from equation (10–15):

$$\eta_T = \frac{bhp}{ihp} \times 100 = \frac{115}{135.3} \times 100 = 84.9\% \qquad Answer$$

The mean effective pressure is obtained from equation (10–9)

$$mep = \frac{Wk_{cycle/cyl}}{cyl \text{ displacement}} = \frac{620 \text{ ft-lbf/cycle}}{(0.06 - 0.004) \text{ ft}^3}$$

$$= 11,071 \text{ lbf/ft}^2 = 76.8 \text{ psi} \qquad Answer$$

The bmep we calculate from equation (10–16)

$$bmep = \frac{bhp}{(\text{displacement})(N)(n)}$$

$$= \frac{115 \text{ hp} \times 33000 \text{ ft-lbf/min-hp}}{(0.054 \text{ ft}^3)(6 \text{ cylinders})(1200 \text{ rpm})}$$

$$= 9750 \text{ lbf/ft}^2 = 67.7 \text{ psi} \qquad Answer$$

The fuel consumption bsfc is calculated from equation (10–17)

$$bsfc = \frac{m}{(bhp)t} = \frac{75 \text{ lbm/hr}}{115 \text{ bhp}}$$

$$= 0.652 \text{ lbm/bhp-hr} \qquad Answer$$

10.9
Diesel-Otto
Comparison

The most common parameter for comparing various engines is the thermodynamic efficiency. Let us derive relationships for the efficiency of the Diesel engine and see how they compare with the Otto cycle. We first use the standard definition of thermodynamic efficiency

$$\eta_T = \frac{Wk_{\text{cycle}}}{Q_{\text{add}}} \times 100$$

and notice that for the Diesel cycle with a perfect gas having constant specific heats

$$Q_{\text{add}} = mc_p(T_3 - T_2) \qquad \textbf{(10–27)}$$

and

$$Wk_{\text{cycle}} = Q_{\text{add}} + Q_{\text{rej}}$$
$$= mc_p(T_3 - T_2) + mc_v(T_1 - T_4)$$

The efficiency can then be written

$$\eta_T = \left[\frac{mc_p(T_3 - T_2) + mc_v(T_1 - T_4)}{mc_p(T_3 - T_2)} \right] \times 100 \qquad \textit{polytropic}$$

$$= \left[1 + \frac{c_v}{c_p} \frac{(T_1 - T_4)}{(T_3 - T_2)} \right] \times 100 = \left[1 - \frac{1}{k}\left(\frac{T_4 - T_1}{T_3 - T_2} \right) \right] \times 100 \qquad \textbf{(10–28)}$$

This can be put in other terms (see problem 10.23) by a little more manipulation to give us

$$\eta_T = \left[1 - \frac{1}{(r_v)^{k-1}} \left(\frac{r_c^k - 1}{k(r_c - 1)} \right) \right] \times 100 \qquad \textbf{(10–29)}$$

For the polytropic processes in the Diesel cycle, equation (10–29) does not hold, but equation (10–28) does apply. A comparison now between the Diesel and Otto cycles shows that at a given compression ratio, the Otto engine is theoretically more efficient [compare equation (10–29) to equation (10–8)] than the Diesel. This is further exemplified by the superimposed Otto cycle on the Diesel cycle in figure 10–23. It is shown that the shaded area represents work realized by the Otto engine and not by the Diesel. The primary reason for developing the Diesel, however, has been that higher compression ratios could be achieved. The fuel in an Otto engine exhibits preignition at elevated compression ratios, but in the Diesel it is just this preignition that caused the fuel to burn. Compression ratios can then be increased to values only limited by material strength and dynamics of the piston-cylinder. Most Diesel engines have thermodynamic efficiencies that are higher than those of Otto engines, because of the increased compression. As a side benefit, the Diesel can generally utilize a less refined fuel than the Otto. In fact, the inventor and first developer of the Diesel, Rudolph Diesel, visualized coal dust as the

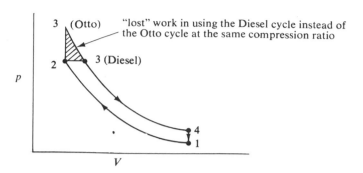

Figure 10–23   *p-V* diagram comparison of Otto and Diesel cycles at the same compression ratio.

primary fuel, but this has since been replaced by liquid fuels. At any rate, the Diesel can generally use a less expensive fuel and use it more efficiently than the Otto engine.

As a review, we list the equations of the Otto and Diesel cycles in table 10–2, which can be used for elementary thermodynamic analyses.

10.10 Engine Design Considerations

In previous sections we have discussed the detailed calculations for predicting the energy transfers of the internal combustion engines, both Otto and Diesel. In addition, the Otto cycle has been discussed from the standpoint of modifications to make its performance more desirable. Here let us consider very briefly some general aspects of IC engine design, presented only to give the reader an overview of the inherent problems and advantages of these engines. There are a number of books, some which are listed in the appendix, which can furnish finer detail. Here we will treat (1) engine balancing; (2) speed limitation; (3) gas flows and manifolding; (4) engine timing; (5) cooling systems; and (6) advantages of the IC engine.

Engine balancing   The piston-cylinder device which reciprocates invariably transfers its mechanical work to a rotating shaft by means of a crank. This mechanism is shown in figure 10–24 with a counterweight superimposed to show its position on the crank. The counterweight is used to prevent extreme imbalance in the crank and rotating shaft when the power stroke of the piston produces a force on the crank. This method of balancing is complimented by reducing the mass of pistons, values, connecting rods, and other moving parts. However, a perfectly balanced reciprocating engine is impossible to achieve.

Speed Limitations   A device which has an imbalance (even a small amount) will be subject to dynamic effects that become extremely

Table 10–2

Equations for IC Engines

Term	Otto	Diesel
Q_{add}	$mc_v\,(T_3 - T_2)$	$\bullet mc_p\,(T_3 - T_2)$
Q_{rej}	$mc_v\,(T_1 - T_4)$	$mc_v\,(T_1 - T_4)$
Wk_{cycle} (ideal)	$\dfrac{mR}{1-k}\,(T_2 - T_1 + T_4 - T_3)$	$\dfrac{mR}{1-k}(T_2 - T_1 + T_4 - T_3) + p_2(V_3 - V_2)$
Wk_{cycle} (polytropic)	$\dfrac{mR}{1-n_{21}}\,(T_2 - T_1) + \dfrac{mR}{1-n_{43}}\,(T_4 - T_3)$	$\dfrac{mR}{1-n_{21}}\,(T_2 - T_1) + \dfrac{mR}{1-n_{43}}\,(T_4 - T_3)$ $+ p_2(V_3 - V_2)$
r_v	V_1/V_2	V_1/V_2
r_c	N A*	V_3/V_2
$\eta\,T$ (ideal)	$1 - \dfrac{1}{(r_v)^{k-1}}$	$1 - \dfrac{1}{(r_v)^{k-1}}\dfrac{r_c^k - 1}{k(r_c - 1)}$
mep	$\dfrac{Wk_{\text{cycle}}}{\text{displacement}} = \dfrac{\text{ihp}}{(\text{displacement} \times N)}$	
bmep	$\dfrac{\text{bhp}}{(\text{displacement} \times N)}$	
η_M	$\dfrac{\text{bhp}}{\text{ihp}} \times 100$	
ihp	$Wk_{\text{cycle}} \times N$	
N	engine speed in rpm or equivalent	

*Not Applicable

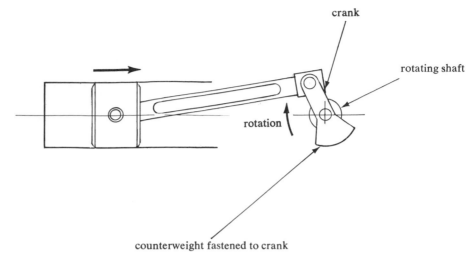

Figure 10-24   Piston-cylinder-crank mechanism of the reciprocating engine of the Otto and/or Diesel cycles.

intolerable at high speeds. Therefore, reciprocating engines have definite inherent speed limitations due to their characteristics of the piston-cylinder crank mechanism. In addition, the Otto and Diesel engines utilize intake and exhaust valves which have response limitations, either in mechanical or hydraulic actuation. That is, valves can physically be opened in a finite time, which will produce an upper limit to engine speeds. Of course, we have not mentioned irreversible effects, which increase with speed and which can become overwhelming in the expansion process where the gases are expected to produce the work of the engine.

The manner in which engine speed affects engine performance is seen in figure 10–25. Here is a typical performance chart which is developed from engine tests. Thermodynamics can give insights into why the parameters behave as they do, but cannot now accurately predict the full shapes of the curves. Notice that the individual curves do not have maximum or minimum values at the same speeds; that is, maximum power is developed at a higher speed than efficiency in the engine of figure 10–25. The optimization process, or selection of the speed which gives the all around best performance, is a problem which must be considered in all engine design or selection methods. We will not pursue this further except to say that the example problems we have considered in this chapter are essentially single speed conditions and the full analysis of engines must include the consideration of a performance chart as illustrated in figure 10–25.

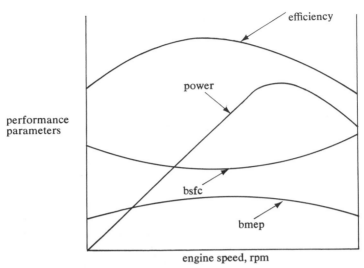

Figure 10–25   Typical performance chart of a reciprocating internal combustion engine.

Gas flows and manifolding Fuel injection is designed to replace carburetors and valves in supplying fuel to cylinders; however, the latter are more common. With the use of carburetors, however, there exists a problem of equal fuel-air mixture to the cylinders as illustrated in figure 10–26. Due to fluid or gas friction and the different distances, cylinders (2) and (3) will obviously get better service than cylinders (1) or (4). The design challenge here is to provide

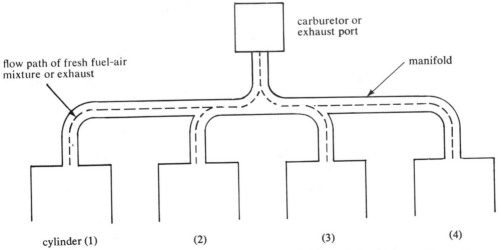

Figure 10–26   Typical configuration of a manifold and the flow pattern through it.

equal flow paths to each cylinder. This applies not only to the carburetor but to the exhaust pipe which tends to retard the flow of exhaust gases from the cylinder.

Engine timing   With the spark ignition engine there exists the problem of synchronizing the spark to the piston position. The spark from the spark plug comes slightly before the piston has fully compressed the gases (at TDC) in general. As speeds are varied, the amount of time between the spark and TDC must also vary, which requires a proper mechanical-electrical coordination of the peripheral engine components such as the distributor, the ignition points, and the carburetor.

Cooling systems   The heat engine must reject heat to the surroundings, and the Otto and Diesel engines are no exceptions. They commonly reject more than three-quarters of the heat added, and if this heat is not taken away from the engine block quickly, the engine will reach extremely high temperatures and permanently damage the materials. Air forced around the engine is the most obvious method of cooling, but this requires a large surface area to be effective. It is generally achieved by putting cooling fins or vanes in the engine block (see figure 10–27), an effective, but expensive process.

cooling vanes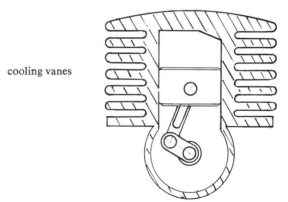

Figure 10–27   Typical cross section of air-cooled engine block.

Water cooling is a common method of removing the rejected heat from the engine block. This is an effective method for large engines as well as small, but it requires additional equipment and thus reduces the reliability of the engine. Radiators, water pumps, and hoses are some of the items readily identified with a water cooling system.

Advantages of the IC reciprocating Engine   We have pointed out a few of the problems of the IC engine, but let us reiterate its advantages. The reciprocating IC engine is an engine which is easily started, can produce its nominal power almost immediately upon starting, can be stopped easily and quickly, and can provide fast changes in speed (acceleration and deceleration).

The Diesel engine is well suited for providing large amounts of power at slow and medium speeds. It is frequently designed with large components to better withstand high compression ratios and thus provide an engine with extended lifetime.

Practice Problems

Section 10.2

10.1. Determine the displacement of an eight-cylinder engine having a bore of 3.25 inches and a stroke of 3 inches.

10.2. Determine the cylinder displacement and engine displacement of an engine having six cylinders with a bore of 102 mm and a stroke of 120 mm.

10.3. Determine the compression ratio of an engine with a bore of 98 mm, a stroke of 100 mm, and a clearance volume of 75 cm³.

10.4. Determine the stroke of an engine with a compression ratio of 8 to 1, a bore of 2.5 inches, and a clearance volume of 2 in³.

10.5. An ideal Otto engine with six cylinders operates on a four-stroke cycle with a compression ratio of 7.5 to 1. Given that p_1 = 14.7 psia, T_1 = 100°F, V_1 = 104 ft³, and the heat added is 1000 Btu/lbm, determine for an air-standard analysis
  (a) p and V and T at the four corners of the cycle.
  (b) T at the four corners.
  (c) Wk_{cycle}.
  (d) Q_{rej}.
  (e) Plot the p-V and T-s diagrams, labeling the coordinate points at (1), (2), (3), and (4).

10.6. A two-stroke, four-cylinder, ideal Otto engine operates at 4800 rpm. Air is taken in at 1×10^6 dyne/cm² and 20°C. The volume before compression is 500 cm³ in each cylinder and 90 cm³ after compression. If 2500 Joule/g air of heat are added, determine
  (a) Compression ratio.
  (b) p, V, and T at the four corners.
  (c) Sketch of p-V and T-s diagrams.
  (d) Entropy change during compression.
  (e) Wk_{cycle}.
  (f) Q_{rej}.
  (g) Power of engine.

10.7. A four-stroke, eight-cylinder Otto engine operates at 250 rpm. The compression process is reversible and polytropic, as is the expansion process. For both $n = 1.5$, and the compression ratio is 6 to 1. The cylinder volume is 2000 cm³ before compression, and at this point the air is at 10^6 dyne/cm² and 35°C. Determine, if the heat addition is 640 cal/g
(a) Wk_{cycle}, or power.
(b) Thermodynamic efficiency.
(c) Heat rejected per cycle.

10.8. A four-stroke, two-cylinder Otto engine has a bore of 24 inches, a stroke of 20 inches, and a compression ratio of 7.2 to 1. The air taken in is at 14.7 psia and 40°F while the compression and expansion processes are reversible and polytropic ($n = 1.49$). If 1400 Btu/lbm air are added and the engine speed is 320 rpm, determine
(a) $\dot{W}k_{cycle}$.
(b) Thermodynamic efficiency.
(c) $\dot{Q}_{rej}$.

Section 10.3

10.9. For the engine of problem 10.5, determine the thermodynamic efficiency.

10.10. For the engine of problem 10.6, determine the thermodynamic efficiency and the mep.

10.11. For the engine of problem 10.7, determine the mep.

10.12. For an eight-cylinder engine which has a stroke of 4 inches, a bore of 3.7. inches, and which delivers 1520 ft-lbf of work per cycle, determine the mep.

10.13. Determine the mep of a six-cylinder engine which has a cylinder displacement of 630 cm³, and which delivers 2000 ft-lbf per cycle.

10.14. An Otto engine has a compression ratio of 6.8 to 1. What is its thermodynamic efficiency?

Sections 10.4 and 10.5

10.15. If the fuel consumption is 1200 lbm/hr and bhp is 2700 hp for the engine of problem 10.8, determine the mechanical efficiency, the thermodynamic efficiency, the bmep, and the bsfc.

10.16. An eight-cylinder four-stroke engine is operating at 6000 rpm when the *p-V* diagram is gotten (figure 10–28). Determine
(a) ihp.
(b) Thermodynamic efficiency.
(c) mep (or imep).

10.17. If the bhp is found to be 450 hp for the engine of problem 10.16 and the fuel consumption is 400 lbm/hr, determine
(a) bmep
(b) bsfc
(c) Mechanical efficiency

10.18. An Otto engine operating on an ideal, four-stroke, air-standard cycle is required to develop 100 hp at 5000 rpm. It must take air in at 14.7 psia and 90°F, and is limited to four cylinders and a compression ratio of 6.5 to 1 and a maximum bore of 2.5 inches. If the proposed fuel to be used can deliver 850 Btu/lbm air, calculate

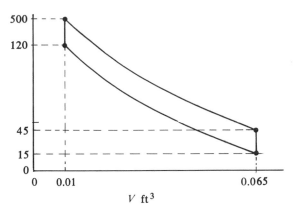

Figure 10–28   Problems 10.16 and 10.17.

      (a) p, v, and T at the four corners. Use gas table B.6.
      (b) wk_{cycle}.
      (c)   Stroke. Use gas table B.6.
      (d) $\dot{Q}_{\text{rej}}$. Use gas table B.6.

10.19. A two-stroke, four-cylinder engine, operating at 6000 rpm produces the p-V diagram shown in figure 10–29. Determine as best as possible
      (a) Power.
      (b) $\dot{Q}_{\text{add}}$.
      (c) Thermodynamic efficiency.
      (d) $\dot{Q}_{\text{rej}}$.

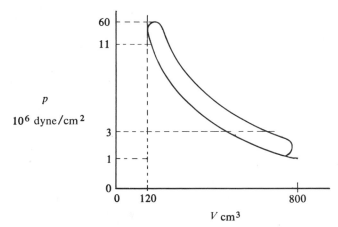

Figure 10–29   Problems 10.19 and 10.20.

10.20. For the engine of problem 10–19, assume the bhp is 180 hp and the fuel consumption is 600 g/min. Determine
      (a) Mechanical efficiency.
      (b) bmep.
      (c) bsfc.

10.21. A two-stroke Diesel engine has six cylinders, a compression ratio of 15 to 1, a cutoff ratio of 3 to 1, and a bore and stroke of 5 inches and 4.9 inches respectively. The air taken in is at 80°F and 15 psia. Determine
(a) p, V, T, at the cycle corners.
(b) Entropy change during process (1–2).
(c) Entropy change during process (2–3).
(d) Wk_{cycle}.
(e) Thermodynamic efficiency.

10.22. If the engine of problem 10.21 consumes 2 lbm/min of fuel when running at 3000 rpm, and delivers 200 bhp, determine
(a) bsfc.
(b) bmep.
(c) Mechanical efficiency.

10.23. From the efficiency equation

$$\eta_T = \frac{Wk_{\text{cycle}}}{Q_{\text{add}}} \times 100$$

derive the relation

$$\eta_T = \left[1 - \frac{1}{(r_v)^{k-1}} \left(\frac{r_c^k - 1}{k(r_c - 1)} \right) \right] \times 100$$

for an ideal Diesel cycle.

10.24. Derive a relation for the thermodynamic efficiency of a Diesel engine having polytropic processes for compression and expansion. Assume the polytropic exponents equal; i.e., $n_{21} = n_{43}$.

10.25. Compare the thermodynamic efficiencies of an Otto engine having a compression ratio of 10.5 to 1, and a Diesel engine having a compression ratio of 15 to 1, and a cutoff ratio of 2.5 to 1. Which is higher?

10.26. A Diesel engine produces 1500 bhp at 280 rpm. It has eight cylinders, 18-inch bore, 18-inch stroke, and a two-stroke cycle. If 12 lbm/min of fuel with a LHV of 18,400 Btu/lbm are used and the mep is 82 psi, determine
(a) Mechanical efficiency.
(b) Thermodynamic efficiency.
(c) bmep.
(d) bsfc.

10.27. There are 80 Btu/cycle of heat supplied to an ideal four-stroke, two-cylinder Diesel engine running at 300 rpm. The following data is given for each cylinder:
$p_1 = 14.7$ psia $T_1 = 90°F$ $V_1 = 1.5$ ft³ $p_2 = 500$ psia
Using the air table B.6, determine
(a) Wk_{cycle}.
(b) mep.
(c) $\dot{Wk}$.
(d) Thermodynamic efficiency.

10.28. There are 0.05 of fuel with an LVH of 18,000 Btu/lbm, and 0.9 lbm of air supplied to a Diesel engine. Given that

$$p_1 = 14.7 \text{ psi} \qquad T_1 = 135°F \qquad r_v = 14$$

determine r_c using the air tables.

10.29. An ideal four-stroke, three-cylinder Diesel engine uses 0.9 lbm of air at 14.7 psia and 100°F. The compression ratio is 15 to 1 and the cutoff ratio is 2.6 to 1. Using the air table (B.6), determine

 (a) Q_{add}.

 (b) Q_{rej}.

 (c) Wk_{cycle}.

 (d) Thermodynamic efficiency.

 (e) mep.

10.30. There are 91 Btu/cycle supplied to an ideal Diesel engine and 0.5 lbm of air at 14.7 psia and 130°F are supplied every cycle. At the end of compression, the pressure is 560 psia. If the engine is running at 900 rpm, determine (using the air tables B.6), assuming a two-stroke cycle,

 (a) r_v.

 (b) r_c.

 (c) Wk_{cycle}.

 (d) $\dot{Wk}_{cycle}$.

 (e) Thermodynamic efficiency.

 (f) mep.

11

Gas Turbines and the Brayton Cycle

The *jet engine*, which extracts energy from a *gas turbine*, has become a popular power-producing device, replacing the reciprocating IC engine wherever higher power per unit engine mass, smoother operation, or increased maintainability is demanded. We will study the thermodynamics of the jet engine and other adaptations of the gas turbine in this chapter.

The *ideal Brayton cycle*, a thermodynamic heat engine cycle, will be defined and compared to the actual operation of the jet engine. To give the reader a better appreciation of the actual engines, the mechanics of the *gas turbine*, the *compressor*, the *combustor*, and *nozzles* and *diffusers* will be presented. These devices are the essential components of the common jet engine and a knowledge of their individual functions helps to make the Brayton cycle more believable.

The perfect gases and their processes, with which we are now familiar, will be used to analyze the Brayton cycle in its ideal configuration. We will see how power is mechanically extracted in a typical jet engine and in stationary power supplies.

By using gas tables, we will remove the perfect gas restriction from the Brayton cycle analysis. Here we will adhere closely to the ideal cycle and its constituting processes, but the generalization of using data from gas tables will provide a truer setting for Brayton cycle calculations.

Regenerative heating, the most common attempt to increase the efficiency of the Brayton cycle, will be considered in the context of a Brayton cycle application. Other variations of the gas turbine power plant, namely the *fan jet*, *ram jet*, and *pulse jet*, will be presented and briefly analyzed.

Finally, the *rocket engine* will be discussed, even though it is neither a gas turbine nor a complete heat engine. An elementary thermodynamic analysis of the solid and liquid rockets will then be made.

11.1
The Ideal Brayton Cycle and the Jet Engine

The ideal Brayton cycle is defined by the following four reversible processes:

(1–2)   Adiabatic compression from state (1) to state (2).
(2–3)   Constant pressure heat addition.
(3–4)   Adiabatic expansion.
(4–1)   Constant pressure heat rejection to state (1).

We can easily plot these four processes on property diagrams, as was done in figure 11–1 in *p-V* and *T-S* diagrams. Traditionally in describing

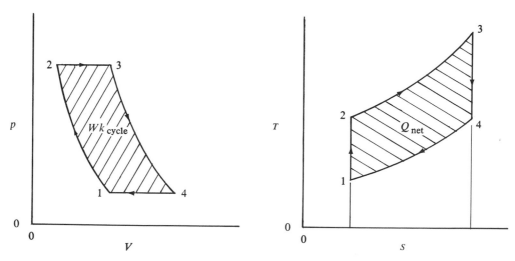

Figure 11–1   Property diagrams of ideal Brayton cycle.

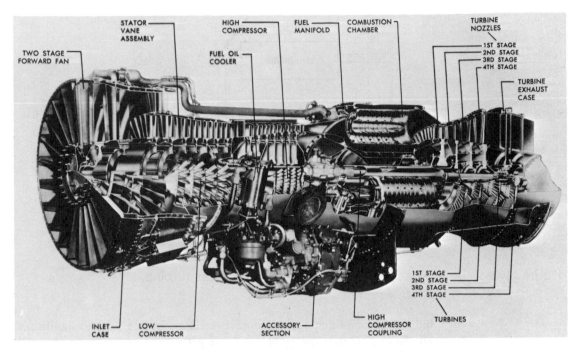

Figure 11–2   A typical aircraft gas turbine engine. Reproduced from "The Aircraft Gas Turbine Engine and Its Operation," August, 1970, ed.; with permission of Pratt & Whitney Aircraft Division.

a real operational engine, the Brayton cycle has been applied to the gas turbine engine. This engine has normally been utilized in powering vehicles (aircraft, trucks, and some experimental configurations) in which it is commonly called a *jet engine;* it has also been used in stationary, small electric power-generating units. In figure 11–2 is shown a typical arrangement of the critical components of the aircraft gas turbine, or jet engine. The major components of the typical gas turbine are listed in a cutaway view in figure 11–2: the fans; the compressor; the combustion chamber or combustor; the turbine and turbine nozzles; and the exhaust section. In the following two sections we will consider in detail the major components: the turbine, the compressor, and the combustor. In section 11.5, we will consider the complete engine and its cycle. Keep in mind that this is only an example and that there are many other engines which have physical characteristics different from those of the gas turbine, but which could be approximated by the Brayton cycle.

In figure 11–3 is shown the schematic diagram of the gas turbine operating on a Brayton cycle. If we apply the steady flow energy equation to the

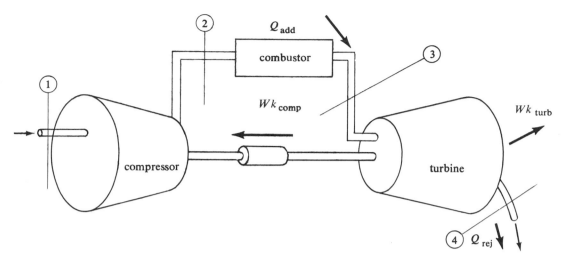

Figure 11-3   System diagram of Brayton cycle gas turbine.

individual components and assume reversible processes and no significant kinetic or potential energy changes, we get

(a)   for the compression:

$$h_2 - h_1 = -wk_{comp} \tag{11-1}$$

(b)   for the combustion:

$$h_3 - h_2 = q_{add} \tag{11-2}$$

(c)   for the turbine expansion:

$$h_4 - h_3 = -wk_{turb} \tag{11-3}$$

For the enclosed area in the p-V diagram of figure 11-1, which is the net work of the cycle, we get

$$wk_{cycle} = wk_{turb} + wk_{comp} \tag{11-4}$$

or

$$wk_{cycle} = h_3 - h_4 + h_1 - h_2 \tag{11-5}$$

The enclosed area in the T-s diagram must be the net heat added to the cycle and

$$q_{net} = q_{add} + q_{rej} \tag{11-6}$$

This also can be identified as

$$q_{net} = wk_{cycle} \tag{11-7}$$

If we assume a perfect gas as the working medium then equations (11–1), (11–2), (11–3), and (11–5) can be further altered. In particular the thermodynamic efficiency

$$\eta_T = \frac{Wk_{\text{cycle}}}{Q_{\text{add}}} \times 100 \tag{11–8}$$

reduces to

$$\eta_T = \left[1 - \frac{1}{(r_p)^{(k-1)/k}}\right] \times 100 \tag{11–9}$$

for perfect gases and reversible conditions (see problem 11.10) where r_p is the pressure ratio given by p_2/p_1 or p_3/p_4.

11.2 The Gas Turbine

The manner by which thermal energy is converted into mechanical energy, thereby providing work in the gas turbine, is shown in the schematic diagram of figure 11–4. Here the high velocity stream of gas (or liquid) is directed against the paddle wheel, thus inducing a rotation of the wheel. This arrangement is called an *impulse turbine* and is one of the two types of turbines. The other type of turbine, the *reaction turbine*, is depicted in figure 11–5; here the fluid is ejected at a high velocity from nozzles attached to the turbine wheel. This causes a reaction that

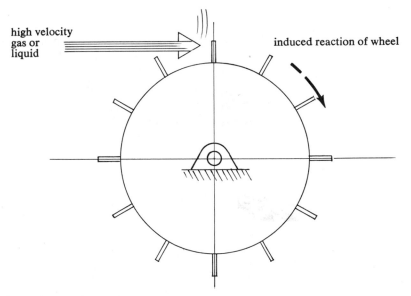

Figure 11–4   Basic principle of impulse turbine.

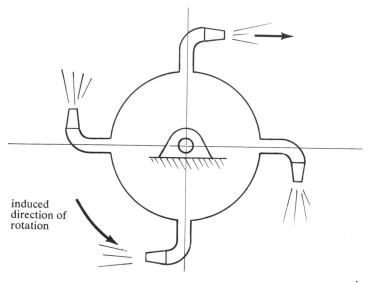

induced
direction of
rotation

Figure 11–5   Reaction turbine.

propels the turbine wheel in the opposite direction of the fluid stream.
Both types of turbines are utilized in practice and if more than one wheel
is used in a turbine (then called a *multistage turbine*), the reaction
principle has many inherent advantages. In this elementary discussion,
however, we will consider only the mechanics of the impulse turbine
with a single stage or wheel.

In the turbine, we want to convert kinetic energy of a fluid into rotational
kinetic energy of a shaft or wheel. This is accomplished as we have
already indicated by paddle wheels or blades in the impulse turbine, and
the most efficient blade shape would be that shown in figure 11–6, the
semicircular blade. Notice that the fluid stream is diverted 180° so that
its acceleration a_s is given by,

$$a_s = \frac{\bar{V}_2 - \bar{V}_1}{\tau}$$ (11–10)

where τ is the time required for a unit mass of fluid to have its velocity
changed from $\bar{V}_1$ to $\bar{V}_2$. The blade therefore impresses, on the fluid, a force
given by Newton's law.

$$F_s = \frac{1}{g_c} m a_s = \text{force on fluid}$$ (11–11)

and specifically,

$$F_s = \frac{1}{g_c} m \left(\frac{\bar{V}_2 - \bar{V}_1}{\tau}\right)$$ (11–12)

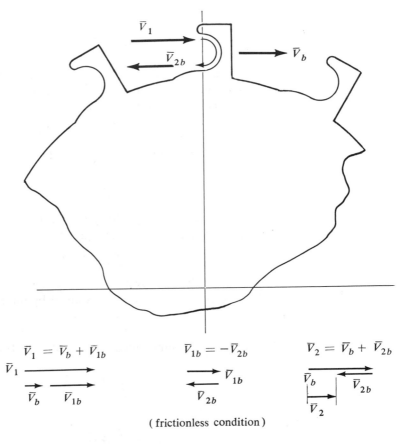

$$\bar{V}_1 = \bar{V}_b + \bar{V}_{1b} \qquad\qquad \bar{V}_{1b} = -\bar{V}_{2b} \qquad\qquad \bar{V}_2 = \bar{V}_b + \bar{V}_{2b}$$

(frictionless condition)

Figure 11–6 Semicircular turbine blade and velocity relationships.

The work gotten from the fluid stream per unit time τ is then

$$Wk = -F_s x = \frac{1}{g_c} mx \left(\frac{\bar{V}_1 - \bar{V}_2}{\tau} \right) \qquad (11\text{–}13)$$

where x is the distance traveled by the blade during the unit time τ. The negative sign is shown to indicate that the work done by the fluid is determined by the force the fluid exerts on the blade $(-F_s)$, and not by the blade force F_s. But

$$\frac{x}{\tau} = \bar{V}_b = \text{blade velocity} \qquad (11\text{–}14)$$

so that,

$$Wk = \frac{1}{g_c} m(\bar{V}_1 - \bar{V}_2)\bar{V}_b \qquad (11\text{–}15)$$

and per unit of fluid mass

$$wk = \frac{1}{g_c}(\bar{V}_1 - \bar{V}_2)\bar{V}_b \tag{11-16}$$

This, of course, is only true for a semicircular blade as shown in figure 11–6. Also, we have that the velocity of incoming fluid relative to the blade, $\bar{V}_{1b}$, is related to the actual velocity $\bar{V}_1$ by the equation

$$\bar{V}_1 = \bar{V}_b + \bar{V}_{1b} \tag{11-17}$$

and shown diagrammatically in figure 11–6. If the fluid is frictionless, then

$$\bar{V}_{1b} = -\bar{V}_{2b} \tag{11-18}$$

where $\bar{V}_{2b}$ is the outgoing fluid velocity relative to the blade and related to the actual velocity $\bar{V}_2$ by the equation

$$\bar{V}_2 = \bar{V}_b + \bar{V}_{2b} \tag{11-19}$$

It can then easily be shown that for the semicircular blade, the magnitude of the outgoing velocity $\bar{V}_2$ is given by the result

$$\bar{V}_2 = 2\bar{V}_b - \bar{V}_1$$

which then can be substituted into equation (11–16) to obtain

$$wk = \frac{1}{g_c}(\bar{V}_1 - 2\bar{V}_b + \bar{V}_1)\bar{V}_b \tag{11-20}$$

or

$$wk = \frac{2}{g_c}(\bar{V}_1\bar{V}_b - \bar{V}_b^2) \tag{11-21}$$

We see that the work is dependent on the blade and the fluid velocities. If we consider a fixed incoming velocity, the maximum work we can get will be determined from a well-known calculus theorem that states, in effect, that the work is greatest when its derivative with respect to its variable $\bar{V}_b$ is zero. Then, using equation (11–21) we get

$$\frac{dwk}{dV_b} = \frac{2}{g_c}(\bar{V}_1 - 2\bar{V}_b) = 0$$

But this implies that, for maximum work,

$$\bar{V}_1 - 2\bar{V}_b = 0$$

or

$$\bar{V}_b = \frac{1}{2}\bar{V}_1 \tag{11-22}$$

That is, the work is greatest when the blade velocity is exactly one-half the

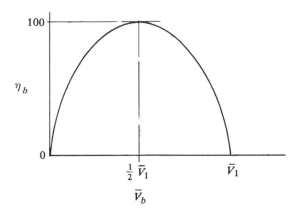

Figure 11-7   Blade efficiency of semicircular blade.

magnitude of the incoming fluid velocity. The blade efficiency we define as the work extracted divided by the incoming fluid kinetic energy. This can be written

$$\eta_b = \left(\frac{1}{g_c}\frac{(\bar{V}_1 - \bar{V}_2)\bar{V}_b}{1/2g_c\ \bar{V}_1^2}\right) \times 100 \qquad (11\text{-}23)$$

and, from equations (11–21) and (11–22) we get for the maximum efficiency

$$\eta_b = \left(\frac{2}{g_c}\frac{(\bar{V}_1\bar{V}_b - \bar{V}_b^2)}{1/2g_c\ \bar{V}_1^2}\right) \times 100 = \left(\frac{2}{g_c}\frac{\bar{V}_1^2/2 - \bar{V}_1^2/4}{1/2g_c\ \bar{V}_1^2}\right) \times 100$$

$$= 100\%$$

If the blade velocity drops below or goes above the value $\frac{1}{2}\bar{V}_1$ the efficiency will be less than 100% and ultimately reach zero efficiency as shown in figure 11–7. Obviously, this blade is ideal at a velocity of one-half the impinging fluid's velocity. There is, however, a difficulty associated with the semicircular blade; the fluid exiting the blade remains in an area which will be occupied by a following blade. This slow-moving fluid then acts as a resistance or barrier to the upcoming blade and produces a drop in the efficiency. In order to have a more practical arrangement for directing fluid past turbine blades, the configuration shown in figure 11–8(a) is used. Incoming fluid is directed at an angle θ into curved blades and subsequently ejected at an angle β. The side view in figure 11–8(b) more clearly shows the blade and fluid interaction, comparable to the simpler semicircular blade in figure 11–6. In figure 11–8(b) we are changing the fluid velocity into directions other than parallel to the blade velocity. It can be shown that for any blade we can write

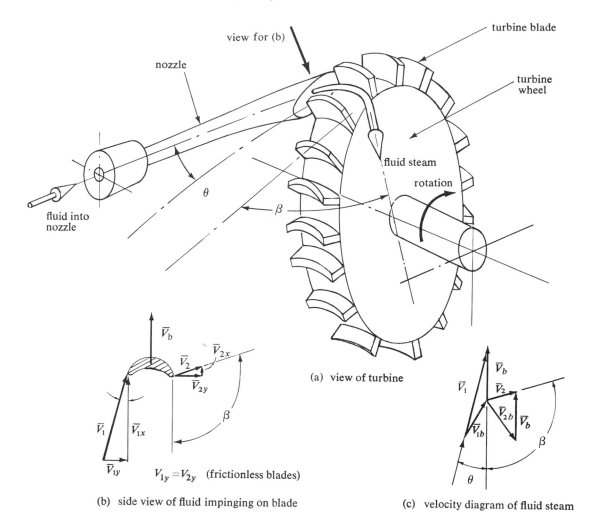

(a) view of turbine

$V_{1y} = V_{2y}$ (frictionless blades)

(b) side view of fluid impinging on blade

(c) velocity diagram of fluid steam

Figure 11–8   Typical impulse turbine.

$$wk = \frac{1}{g_c} (\bar{V}_{1x} - \bar{V}_{2x})\bar{V}_b \qquad (11–24)$$

where the $1x$ and $2x$ terms are just components of the velocity vectors:

$$\bar{V}_{1x} = \bar{V}_1 \cos \theta \qquad (11–25)$$
$$\bar{V}_{2x} = \bar{V}_2 \cos \beta \qquad (11–26)$$

We have seen that the velocities relative to the blade are given by the vector equations (11–17) and (11–19) for the semicircular configuration. These relations are correct for any blade shape as long as we interpret the addition signs as vector additions rather than simple algebraic additions. These operations are shown graphically in figure 11–8(c), a velocity diagram. Here the maximum work is extracted when $\bar{V}_{1b}$ and $\bar{V}_{2b}$ are tangent to the blade profiles at the respective positions. In addition, for a frictionless blade, regardless of its shape, equation (11–18) holds

$$|\bar{V}_{1b}| = |\bar{V}_{2b}|$$

From this simple relation, the terms $\bar{V}_2$ and $\bar{V}_{2x}$ can be found from known values for $\bar{V}_1$ and $\bar{V}_b$.

If fluid or blade velocities become excessive in attempting to produce power, additional turbine wheels can be used to cascade the fluid through a series of adjacent blades. This reduces the velocities in any one blade, yet provides work from all the wheels. This device, called a *multistage turbine*, is typified by the configuration shown in figure 11–2. Notice also that the wheel diameter increases for succeeding stages — this is a normal procedure and is done for kinetic reasons.

The fluid directed against the blades of a turbine attains its high velocity through a nozzle. In figure 11–8 is shown a typical nozzle, and in figure 11–9 we indicate the nozzle in relation to the blade and exhaust. If we consider only the nozzle as a system, then, using the steady flow equation of the first law, we have

$$\frac{\bar{V}_1^2 - \bar{V}_0^2}{2g_c} + h_1 - h_0 = q - wk_{os} \qquad (11\text{--}27)$$

but here the work is zero and if we assume adiabatic walls then

$$\frac{\bar{V}_1^2 - \bar{V}_0^2}{2g_c} + h_1 - h_0 = 0 \qquad (11\text{--}28)$$

for the nozzle. For the whole turbine stage of nozzle, blade, and exhaust we get

$$\frac{\bar{V}_3^2 - \bar{V}_0^2}{2g_c} + h_3 - h_0 = q - wk_{os}$$

and then the work wk_{os} is that done by the fluid in rotating the turbine wheel. If everything is adiabatic then

$$\frac{\bar{V}_3^2 - \bar{V}_0^2}{2g_c} + h_3 - h_0 = -wk_{os} \qquad (11\text{--}29)$$

and we can simplify this by noting that $\bar{V}_0$ and $\bar{V}_3$ are negligible. Note that $\bar{V}_1$ is not negligible and is the whole reason for extracting work. Then

$$h_2 - h_1 = wk_{os} \qquad (11\text{--}30)$$

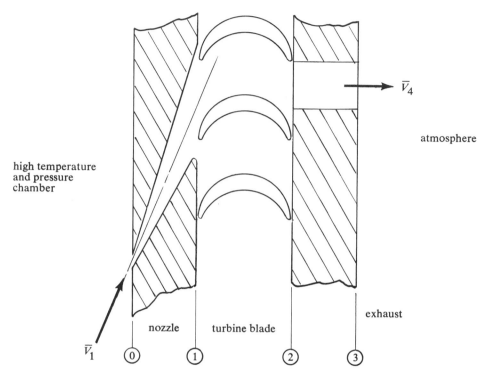

Figure 11–9   Typical turbine stage.

which is the work produced in a turbine if the process is frictionless and adiabatic, and if kinetic energy changes are neglected. The power is then found from

$$\dot{m}(h_2 - h_1) = -\dot{w}k_{os} \qquad \text{(11–31)}$$

which is a common result.

Example 11.1   A reversible, adiabatic turbine receives 3.0 lbm/s air at 1200°F and 300 psia and then exhausts the air to 14.7 psia. If the blade and nozzle profiles are those shown in figure 11–10 and the exhaust air temperature is 400°F, determine the turbine work per unit mass of fluid, the power extracted, the velocity of the air impinging on the blades, and the blade velocity.

Solution   We can determine the work of the system, the turbine, from equation (11–29)

$$-wk_{os} = h_3 - h_0 + \frac{\bar{V}_3^2 - \bar{V}_0^2}{2g_c}$$

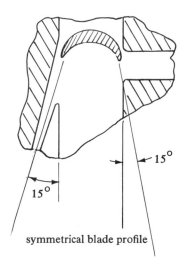

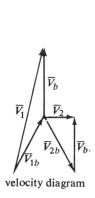

velocity diagram

Figure 11–10   Diagram for example 11.1.

If we neglect the velocity of the air entering and leaving the turbine and assume constant specific heat values, we then have

$$-wk_{os} = c_p(T_3 - T_0)$$

Then

$$-wk_{os} = 0.24 \text{ Btu/lbm°R } (400 - 1200)°R$$
$$= -192 \text{ Btu/lbm}$$

or

$$wk_{os} = 192 \text{ Btu/lbm} \qquad\qquad Answer$$

The power is obtained from

$$\dot{W}k_{os} = \dot{m}wk_{os} \qquad\qquad \textbf{(11–32)}$$

which yields

$$\dot{W}k_{os} = 3 \text{ lbm/s } (192 \text{ Btu/lbm}) = 576 \text{ Btu/s}$$

or

$$\dot{W}k_{os} = 815 \text{ hp} \qquad\qquad Answer$$

The blade velocity can be determined by first recalling equation (11–24)

$$wk = \frac{1}{g_c} (\bar{V}_{1x} - \bar{V}_{2x})V_b$$

and noting that, from equation (11–26)

$$\bar{V}_{1x} = \bar{V}_1 \cos 15°$$
$$V_{2x} = V_2 \cos 90° = 0$$

From the velocity diagram of figure 11–10 we see that

$$|\bar{V}_{1b}| = |\bar{V}_{2b}|$$

so that

$$\bar{V}_2 = \bar{V}_{1y} = \bar{V}_1 \sin 15°$$

Also, from this same diagram

$$2\,\bar{V}_b = \bar{V}_1 \cos 15°$$

or

$$\bar{V}_b = \frac{V_1 \cos 15°}{2} \tag{11–33}$$

Using these trigonometric results then in equation (11–24) we get

$$wk = \frac{1}{g_c} (\bar{V}_1 \cos 15°)\left(\frac{\bar{V}_1 \cos 15°}{2}\right)$$

but wk is equal to 192 Btu/lbm so that

$$192 \text{ Btu/lbm} = (1 \text{ Btu/778 ft-lbf}) \, (1 \text{ s}^2\text{-lbf/322 ft-lbm})\left(\frac{1}{2}\right)(\cos 15°)^2(\bar{V}_1)^2$$

yielding, for $\bar{V}_1$

$$\bar{V}_1 = \frac{\sqrt{192 \text{ Btu/lbm} \times 32.2 \text{ ft-lbm/s}^2\text{-lbf} \times 2 \times 778 \text{ ft-lbf/Btu}}}{\cos 15°}$$

$$= 3211 \text{ ft/s} \qquad\qquad Answer$$

Since this represents a velocity greater than sound,* there would be phenomena occurring in the turbine which we have not accounted for. Let us merely note here that we would probably recommend a series of turbine stages in order that nozzle velocities may be reduced. At any rate the blade velocity expected from the above situation can be determined from equation (11–33)

$$\bar{V}_b = \frac{\bar{V}_1 \cos 15°}{2} = \frac{3211 \text{ ft/s} \times 0.966}{2}$$

$$= 1551 \text{ ft/s} \qquad\qquad Answer$$

---

*Velocity of sound is near 1200 ft/s for common earth atmospheric conditions.

**11.3
Combustors
and
Compressors**

The high temperature gases furnished to the gas turbine nozzle are produced in the combustor. A combustor is merely a chamber at a constant pressure which allows for the burning or combustion of fuel and air. It can be visualized as a pipe with fuel injectors situated as shown in figure 11–11. The flame wall is a continuous, steady state chemical reaction of fuel and air, producing the hot gases for the turbine. We can consider then the flame wall as a heat addition process and the whole combustor can be represented by a system diagram shown in figure 11–12. Notice in this figure that to conserve mass we must have that

$$\dot{m}_2 = \dot{m}_1 + \dot{m}_f \tag{11–34}$$

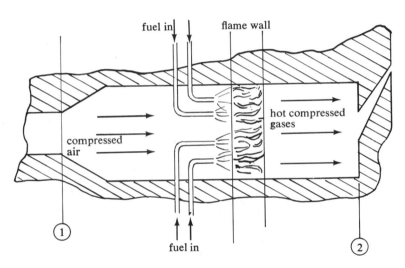

Figure 11–11   Typical combustor.

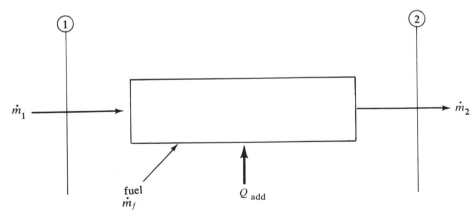

Figure 11–12   System diagram of combustor.

The fuel-air ratio, $\dot{m}_f/\dot{m}_1$, is generally about 1 to 30, but can vary greatly from this because of the fuel, air conditions, or operating requirements. The first law written for the combustor is

$$\dot{m}_1 h_1 + \dot{m}_f h_f - \dot{m}_2 h_2 = \dot{Q}_{added} \qquad (11-35)$$

assuming that no kinetic or potential energy changes occur. The work or power is obviously zero in the combustor and if we further neglect the fuel mass, equation (11–35) can be written

$$\dot{m}_2(h_2 - h_1) = \dot{Q}_{added} \qquad (11-36)$$

or

$$h_2 - h_1 = q_{added} \qquad (11-37)$$

where q is based on unit mass of air flowing through the combustor.

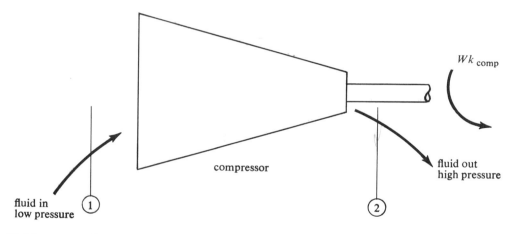

Figure 11–13   System diagram for compressor.

The gases flowing through the combustor are at an essentially constant high pressure. To get the air or gases up to a high pressure as they enter the combustor, a pump or compressor must be used. We will see in the ram jet that other methods exist to compress the gas, but characteristically in the pump or compressor, gases are received at a low velocity, are accelerated through blades or fans to a high velocity, and finally restricted to increase the gas density and pressure while reducing the velocity to a low value again. There are many varied configurations to produce compressed gases, but here we will consider only the typical one used with a gas turbine. It is essentially a reversed turbine which has been considered in some detail already. Gases are taken in at low pressure, as indicated in the system diagram of figure 11–13, and are compressed

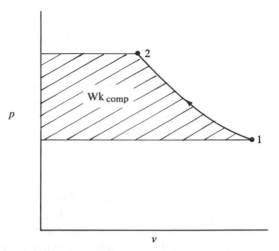

Figure 11–14   Diagram of compressor.

through a series of turbine-type blades and nozzles to get a high pressure gas. The work to drive the compressor is normally supplied by a shaft from the gas turbine itself. The first law of thermodynamics can take the form

$$\frac{\bar{V}_2^2 - \bar{V}_1^2}{2g_c} + h_2 - h_1 = q - wk_{\text{comp}} \tag{11–38}$$

for the compressor and if the walls are assumed to be adiabatic, we can eliminate q. The velocity terms are also neglected in some cases, although this can represent a significant error for high speed jet engine aircraft. At any rate, if we do make these simplifying assumptions, then

$$h_2 - h_1 = -wk_{\text{comp}} \tag{11–39}$$

and we can then also determine work, if we have a reversible compressor, from the area to the left of a curve in a p-v diagram. This is seen in figure 11–14 where the area is found from equation (6–49)

$$-wk_{\text{comp}} = \frac{k}{1-k}(p_2 v_2 - p_1 v_1)$$

By using the polytropic relation for work

$$-wk_{\text{comp}} = \frac{n}{1-n}(p_2 v_2 - p_1 v_1)$$

we can consider *nonadiabatic* compressors as well, but then we must use equation (11–38) instead of (11–39) for any energy balances.

**11.4
Nozzles and
Diffusers**

Fluids such as gases and liquids are carriers of energy and during their flow through pipes, conduits, channels, or other conveyors, it is frequently desired to convert the energy to other forms. For instance, high temperature gases (containing internal energy) may be accelerated to increase the kinetic energy by forcing the gases through a restriction called a *nozzle*. The kinetic energy increase would be done at the expense of internal energy or enthalpy. Similarly, a high pressure liquid may be accelerated by passing it through a nozzle, as illustrated in figure 11–15. Here the kinetic energy is increased at the expense of a pressure (or enthalpy) drop in the fluid. These two examples are typical of the use of *converging nozzles*, nozzles which have a decreasing cross-sectional area downstream in the flow. There are, however, other types of nozzles: *diverging* nozzles or *diffusers*, and *converging-diverging* nozzles or *deLaval* nozzles. Diffusers are commonly used to decelerate high velocity gases or liquids from regions of low pressure to those at a higher pressure. That is, diffusers compress fluids at the expense of kinetic energy. A typical one is shown in figure 11–16.

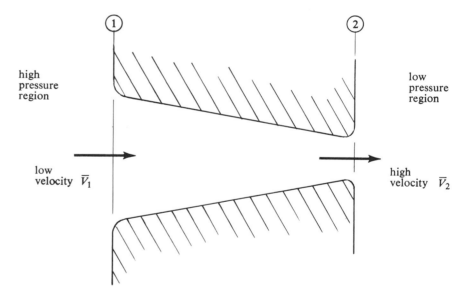

Figure 11–15   Typical converging nozzle.

Fluids which travel through nozzles at velocities less than sonic (less than the speed of sound) behave rather predictably. No uncommon situations arise and their use for converging nozzles and diffusers can be intuitively seen. However, gases traveling at sonic or supersonic velocities are subject to complicating events such as shock waves and accelerations in diffuser sections. As a consequence of supersonic flow of compressible

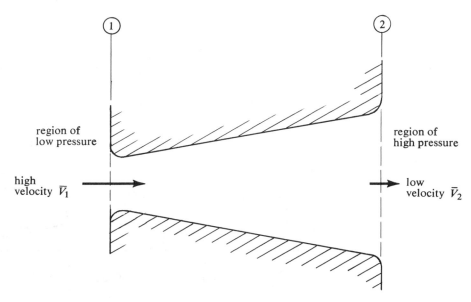

region of
low pressure

region of
high pressure

high
velocity $\bar{V}_1$

low
velocity $\bar{V}_2$

Figure 11–16   Typical diffuser.

gases, deLaval nozzles are used to accelerate these fluids. In figure 11–17 is shown the normal configuration of a deLaval nozzle and typical velocity and pressure curves in the passage. Notice that for subsonic flow (*a*) the fluid is accelerated very slightly. In fact, if the section between the throat (smallest cross-sectional area) and station (2) were removed, we would have a converging nozzle with much better accelerating characteristics.

For supersonic flow (*b*) we see that this velocity increases throughout the fluid flow in the nozzle. Normally, the velocity of the fluid at the throat of a nozzle will be sonic if the velocities in the divergent section are supersonic. The solid curve (I) for velocity represents the supersonic flow condition where the gases leave the nozzle at a high supersonic velocity. At some place external to the nozzle, a shock wave will be present which quickly decelerates the gases to subsonic speed. The discontinuous curve (II) represents a condition where a shock wave occurs in the diverging section of the nozzle. The shock wave can be seen to be a jump discontinuity in pressure and velocity and is normally described as an irreversible adiabatic process. The velocity downstream of the shock wave is subsonic and decreases to the exit in the normal fashion of subsonic diffusers. A more complete description and analysis of nozzle flows of compressible or incompressible fluids can be found in many good fluid mechanics or gas dynamics texts.

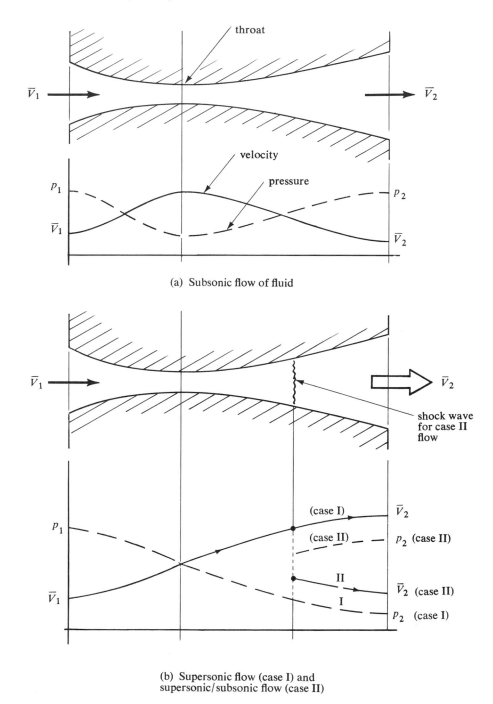

(a) Subsonic flow of fluid

(b) Supersonic flow (case I) and
supersonic/subsonic flow (case II)

Figure 11–17   Flow through a deLaval nozzle.

Let us now review the meaning of nozzles. In table 11–1 we see listed the energy conversion for the three types of nozzles we have considered: converging, diverging, and deLaval.

Table 11–1
The Energy Conversions in Nozzles

Nozzle Type	Energy Converted		Type of flow
	from	*to*	
Converging	Internal or Enthalpy	Kinetic	Subsonic or Supersonic
Diverging, Diffuser	Kinetic	Internal or Enthalpy	Subsonic or Supersonic
DeLaval	Internal or Enthalpy	Kinetic	Supersonic

We have seen before that gas turbines are visualized as devices converting the internal energy of a high temperature/pressure gas into rotational kinetic energy by means of turbine blades. The nozzle, however, is required to convert gas energy into kinetic energy and the turbine merely converts the kinetic energy to rotational kinetic energy. In steam turbine operations, the nozzle serves the same useful purpose as it does for gas turbine. In propelling high speed aircraft, the nozzle accelerates the exhaust gases, thus causing a forward thrust in the aircraft.

Let us now analyze the nozzle from a thermodynamic viewpoint. The nozzles we have considered are normally operated under steady flow conditions so that the equation

$$\rho_1 A_1 \bar{V}_1 = \rho_2 A_2 \bar{V}_2 \tag{11–40}$$

will then be descriptive of the nozzle.

The work done on or by any of the nozzles is zero, and commonly the flow is assumed to be adiabatic. Then the first law energy equation for steady flow becomes, per unit mass,

$$\frac{1}{2g_c}(\bar{V}_2^2 - \bar{V}_1^2) + \frac{g}{g_c}(z_2 - z_1) + h_2 - h_1 = 0$$

for the nozzle. Also potential energy changes are negligible ($z_2 \approx z_1$) and we have

$$\frac{1}{2g_c} (\bar{V}_2^2 - \bar{V}_1^2) = h_1 - h_2 \qquad (11\text{–}41)$$

Frequently a stagnation state is defined for the inlet nozzle condition. By *stagnation* is meant that the flow is replaced by a static or stagnating state so that the stagnation enthalpy h^* at state (1) is

$$h^*_1 = h_1 + \frac{1}{2g_c} \bar{V}_1^2 \qquad (11\text{–}42)$$

and the stagnation temperature T^* at state (1) is, for gases having constant specific heats

$$T^*_1 = T_1 + \frac{1}{2c_pg_c} \bar{V}_1^2 \qquad (11\text{–}43)$$

so that

$$h^*_1 = c_pT^*_1 \qquad (11\text{–}44)$$

Then equation (11–41) can be written as

$$\frac{1}{2g_c} \bar{V}_2^2 = h_1 + \frac{1}{2g_c} \bar{V}_1^2 - h_2 = h^*_1 - h_2$$

or

$$\bar{V}_2 = \sqrt{2g_c(h^*_1 - h_2)} \qquad (11\text{–}45)$$

This can be reduced to

$$\bar{V}_2 = \sqrt{2g_c c_p(T^*_1 - T_2)} \qquad (11\text{–}46)$$

for gases having constant specific heats.

Since nozzles are assumed to be adiabatic and if they are also reversible, the flow of fluid would be conducted at constant entropy. On a *T-s* diagram, shown in figure 11–18, the reversible adiabatic flow through a converging nozzle would be represented by path (1–2). We say this is 100% efficient and note that an expansion to the same pressure p_2 conducted in an irreversible manner, as in path (1–2′) would reduce the nozzle efficiency η_N. We define nozzle efficiency by the equation

$$\eta_n = \left(\frac{h^*_1 - h_{2'}}{h^*_1 - h_2}\right) \times 100 \qquad (11\text{–}47)$$

where $h_{2'}$ is the actual enthalpy of the exiting fluid. For gases having constant specific heats, equation (11–47) becomes

$$\eta_n = \left(\frac{T^*_1 - T_{2'}}{T^*_1 - T_2}\right) \times 100 \qquad (11\text{–}48)$$

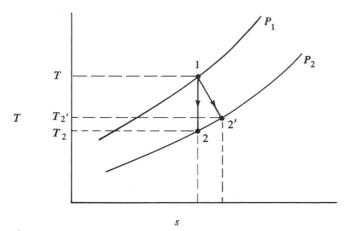

Figure 11–18   *T-s* diagram of flow through a converging nozzle.

Example 11.2   Air at a stagnation state of 500°F and 160 psia enters a converging nozzle. Determine the exit velocity if the nozzle is reversible and the pressure on the exit side is 14.7 psia.

Solution   We may use equation (11–46) since our system, a converging nozzle, is conducting air which we assume has a constant specific heat over the temperature change; we will use $c_p = 0.24$ Btu/lbm°R. The exit temperature we will calculate from the perfect gas relationships and the reversible adiabatic process equation $pv^k = C$. From the equation

$$\frac{p_1}{p_2} = \left(\frac{T_1}{T_2}\right)^{k/k-1}$$

we solve for T_2

$$T_2 = T_1\left(\frac{p_2}{p_1}\right)^{(k-1)/k} = T_1{}^*\left(\frac{p_2}{p_1{}^*}\right)^{(k-1)/k}$$

$$= (500 + 460)\left(\frac{14.7}{160}\right)^{0.4/1.4} = 485°R$$

From equation (11–46) we obtain

$$\bar{V}_2 = \sqrt{2(32.2 \text{ ft-lbm/s}^2\text{-lbf})(0.24 \text{ Btu/lbm°R})(960 - 485°R)}$$

$$= \sqrt{7350 \text{ ft-Btu/s}^2\text{-lbf} \times 778 \text{ ft-lbf/Btu}}$$

$$= 2390 \text{ ft/s} \hspace{4cm} Answer$$

Example 11.3   Air enters a diffuser at 1500 ft/s, 600°R, and 15 psia. If 3 lbm/s of air flows through the diffuser, determine the entrance area of the diffuser, the pressure at the exit, and the mach number of the air at the entrance. Assume the velocity at the exit is negligible and the diffuser efficiency is 90%.

Solution

The entrance area can be calculated from the continuity equation (11–40)

$$\rho_1 A_1 \bar{V}_1 = \dot{m}_1$$

or

$$A_1 = \frac{\dot{m}_1}{\bar{V}_1 \rho_1}$$

We have that

$$v_1 = \frac{RT_1}{p_1} = \frac{1}{\rho_1}$$

or

$$\rho_1 = \frac{p_1}{RT_1} = (15 \text{ lbf/in}^2)/(53.3 \text{ ft-lbf/lbm°R})(600°R)$$

$$= \frac{15 \times 144}{53.3 \times 600} \text{ lbm/ft}^3 = 0.0675 \text{ lbm/ft}^3$$

Then

$$A_1 = \frac{(3 \text{ lbm/s})}{(1500 \text{ ft/s})(0.0676 \text{ lbm/ft}^3)}$$

$$= 0.0296 \text{ ft}^2 = 4.26 \text{ in}^2 \qquad\qquad Answer$$

If we assume that the diffuser, shown in figure 11–19, is adiabatic then we may apply equation (11–35) in the form

$$\frac{1}{2g_c} \bar{V}_1^2 = h_2 - h_1 = c_p(T_{2'} - T_1)$$

Then

$$T_{2'} = T_1 + \frac{1}{2c_p g_c} \bar{V}_1^2$$

$$= 600°R + \frac{1}{2(0.24 \text{ Btu/lbm°R})(32.2 \text{ ft-lbm/s}^2\text{-lbf})} \frac{(1500)^2 \text{ ft}^2/\text{s}^2}{(778 \text{ ft-lbf/Btu})}$$

$$= 787°R$$

Notice now that since the diffuser is not reversible, but rather is 90% efficient, we can see on the *T-s* diagram of the process (figure 11–20) that the actual temperature $T_{2'}$ is higher than the reversible adiabatic T_2. That is, we define the diffuser efficiency in a manner similar to the nozzle efficiency

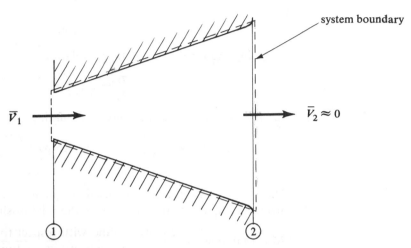

Figure 11–19   Adiabatic diffuser.

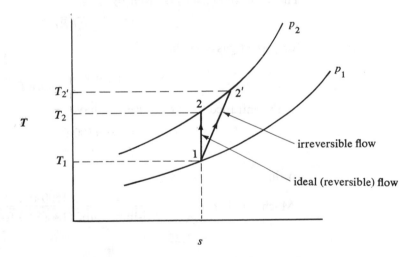

Figure 11–20   *T-s* diagram for diffuser flow in example 11.3.

$$\eta_D = \left(\frac{T_2 - T_1}{T_{2'} - T_1}\right) \times 100 \qquad \textbf{(11–49)}$$

where the temperatures at state (2) are effectively the stagnation temperatures and where the gas has constant specific heat values. The primed subscript 2′ represents the actual temperature at state (2′) and the unprimed, the ideal temperature. Then

$$90\% = \left(\frac{T_2 - 600}{787 - 600}\right) \times 100$$

or

$$T_2 = 768°R$$

and using the reversible isentropic relation

$$\frac{p_1}{p_2} = \left(\frac{T_1}{T_2}\right)^{k/(k-1)}$$

we may obtain the exit pressure. Calculating

$$p_2 = p_1 \left(\frac{T_2}{T_1}\right)^{k/(k-1)} = (15 \text{ psia})\left(\frac{768}{600}\right)^{1.4/0.4}$$

$$= 35.6 \text{ psia} \hspace{3cm} Answer$$

The *mach number* is a term used frequently in gas dynamic or fluid mechanics analysis. It is defined by the relationship

$$\text{Mach number} = \frac{\text{velocity of fluid with respect to the system boundary}}{\text{sonic velocity in the fluid at that point}}$$

The sonic velocity V_s is given by

$$V_s = \sqrt{(g_c k R T)}$$

for perfect gases so that

$$\text{Mach number} = \frac{V}{\sqrt{g_c k R T}} \hspace{2cm} \textbf{(11–50)}$$

At the entrance to our diffuser we have

$$V_1 = 1500 \text{ ft/s}$$
$$T_1 = 600°R$$

so that

$$\text{Mach number} = \frac{1500 \text{ ft/s}}{\sqrt{(32.2 \text{ ft-lbm/s}^2\text{-lbf})(1.4)(53.3 \text{ ft-lbf/lbm°R})(600°R)}}$$

$$= 1.25 \hspace{4cm} Answer$$

Notice that a mach number equal to unity (1) would describe sonic velocity conditions. In our problem, the velocity is greater than sonic so it is called *supersonic*. We have, in general, that

(implies)

$$M = 1 \longrightarrow \text{sonic velocity}$$
$$M > 1 \longrightarrow \text{supersonic}$$
$$M < 1 \longrightarrow \text{subsonic}$$

Also, observe that mach number is dependent upon the air or fluid temperature as well as upon the velocity, and the condition of flow (whether subsonic, sonic, or supersonic) is not only determined by velocity but by the fluid properties of temperature, pressure, or volume as well.

**11.5
Perfect Gas
Analysis of
Brayton Cycle**

Two applications of the Brayton cycle will be treated here. In each case the gas turbine will be used and the medium will be considered a perfect gas.

Example 11.4

An aircraft is propelled by a reversible gas turbine, or turbo-jet engine, shown in figure 11–21. The aircraft is flying at 40,000 ft altitude under standard day atmospheric conditions at 500 mph. Assume the compressor has a pressure ratio of 10 to 1, the fuel-air ratio is 1 to 35, and the fuel is kerosene. Assume also that the kerosene has the same LHV as octane (C_8H_{18}) and burns completely, releasing the full LHV in the combustor. If the engine operates on an ideal Brayton cycle and the gases are ideal, determine the values of p, v, T, and s for the Brayton cycle corners. Then determine the thermodynamic efficiency and the mass flow rate of air required to power the aircraft if the wind resistance (drag) is 600 lbf and the flight is level.

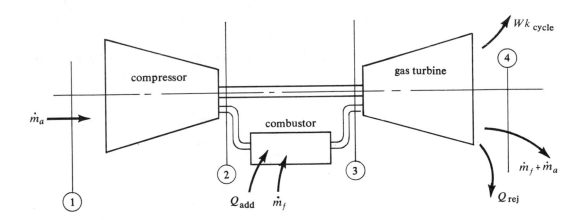

Figure 11–21   Turbo-jet engine.

Solution

In figure 11–22 are shown the p-v and T-s diagrams for the jet engine. The standard day atmospheric condition fixes the properties at state (1). From table B.16 they are $p_1 = 5.56$ inches Hg; $\rho_1 = 0.0189$ lbm/ft³; and $T_1 = 389.97°R$. The pressure at state (2) is obtained from the compressor pressure ratio $p_2/p_1 = 10$, which gives us

$$p_2 = 10\, p_1 = 55.6 \text{ in Hg}$$

Also

$$p_3 = p_2 = 55.6 \text{ in Hg} \times 0.491 \text{ psi/in Hg} = 27.2 \text{ psia}$$

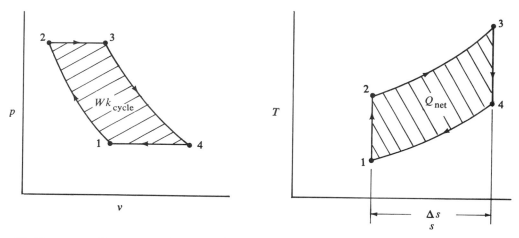

Figure 11–22   Property diagrams for reversible jet engine.

and

$$p_4 = p_1 = 2.72 \text{ psia}$$

Since the engine is operating in accordance with an ideal Brayton cycle, we can write

$$p_1 v_1{}^k = p_2 v_2{}^k$$

and

$$\frac{p_1}{p_2} = \left(\frac{v_2}{v_1}\right)^k$$

or

$$v_2 = \left(\frac{p_1}{p_2}\right)^{1/k} v_1$$

But

$$v_1 = \frac{1}{\rho_1} = 1 \text{ ft}^3/0.0189 \text{ lbm} = 52.9 \text{ ft}^3/\text{lbm}$$

so that

$$v_2 = 52.9 \text{ ft}^3/\text{lbm} \left(\frac{2.73}{27.3}\right)^{10/1.4} = 52.9 \left(\frac{1}{10}\right)^{0.713}$$

$$= 10.2 \text{ ft}^3/\text{lbm} \qquad\qquad\qquad Answer$$

Using the perfect gas relation

$$T_2 = \frac{p_2 v_2}{R}$$

$$T_2 = \frac{(27.2 \text{ lbf/in}^2)(10.2 \text{ ft}^3/\text{lbm})(144 \text{ in}^2/\text{ft}^2)}{(53.3 \text{ ft-lbf/lbm}°R)}$$

$$= 750°R \qquad\qquad Answer$$

To determine the properties at state (3), we now observe that process (2–3) is a constant pressure heat addition in the combustor. Using equation (11–35)

$$\dot{m}_2 h_3 + \dot{m}_f h_f - \dot{m}_2 h_3 = Q_{\text{added}}$$

we observe that the heat added is just the LHV of the fuel. For kerosene, as for octane, this has a value of 19,256 Btu/lbm fuel. The enthalpy of the liquid fuel h_f can be calculated from an equation from *Gas Tables* by Keenan and Kaye*

$$h_f = (0.5\,T - 287) \text{ Btu/lbm fuel} \qquad\qquad \textbf{(11–51)}$$

This gives us, assuming in our problem a fuel temperature of 40°F or 500°R, that

$$h_f = (0.5)(500) - 287 = -37 \text{ Btu/lbm fuel}$$

Then

$$\dot{m}_2 h_3 + \dot{m}_f(-37) - \dot{m}_2 h_3 = \dot{m}_f(19{,}256)$$

or

$$\left(\frac{\dot{m}_2}{\dot{m}_f}\right) h_3 - 37 - \left(\frac{\dot{m}_2}{\dot{m}_f}\right) h_3 = 19{,}256$$

But

$$\frac{\dot{m}_2}{\dot{m}_f} = 35$$

$$\frac{\dot{m}_3}{\dot{m}_f} = \frac{(\dot{m}_2 + \dot{m}_f)}{\dot{m}_f} = \left(\frac{\dot{m}_2}{\dot{m}_f} + 1\right) = 36$$

yielding

$$\left(\frac{\dot{m}_2}{\dot{m}_f} + 1\right) h_3 - 37 - \frac{\dot{m}_2}{\dot{m}_f} h_2 = 19{,}256$$

or

$$36\,h_3 - 37 \text{ Btu/lbm} - 35\,h_2 = 19{,}256 \text{ Btu/lbm}$$

If we assume the same values of specific heat for both air and hot gases, then we can write

---

*From Kennan and Kaye, *Gas Tables* (New York, 1948) by permission of John Wiley & Sons, Inc.

$$36 \, c_p T_3 - 37 - 35 \, c_p T_2 = 19,256$$

$$36(0.24 \text{ Btu/lbm}°\text{R})T_3 - 37 - 35(0.24 \text{ Btu/lbm}°\text{R})(750°\text{R})$$
$$= 19,256 \text{ Btu/lbm}$$

giving us

$$T_3 = (19,256 + 35 \times 0.24 \times 750 + 37) \frac{1}{36 \times 0.24}$$

$$= 2962°\text{R} \qquad\qquad\qquad \textit{Answer}$$

From the perfect gas relationship

$$v_3 = \frac{RT_3}{p_3} = \frac{53.3 \times 2962}{144 \times 27.2} = 40.3 \text{ ft}^3/\text{lbm} \qquad \textit{Answer}$$

Using isentropic expansion equations for process (3–4), we have

$$\frac{p_4}{p_3} = \left(\frac{v_3}{v_4}\right)^k$$

or

$$v_4 = \left(\frac{p_3}{p_4}\right)^{1/k} v_3 = (10)^{1/1.4}(40.3)$$

$$= 209 \text{ ft}^3/\text{lbm} \qquad\qquad\qquad \textit{Answer}$$

The temperature at state (4) we can determine from the perfect gas relation

$$T_4 = \frac{p_4 v_4}{R} = \frac{2.72 \times 144 \times 209}{53.3} = 1536°\text{R} \qquad \textit{Answer}$$

The entropy change is obtained from process (2–3) or process (4–1). Both of these processes are isobaric and we can write, from equation (7–21)

$$\Delta s = s_4 - s_1 = s_3 - s_2 = c_p \ln \frac{T_4}{T_1} = c_p \ln \frac{T_3}{T_2}$$

$$= 0.24 \text{ Btu/lbm}°\text{R} \ln \frac{1536}{389.97} = 0.329 \text{ Btu/lbm}°\text{R} \quad \textit{Answer}$$

The thermodynamic efficiency we can calculate from either equation (11–8) or (11–9). From the latter

$$\eta_T = \left(1 - \frac{1}{10^{0.4/1.4}}\right) \times 100 = 48.2\% \qquad \textit{Answer}$$

The power required to propel the aircraft can be determined from

$$\dot{w}k = F\bar{V} \qquad\qquad\qquad \textbf{(11–52)}$$

where F is the force (in this case the drag), and $\bar{V}$ is the velocity. We get then

$$\dot{W}k = (600 \text{ lbf})\left(500 \times \frac{5280}{3600} \text{ ft/s}\right)$$

$$= 440,000 \text{ ft-lbf/s}$$

$$= 440,000 \times \frac{1}{550} \text{ hp} = 800 \text{ hp}$$

The mass flow rate of air required can be obtained from the relationship

$$\dot{W}k = \dot{m}wk_{\text{cycle}} = 800 \text{ hp}$$

The work of the Brayton cycle can be obtained from equations (11–6) and (11–7)

$$wk_{\text{cycle}} = q_{\text{net}} = q_{\text{add}} + q_{\text{rej}}$$

We have

$$q_{\text{add}} = 19,256 \text{ Btu/lbm fuel}$$

$$= \frac{1}{35} \times 19,256 \text{ Btu/lbm air} = 550 \text{ Btu/lbm air}$$

The heat rejected is just process (4–1), occurring in the surroundings

$$q_{\text{rej}} = h_1 - h_4 = c_p(T_1 - T_4)$$
$$= 0.24 \text{ Btu/lbm°R } (389.97°R - 1528°R)$$
$$= -273 \text{ Btu/lbm air}$$

Then

$$\dot{m}_{\text{air}} = \frac{800 \text{ hp}}{273 \text{ Btu/lbm air}} = \frac{440,000 \text{ ft-lbf/s}}{277 \times 778 \text{ ft-lbf/lbm air}}$$

$$= 2.07 \text{ lbm/s} \qquad \qquad \textit{Answer}$$

An interesting mechanics problem is embodied in the problem just considered. The gas turbine drives the compressor through a shaft, thus providing the high pressure gases. But the question of how the aircraft is propelled by such a device is not answered. The turbine can rotate but cannot exert a direct force on the aircraft. In fact, the manner of driving the vehicle through air is the mechanics problem referred to. The answer is found in an application of Newton's law; namely, as shown in figure 11–23, the air is increased in velocity from $\bar{V}_1$ (which would be the negative aircraft velocity) to $\bar{V}_4$. This is an acceleration, $\bar{V}_4 - \bar{V}_1/\tau$, and from the relation $F = ma$ we get

$$F = \frac{m}{g_c}\left(\frac{\bar{V}_4 - \bar{V}_1}{\tau}\right)$$

But every force has an opposing force and we call this force F_I the impulse which actually drives the aircraft. In addition, the difference in

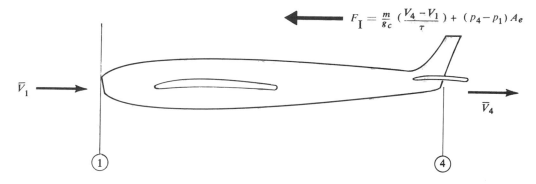

$$F_I = \frac{m}{g_c} \left(\frac{V_4 - V_1}{\tau} \right) + (p_4 - p_1) A_e$$

Figure 11–23   Reaction principle of jet aircraft propulsion.

pressure between the entrance at atmospheric pressure and the exit can cause a force given by $(p_4 - p_1)A_e$ where A_e is the area of the exhaust nozzle. Adding this to the impulse force of the above equation, we then get

$$F_I = \frac{m}{g_c} \left(\frac{\bar{V}_4 - \bar{V}_1}{\tau} \right) + (p_4 - p_1)A_e \qquad \textbf{(11–53)}$$

Frequently the specific impulse, defined by

$$I_{sp} = \frac{F_I}{\dot{m}} = \frac{\bar{V}_4 - \bar{V}_1}{g_c} + \frac{(p_4 - p_1)A_e}{\dot{m}} \qquad \textbf{(11–54)}$$

is referred to in the literature. This quantity provides a quick measure of the propulsive capability of a particular design.

Let us now consider a gas turbine with some irreversibilities included.

Example 11.5   A closed turbine engine operates with 20 lbm/s of air as the working medium operates on an air-standard Brayton cycle. The air-standard Brayton cycle is defined as a Brayton cycle in which the properties of all the gases are assumed to be like those of air. In addition, the mass of the fuel added in the combustor is generally neglected in the air-standard cycle. For this example problem, the engine is subject to the following restrictions: the compressor is 90% efficient (assuming reversible adiabatic to be ideal); the turbine is 88% efficient; and the combustor utilizes 70% of the LHV of the fuel-kerosene. The combustor allows the kerosene to burn and the heat is then transferred by conduction to the confined constant pressure air in the closed cycle. Assume 1 lbm of kerosene is used for 40 lbm of air flowing through the combustor. The engine is shown in figure 11–24 where it can be seen that the power

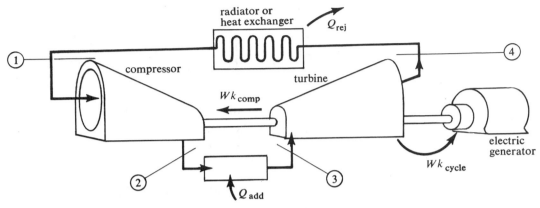

Figure 11–24   Closed air-standard Brayton cycle.

produced is used to provide electric power through a generator. The following data apply: $p_1 = 14.0$ psia, $t_1 = 100°F$, and $p_2 = 100$ psia. Determine the critical property values on p-v and T-s diagrams and determine the power produced. Also calculate the cycle efficiency and the heat rejected.

Solution

Since we are faced with an irreversible process, the process equations developed in chapter 6 will not be of full assistance. First we note that $p_1 = p_4$ and $p_3 = p_2$. Also, from the perfect gas relation

$$v_1 = \frac{RT_1}{p_1} = \frac{53.3 \times 560}{14 \times 144} = 14.81 \text{ lbm/ft}^3$$

and we have determined state (1). The compressor efficiency we write as

$$\eta_c = \frac{\text{ideal } Wk}{\text{actual } Wk} \times 100 = \frac{h_2 - h_1}{h_{2'} - h_1} \times 100 \qquad \textbf{(11–55)}$$

where $h_{2'}$ is the enthalpy value of the gas if it were compressed reversibly and adiabatically to the given pressure p_2. We can then write, for this engine, assuming constant specific heat values that

$$90\% = \frac{c_p(T_2 - T_1)}{c_p(T_{2'} - T_1)} \times 100$$

Using the reversible adiabatic process as the idealized compression

$$\frac{T_2}{T_1} = \left(\frac{p_2}{p_1}\right)^{(k-1)/k}$$

or

$$T_2 = (560°R)\left(\frac{100}{14}\right)^{0.4/1.4} = 982°R$$

We then easily calculate the true temperature at state (2) from equation (11–55)

$$0.90 = \frac{T_2 - T_1}{T_{2'} - T_1} = \frac{982 - 560}{T_{2'} - 560} = \frac{422}{T_{2'} - 560}$$

or

$$T_{2'} = 469 + 560 = 1029°R$$

Now, determining the specific volume, we have

$$v_2 = \frac{RT_{2'}}{p_2} = \frac{53.3 \times 1029}{100 \times 144} = 3.81 \text{ ft}^3/\text{lbm}$$

To obtain the temperature at state (3), we use the first law steady flow equation which reduces to

$$h_3 - h_{2'} = q_{add}$$

or

$$c_p(T_3 - T_{2'}) = q_{add}$$

The heat added is just 70% of the LHV of kerosene

$$19{,}256 \text{ Btu/lbm fuel} \times 0.70 = 13{,}479 \text{ Btu/lbm fuel}$$

and per unit mass of air this is

$$q_{add} = 13{,}479 \text{ Btu/lbm fuel} \times 1 \text{ lbm fuel}/40 \text{ lbm air} = 337 \text{ Btu/lbm}$$

Then

$$c_p(T_3 - T_2) = 337 \text{ Btu/lbm}$$

$$0.24 \text{ Btu/lbm°R } (T_3 - 1029) = 337 \text{ Btu/lbm}$$

$$T_3 = \frac{337}{0.24}°R + 1029 = 2433°R$$

Also,

$$v_3 = \frac{RT_3}{p_3} = \frac{53.3 \times 2433}{100 \times 144} = 9.0 \text{ ft}^3/\text{lbm}$$

The turbine efficiency we can write as

$$\eta_{turb} = \left(\frac{\text{actual work}}{\text{ideal work}}\right) \times 100 = \left(\frac{h_3 - h_{4'}}{h_3 - h_4}\right) \times 100 \quad \textbf{(11–56)}$$

where h_4 is the enthalpy if the gases were expanded reversibly and adiabatically through the turbine. Then

$$88\% = \left(\frac{h_3 - h_{4'}}{h_3 - h_4}\right) \times 100 \quad \text{and} \quad 0.88 = \frac{c_p(T_3 - T_{4'})}{c_p(T_3 - T_4)} = \frac{T_3 - T_{4'}}{T_3 - T_4}$$

We obtain T_4 from the ideal cycle relation

$$\frac{T_4}{T_3} = \frac{T_1}{T_2}$$

so that

$$T_4 = (2433°\text{R})\left(\frac{560°\text{R}}{982°\text{R}}\right) = 1387°\text{R}$$

and

$$0.88 = \frac{2433°\text{R} - T_{4'}}{2433 - 1387}$$

or

$$T_{4'} = 1513°\text{R}$$

We can quickly calculate $v_4 = 40.0$ ft^3/lbm and plot the *p-v* and *T-s* diagrams as shown in figure 11–25. To calculate the entropy changes we may use the constant pressure reversible process equation

$$\Delta s = c_p \ln \frac{T_f}{T_i}$$

and, specifically

$$s_{2'} - s_1 = s_{2'} - s_2 = c_p \ln \frac{T_{2'}}{T_2} = 0.24 \ln \frac{1029}{982} = 0.0112 \text{ Btu/lbm°R}$$

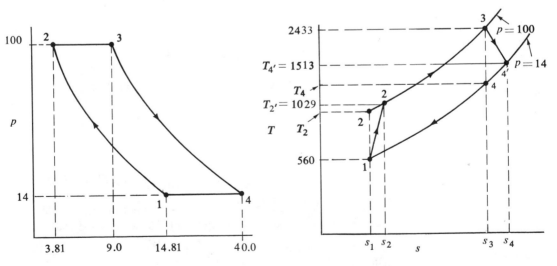

Figure 11–25   Property diagram for nonideal gas turbine cycle (not drawn to scale).

$$s_3 - s_{2'} = c_p \ln \frac{T_3}{T_{2'}} = 0.24 \ln \frac{2433}{1029} = 0.207 \text{ Btu/lbm}°\text{R}$$

$$s_4 - s_1 = c_p \ln \frac{T_4}{T_1} = 0.24 \ln \frac{1513}{560} = 0.239 \text{ Btu/lbm}°\text{R}$$

Notice in the *T-s* diagram that the irreversible cycle has a higher heat rejection than the ideal Brayton cycle, both operating between the same pressures. At the same time the net heat added is approximately the same.

The power produced by the cycle and supplied to the electric generator is obtained

$$\dot{W}k = \dot{m}wk_{\text{cycle}}$$

where

$$\dot{m} = 20 \text{ lbm/s}$$

and

$$
\begin{aligned}
wk_{\text{cycle}} &= wk_{\text{turb}} + wk_{\text{comp}} \\
&= h_3 - h_4 + h_1 - h_2 \\
&= c_p(T_3 - T_4 + T_1 - T_2) \\
&= 0.24 \, (2433 - 1513 + 560 - 1029) \text{ Btu/lbm} \times 778 \text{ ft-lbf/Btu} \\
&= 84{,}211 \text{ ft-lbf/lbm}
\end{aligned}
$$

Then

$$\dot{W}k = 20 \text{ lbm/s} \times 84{,}211 \text{ ft-lbf/lbm} = 1{,}684{,}220 \text{ ft-lbf/s}$$

or

$$\dot{W}k = 1{,}684{,}220 \times \frac{1}{550} \text{ hp} = 3{,}062 \text{ hp} \qquad \textit{Answer}$$

The cycle efficiency we determine from equation (8–5)

$$\eta_T = \frac{wk_{\text{cycle}}}{q_{\text{add}}} \times 100$$

which yields

$$\eta_T = \frac{84{,}211 \text{ ft-lbf/lbm}}{778 \text{ ft-lbf/Btu} \times 337 \text{ Btu/lbm}} \times 100 = 32.1\% \quad \textit{Answer}$$

Even though we have irreversible sources of heat in this problem, from a system diagram, as sketched in figure 11–26, we can see that

$$\dot{Q}_{\text{rej}} = \dot{Q}_{\text{add}} - \dot{W}k$$

which gives us

$$
\begin{aligned}
\dot{Q}_{\text{rej}} &= 20 \text{ lbm/s} \times 337 \text{ Btu/lbm} \times 1.41 \text{ hp-s/Btu} - 3{,}062 \text{ hp} \\
&= 6{,}441 \text{ hp} = 4{,}568 \text{ Btu/s} \qquad \textit{Answer}
\end{aligned}
$$

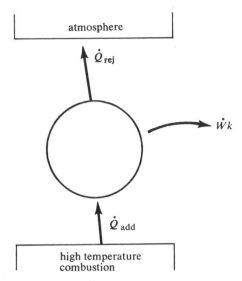

atmosphere

$\dot{Q}_{rej}$

$\dot{W}k$

$\dot{Q}_{add}$

high temperature
combustion

Figure 11-26   System diagram of gas turbine.

**11.6
Gas Tables and
Brayton Cycle
Calculations**

The gas turbine can be analyzed with the aid of gas table data; this will increase the accuracy of predicted results or property values over the perfect gas analysis as given previously. The following example problem should aid in clarifying this method. A reading of section 10.4 should help in clarifying the use of gas tables.

Example 11.6

For the turbo-jet engine of problem 11.4, determine the property values at the four corners of the ideal Brayton cycle. (Use the air tables B.6.) Using these results, obtain the required air flow for level flight and for a 10° climb angle assuming a drag of 600 lbf, and an aircraft weight of 40,000 lbf.

Solution

The initial conditions are still the same as for problem 11.4: $p_1 = 2.73$ psia; $\rho_1 = 0.0189$ lbm/ft³; and $T_1 = 389.97°R$. From the following calculations we determine p_{r1}.

From table B.6 we have

$$p_r = 0.4858 \quad \text{at} \quad T = 400°R$$
$$p_r = 0.3048 \quad \text{at} \quad T = 350°R$$

Then

$$\frac{400°R - 389.97°R}{400°R - 350°R} = \frac{0.4858 - p_{r1}}{0.4858 - 0.3048}$$

$$p_{r1} = -\left(\frac{(10.03)}{(50)}\right)(0.1810) + 0.4858 = 0.4495$$

Then

$$\left(\frac{p_2}{p_1}\right) = 10 = \left(\frac{p_{r2}}{p_{r1}}\right)$$

and

$$p_{r2} = 10 \times p_{r1} = 4.495$$

which allows us to determine T_2. We again interpolate, here between

$$p_r = 5.526 \text{ at } T = 800°R$$
$$p_r = 4.396 \text{ at } T = 750°R$$

Then

$$\frac{750 - T_2}{750 - 800} = \frac{4.396 - 4.495}{4.396 - 5.526}$$

so that

$$-T_2 = \left(\frac{0.099}{1.130}\right)(50) - 750$$

or

$$T_2 = 746°R$$

The pressure at state (2) is 27.3 psia since $p_2/p_1 = 10$ and $p_2 = 10 \, p_1$. For the combustor we use equation (11–35)

$$\dot{m}_3 h_3 + \dot{m}_f h_f - \dot{m}_2 h_2 = \dot{Q}_{add}$$
$$h_f = -37 \text{ Btu/lbm fuel} \qquad \text{See equation (11–51)}.$$

Here

$$\frac{\dot{m}_f}{\dot{m}_2} = 35 \, q_{add} = 19{,}256 \text{ Btu/lbm fuel}$$

and

$$h_2 = 178.7 \text{ Btu/lbm air} \qquad \text{(from table B.6)}$$

We then write the energy equation (11–35) for the combustor

$$\left(\frac{\dot{m}_2 + \dot{m}_f}{\dot{m}_f}\right) h_3 + h_f - \left(\frac{\dot{m}_2}{\dot{m}_f}\right) h_2 = q_{add}$$

or

$$36 \, h_3 - 37 - 35(178.7) = 19{,}256$$

and then
$$h_3 = \left(\frac{1}{36}\right)(19256 + 6300 + 37)$$
$$= 710.0 \text{ Btu/lbm}$$

Now we can find T_3 from table B.6. Interpolating between 2600°R and 3000°R we obtain
$$\frac{3000 - T_3}{3000 - 2600} = \frac{790.68 - 710.0}{790.68 - 674.49}$$
$$T_3 = 3000 - \left(\frac{80.68}{116.19}\right)(400) = 2722.2°\text{R}$$

At state (3) we calculate the relative pressure by using the interpolation
$$p_{r3} = p_{r3000°\text{R}} - \left(\frac{80.68}{116.19}\right)(p_{r3000°\text{R}} - p_{r2600°\text{R}})$$

Substituting values into this from table B.6 we obtain
$$p_{r3} = 941.4 - (0.694)(427.9) = 644.4$$

which gives us
$$p_{r4} = \left(\frac{1}{10}\right)(p_{r3}) = 64.44$$

At this relative pressure we get, from table B.6
$$T_4 = 1556.2°\text{R}$$

In figure 11–27 are plotted the property diagrams for the cycle. The specific volumes are obtained from the following calculations:
$$v_1 = \frac{1}{\rho_1} = \frac{1}{0.0189} = 53 \text{ ft}^3/\text{lbm}$$
$$v_2 = v_1 \left(\frac{v_{r2}}{v_{r1}}\right) = (53)\left(\frac{64.1}{331.8}\right) = 10.24 \text{ ft}^3/\text{lbm}$$
$$v_3 = v_2 \left(\frac{T_3}{T_2}\right) = 10.24 \left(\frac{2722.2}{746}\right) = 37.36 \text{ ft}^3/\text{lbm}$$

and
$$v_4 = v_3 \frac{v_{r4}}{v_{r3}} = 37.36 \left(\frac{8.958}{1.664}\right) = 201 \text{ ft}^2/\text{lbm}$$

The entropy change is obtained from
$$\Delta s = s_4 - s_1 = \phi_4 - \phi_1 = 0.863 - 0.523$$
$$= 0.340 \text{ Btu/lbm°R}$$

To determine the air flow required, we must calculate the cyclic work from the equation

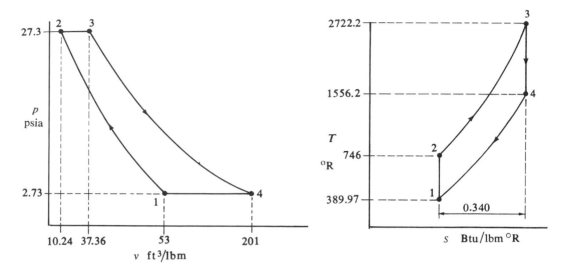

Figure 11–27   Property diagram of Brayton cycle (not drawn to scale).

$$wk_{\text{cycle}} = wk_{\text{turb}} + wk_{\text{comp}}$$
$$= h_3 - h_4 + h_1 - h_2$$

The enthalpy values we obtain from table B.6:

$$h_1 = 93.2 \text{ Btu/lbm}$$
$$h_2 = 178.7 \text{ Btu/lbm}$$
$$h_3 = 711.0 \text{ Btu/lbm}$$
$$h_4 = 384.1 \text{ Btu/lbm}$$

and then

$$wk_{\text{cycle}} = 711.0 - 384.1 + 93.2 - 178.7$$
$$= 241.4 \text{ Btu/lbm air}$$

The power required for level flight at 500 mph we found to be 800 hp so that

$$800 \text{ hp} = \dot{m}wk_{\text{cycle}} = \dot{m}(241.4) \text{ Btu/lbm}$$

This gives us

$$\dot{m} = \frac{800 \text{ hp}}{241.4 \text{ Btu/lbm} \times 1.41} \text{ hp-s/Btu}$$
$$= 2.35 \text{ lbm/s} \qquad\qquad \textit{Answer}$$

This is slightly more than 2.04 lbm/s of air required when we assumed perfect gas conditions in problem 11.4. The trend is altogether to be expected.

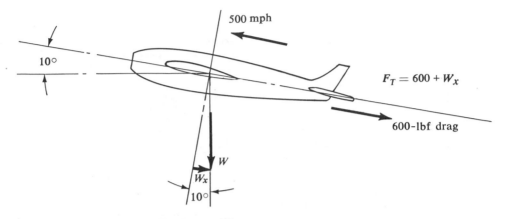

Figure 11-28   Forces on aircraft in climb.

For the aircraft climbing at 10 degrees the power required is determined by adding the drag and the component of the weight to be overcome. From figure 11–28 we see that this gives a required thrust force of

$$F = 600 \text{ lbf} + W_x = 600 + 40{,}000 \sin 10°$$
$$= 7{,}550 \text{ lbf}$$

The power required at 500 mph is then

$$\dot{W}k = 7{,}550 \text{ lbf} \times 500 \times \frac{5280}{3600} \text{ ft/s}$$

$$= 5{,}536{,}666 \text{ ft-lbf/s}$$

and the air flow is quickly determined;

$$\dot{m} = \frac{5{,}536{,}666 \text{ ft-lbf/s}}{241.4 \text{ Btu/lbm} \times 778 \text{ ft-lbf/Btu}} = 29.5 \text{ lbm/s} \quad \textit{Answer}$$

We thus see that an appreciable increase in air is required to maneuver the aircraft up. The engine we have considered here in driving an aircraft, the turbo-jet, is an open cycle type of engine and is an "air breather." That is, it requires a large atmosphere of air to function properly. For this reason, the jet aircraft cannot operate at high altitudes or in space. In addition, it is not reusing the gases from the exhaust as was done in the closed cycle of problem 11.5. We will see a scheme to reduce this wastefulness in the next section.

**11.7
Regenerative
Heating and
Engine Design
Considerations**

The open cycle Brayton engine exhausts a high temperature gas which must also have a high velocity if it is to propel the aircraft. In applications where the power may be extracted from the turbine shaft, such as was done in problem 11.5 and could be done in a ground vehicle, the hot

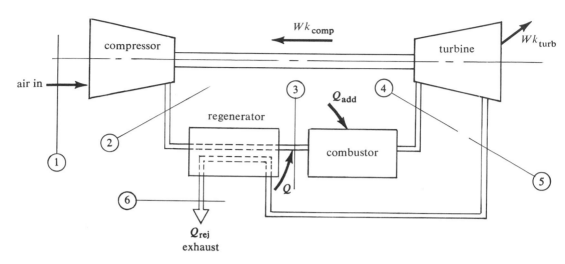

Figure 11–29   Regenerative heating Brayton cycle.

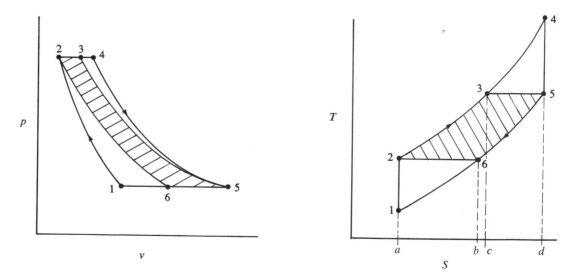

Figure 11–30   Regenerative heat Brayton cycle property diagrams.

exhaust gases are a total loss. To reduce this waste of energy and increase efficiency, a regenerative heater can be attached to the normal gas
turbine cycle, as diagrammed in figure 11–29. The technique merely
transfers some of the energy from the exhaust to the fresh gases and the
ideal Brayton cycle property diagrams then take the form shown in
figure 11–30. In each diagram, the shaded area represents the work or

heat added by virtue of the regeneration process. From an application of the steady flow energy to the components we get

(a) for the compressor:

$$h_2 - h_1 = -wk_{comp} \qquad (11\text{–}57)$$

(b) for the regenerator:

$$h_5 - h_6 = h_3 - h_2 \qquad (11\text{–}58)$$

(c) for the combustor:

$$h_4 - h_3 = q_{add} \qquad (11\text{–}59)$$

(d) for the turbine:

$$h_4 - h_5 = -wk_{turb} \qquad (11\text{–}60)$$

The thermodynamic efficiency can then be written

$$\eta_T = \left(\frac{h_4 - h_5 + h_1 - h_2}{h_4 - h_3} \right) \times 100 \qquad (11\text{–}61)$$

Now, if we assume a perfect gas medium with constant specific heats, we can make the substitution $h = c_p T$ and get

$$\eta_T = \left(\frac{T_4 - T_5 + T_1 - T_2}{T_4 - T_3} \right) \times 100$$

or if we have $T_3 = T_5$

$$\eta_T = \left(1 - \frac{T_2 - T_1}{T_4 - T_5} \right) \times 100 \qquad (11\text{–}62)$$

This can be revised further by some algebra:

$$\eta_T = \left(1 - \frac{T_1}{T_4} \frac{(T_2/T_1 - 1)}{(1 - T_5/T_4)} \right) \times 100 = \left[\left(1 - \frac{T_1}{T_4} r_p^{\,k-1/k} \right) \right] \times 100 \qquad (11\text{–}63)$$

where

$$r_p = \frac{p_2}{p_1} = \frac{p_4}{p_5} \qquad \text{and} \qquad \frac{T_2}{T_1} = \frac{T_4}{T_5}$$

This interesting result shows that an increase in compression or pressure ratio will decrease the efficiency of a regenerative Brayton cycle engine. This is directly opposite to the result of the simple Brayton cycle, as given by equation (11–9), where increases in pressure ratios increase the cycle efficiency. This should provide enough indication that any given engine needs to be analyzed from a fresh outlook. Keep the basic thermodynamic tools as foundations, and from this base make assumptions and restrictive developments in context with the actual situation.

Example 11.7   A reversible, regenerative gas turbine uses 2.0 lbm/s of air. Its operating pressure ratio is 15 to 1, and the compressor inlet conditions are 14.7 psia and 80°F. If the turbine exhaust is at a temperature of 1260°R, determine the following from an air-standard analysis assuming perfect gas behavior: (a) cycle thermodynamic efficiency, (b) power developed and, (c) heat added and rejected.

Solution

(a)   We calculate the efficiency from equation (11–63)

$$\eta_T = \left(1 - \frac{T_1}{T_4} (r_p)^{(k-1)/k}\right) \times 100$$

First, however we need to determine T_4:

$$\frac{T_4}{T_5} = \left(\frac{p_4}{p_5}\right)^{(k-1)/k} = (r_p)^{(k-1)/k}$$

and we can then substitute this into the efficiency equations to obtain

$$\eta_T = \left(1 - \frac{T_1}{T_5}\right) \times 100$$

which yields

$$\eta_T = \left(1 - \frac{540}{1260}\right) \times 100 = 57.1\% \qquad \textit{Answer}$$

Notice that the efficiency of a simple Brayton cycle gas turbine operating with the same pressure ratio is

$$\eta_T = \left(1 - \frac{1}{(r_p)^{(k-1)/k}}\right) \times 100 = \left(1 - \frac{1}{15^{0.286}}\right) \times 100 = 53.9\%$$

We therefore gain 3.2% efficiency by using the reheater on this engine.

(b)   The power developed can be obtained from the relation $\dot{W}k = \dot{m}wk_{\text{cycle}}$. For air-standard analyses

$$wk_{\text{cycle}} = c_p(T_4 - T_5 + T_1 - T_2)$$

and

$$T_4 = T_5 \left(\frac{p_4}{p_5}\right)^{(k-1)/k} = (1260°R)(15)^{0.286} = 2734°R$$

and

$$T_2 = T_1 \left(\frac{p_2}{p_1}\right)^{(k-1)/k} = (540°R)(15)^{0.286} = 1172°R$$

Then

$$wk_{\text{cycle}} = 0.24 \text{ Btu/lbm°R } (2734 - 1260 + 540 - 1172)°R$$
$$= 202.1 \text{ Btu/lbm}$$

The delivered power is then obtained

$$\dot{W}k_{\text{cycle}} = 2 \times 202.1 \text{ Btu/s} \times 1.41 \text{ hp-s/Btu}$$
$$= 570 \text{ hp}$$

(c) We can calculate the heat added in at least two ways:

$$q_{\text{add}} = \left(\frac{wk_{\text{cycle}}}{\eta_T}\right) \times 100$$

or

$$q_{\text{add}} = c_p(T_4 - T_5)$$

From the first relationship

$$q_{\text{add}} = \frac{202.1 \text{ Btu/lbm}}{57.1} \times 100 = 354 \text{ Btu/lbm} \qquad \textit{Answer}$$

and from the second

$$q_{\text{add}} = 0.24 \text{ Btu/lbm } (2734 - 1260) = 354 \text{ Btu/lbm} \qquad \textit{Answer}$$

which verifies our results.

The heat rejected can be determined from

$$q_{\text{rej}} = wk_{\text{cycle}} - q_{\text{add}}$$
$$= 202.1 \text{ Btu/lbm} - 354 \text{ Btu/lbm}$$
$$q_{\text{rej}} = -151.9 \text{ Btu/lbm} \qquad \textit{Answer}$$

The gas turbine, operating in accordance with the Brayton cycle, is a device which is capable of high performance. Operating efficiences of 40% to 50% have been achieved, and it is capable of high bursts of power (with an accompanied decrease in efficiency). Mechanically it is quite simple to operate and maintain. It is inherently stable, and by rotating symmetrically about a single axis, it can be easily balanced. Its main detriment is its limitation on a high temperature and consequent efficiency limitation; that is, T_3 in the simple cycle or T_4 in the regenerative cycle cannot exceed maximum temperatures that materials can constantly withstand. The cycle is in steady state and the upper temperature is retained for long periods of time — not like in the Otto cycle. (See chapter 10.) The developing of materials which can withstand ever higher temperature, of course, helps the push for higher gas turbine efficiency, but a limit is still there. Additionally, a gas turbine can represent a relatively high initial investment in construction, which can be considered a drawback, but more constant criticism is toward its slow acceleration characteristics for ground vehicles. While the gas turbine is a device which functions best at one constant, optimum speed (as is true of most all engines), the lack of acceptable acceleration in propelling surface vehicles has not been proven and in fact, the exact opposite has frequently been demonstrated.

Two techniques which provide the turbo-jet engine with quick bursts of power and increased performance for short durations are *after-burning* and *water-injection*. The former involves an additional combustion of fuel immediately behind the turbine, which increases the temperature of the already hot exhaust gases and induces an increased exit velocity. As we have seen from equation (11–53), the force on the aircraft is increased and thus drives it faster.

Water-injection is a scheme whereby water is sprayed into the exhaust from the turbine and, while this does not increase the gas temperature, it increases the mass of the expelled gases. This requires equation (11–53)

$$F_{\text{I}} = \frac{m}{g_c} \left(\frac{\bar{V}_4 - \bar{V}_1}{\tau} \right) + (p_4 - p_1)A_e$$

to be read as

$$F_{\text{I}} = \frac{1}{g_c} \left(\frac{m_4 \bar{V}_4 - m_1 \bar{V}_1}{\tau} \right) + (p_4 - p_1)A_e$$

which then can be seen as an increase in the thrust force F_{I} due to m_4 being greater than m_1.

11.8
The Ram Jet

We have considered the gas turbine as the prime engine described by the Brayton cycle. As specific examples we looked at the turbo-jet engine and the closed cycle gas turbine. Now let us look at the case of a vehicle traveling at such a high speed in air that the turbo-jet compressor is not needed. That is, at high velocities the entering air will be already traveling faster than a turbine driven compressor can induce. Then, we can neglect the compressor and just restrict the air to increase its pressure. This high pressure air is then allowed to combust with fuel and is subsequently expelled as hot exhaust gases. Since the turbine merely drove the compressor in the turbo-jet engine, we can also neglect this component and direct the exhaust through a nozzle to increase the velocity for impulse. This whole series of processes describes the workings of the ram jet engine. It derives its name from the fact that air is "rammed" into the engine from the high velocity of the air relative to the engine. In figure 11–31 is shown the basic schematic of the engine and the mechanics of it are depicted in figure 11–32. The shock waves shown as lines in this figure serve to decrease the relative velocity of the

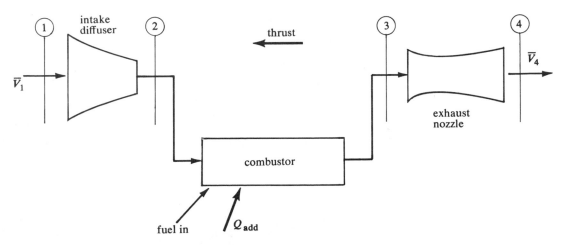

Figure 11–31   Bosie ram jet components.

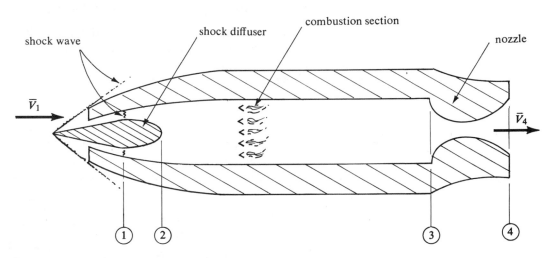

Figure 11–32   Typical ram jet engine.

air entering the engine and consequently increase the pressure. Shock waves are a normal occurrence in supersonic speeds, which are the operating regime of the ram jet.

The thermodynamic analysis of the engine is identical to that of the turbo-jet; that is, for an air-standard analysis we have

$$q_{\text{add}} = c_p(T_3 - T_2) \tag{11–64}$$

$$q_{\text{rej}} = c_p(T_1 - T_4) \tag{11–65}$$

and

$$wk_{cycle} = q_{add} + q_{rej} = c_p(T_3 - T_2 + T_1 - T_4) \qquad \textbf{(11–66)}$$

The efficiency is gotten from

$$\eta_T = \left(1 - \frac{1}{(r_p)^{(k-1)/k}}\right) \times 100 \qquad \textbf{(11–67)}$$

In equations (11–64), (11–65), (11–66), above all the temperatures should properly be replaced by total temperature T^* defined by equation (11–43)

$$T^* = T + \frac{\overline{V}^2}{2g_c c_p}$$

Total temperature obviously includes the kinetic energy of the gases. In the ram jet in particular the kinetic energy effects cannot generally be ignored.

The ram jet has the following advantages over a turbo-jet engine:

(a) Higher operating temperature due to a simplified construction. This implies an increased efficiency.
(b) More power per unit engine weight.
(c) Higher operating speeds.
(d) Less maintenance.

Along with these advantages, it must be said that the ram jet is not self-starting; that is, it must be traveling at sufficient speeds before it can function. It also has an inherent problem of sustaining combustion due to the high air speed through the chamber, and this problem has not been completely solved.

11.9 The Pulse Jet

One of the first engines to successfully power a missile from takeoff at zero velocity to a high velocity was the pulse jet. The engine is not currently being used in practical applications, but it does represent an ingenious method of operating a heat engine.

The pulse jet engine, shown in figure 11–33, is started by firing a fuel air mixture with an electrical spark. This increases the pressure and temperature in the combustion chamber (process 2–3) isometrically. Then the hot gases are expelled through the exhaust nozzle, thus imparting an impulse force to the vehicle. Immediately following this process, the air is drawn into the inlet diffuser and then through the flapper valves into the combustion chamber. The flapper valves allow air to flow into

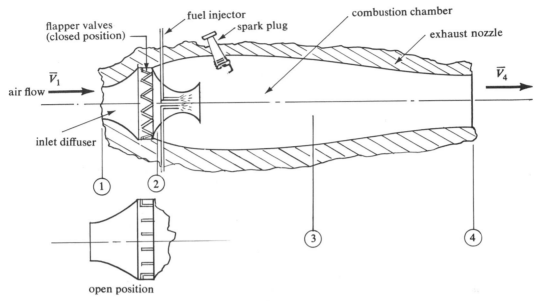

flapper valves (closed position)

fuel injector

spark plug

combustion chamber

exhaust nozzle

$\overline{V}_1$

air flow

inlet diffuser

$\overline{V}_4$

① ② ③ ④

open position

Figure 11–33   Cross section of pulse jet.

the chamber, but they close if air tends to flow back. In figure 11–33 are shown the two normal positions of the flapper valves; the closed position during combustion (process 2–3) and expansion (process 3–4), and the open position during compression (process 1–2). The engine has a distinct cycle during which the flapper valves open and close to allow for air intake and combustion respectively. Depending on the engine's size, the cycle will proceed at a speed inherent to the design. That is, no external controls are required to actuate the flapper valves as they are moved by differential pressures between the diffuser and the combustion chamber. One of the earliest and the most famous applications of the pulse jet was the German V-1 rocket, a pilotless warcraft, used in World War II, which operated at approximately 40 cycles/s. Because of this particular frequency, it produced a loud buzzing sound when in flight, and for this reason, the V-1 was referred to as the "buzz bomb." Other engines of smaller size than the V-1 power plant have been built with operating speeds of 250 to 300 cycles/s.

The thermodynamic cycle of the ideal pulse jet is described by the *p-V* diagram of figure 11–34; however, the deviation from this is great in actual engines. As can be seen, the cycle is a hybrid of the Brayton and Otto cycles. Today, however, there are no serious efforts at adapting the pulse jet to produce actual engines, even though it is relatively inexpensive, can produce power at zero velocity, and increases its effectiveness with increased vehicle velocity.

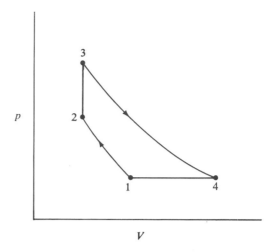

Figure 11–34   *p-V* diagram of ideal pulse jet.

11.10 Rockets

The rocket engine is included here because it is frequently compared to the jet engine as a means of producing power. It has some distinct advantages over the gas turbine engine:

(a) It can develop maximum power at zero velocity.
(b) It is not "air-breathing" as are all other gas turbine or jet engines.
(c) It can develop much higher thrust-to-weight ratios.

These characteristics have made the rocket engine the primary means of power for travel beyond the earth's atmosphere. The rocket engine has a distinct advantage when compared to other heat engines, however, since the rocket is not a true cyclic device, that is, it is not a heat engine, but is rather a device executing processes which *never* return the engine to its initial state. This means the rocket need not be limited to the second law of thermodynamics — it may produce work with only one heat exchange. This is precisely what is done, and in figure 11–35 is shown the cross section of a typical engine to indicate the physical processes being conducted. There are only two: a combustion of a fuel (either solid or liquid) normally done at constant pressure; and an expansion of the hot exhaust gases through a nozzle. Ideally, this expansion is reversible and adiabatic. In figure 11–36 are drawn the process curves of these two ideal processes. As can be seen, the work available to propel the rocket is the area to the left of the *p-v* curve and the heat added is the area under the *T-s* curve. The major part of the rocket is the nozzle, which accelerates the hot gases; and we will look at an example problem concerned with just this. First, however, let us write the general equation for thrust

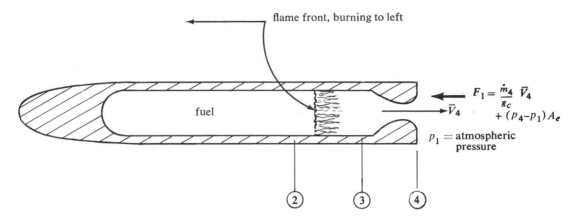

Figure 11–35   Cross section of typical rocket.

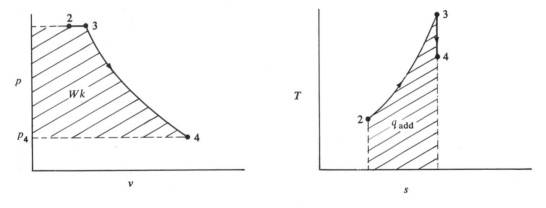

Figure 11–36   Property diagrams for rocket engine.

induced, F_I from equation (11–53)

$$F_I = \frac{m}{g_c}\left(\frac{\bar{V}_4 - \bar{V}_1}{\tau}\right) + (p_4 - p_1)A_e$$

In the rocket, however, no mass is entering the system so $\bar{V}_1 = 0$ yielding

$$F_I = \frac{m}{g_c\tau}(\bar{V}_4) + (p_4 - p_1)A_e \qquad \textbf{(11–68)}$$

The specific impulse of the turbine is then written

$$I_{sp} = \frac{F_I}{m} = \frac{\bar{V}_4}{g_c} + \left(\frac{p_4 - p_3}{\dot{m}}\right)A_e \qquad \textbf{(11–69)}$$

Example 11.8   A rocket burns 200 lbm/s of fuel and oxidizer at 2700°F and 300 psia. The nozzle exhaust area is 2 ft², and the pressure is assumed to be atmospheric (14.7 psia) at the exit. Determine the exhaust velocity, the specific impulse, impulse, and power produced by the engine if the nozzle is reversible and adiabatic, and we assume air-standard conditions.

Solution   For a reversible adiabatic nozzle we write the steady flow energy equation

$$h_4 - h_3 + \frac{\bar{V}_4^2 - \bar{V}_3^2}{2g_c} = 0$$

Since $\bar{V}_3 = 0$ we then have

$$\bar{V}_4 = \sqrt{2g_c(h_3 - h_4)}$$

Invoking air-standard conditions

$$h_3 - h_4 = c_p(T_3 - T_4)$$

Then, for reversible adiabatic processes

$$T_4 = T_3 \left(\frac{p_4}{p_3}\right)^{(k-1)/k} = (3160°R)\left(\frac{14.7}{300}\right)^{0.286}$$

$$= 1334°R$$

and the velocity can then be calculated:

$$\bar{V}_4 = \sqrt{2 \times 32.2 \text{ ft lbm/s}^2 \text{ lbf} \times 0.24 \text{ Btu/lbm°R}}$$
$$\times 778 \text{ ft-lbf/lbm } (3160 - 1334)°R$$
$$= 4685 \text{ ft/s} \qquad\qquad\qquad\qquad\qquad\qquad Answer$$

The specific impulse is then obtained from equation (11–69), assuming $p_4 = p_3$

$$I_{sp} = \frac{4685 \text{ ft/s}}{32.2 \text{ ft-lbm/lbf-s}^2} + 0 = 145 \text{ lbf-s/lbm} \qquad Answer$$

The impulse F_I can be quickly calculated from equation (11–68)

$$F_I = \frac{\dot{m}}{g_c}(\bar{V}_4) = \dot{m}I_{sp} = 200 \text{ lbm/s} \times 145 \text{ lbf-s/lbm}$$

$$= 29{,}000 \text{ lbf} \qquad\qquad\qquad\qquad\qquad\qquad Answer$$

Using our common equations, we obtain the power from the equation

$$\dot{W}k = \dot{m}wk$$

where

$$wk = c_p(T_3 - T_4)$$

Then

$$\dot{W}k = 200 \text{ lbm/s} \times 0.24 \text{ Btu/lbm}°\text{R} (3160 - 1334°\text{R})$$

or

$$\dot{W}k = 87648 \text{ Btu/s} = 123,584 \text{ hp} \qquad\qquad Answer$$

Practice Problems

Problems designated with an asterisk * preceding the number should only be attempted by those having a background which includes the knowledge of integral calculus.

Section 11.1

11.1. Determine the thermodynamic efficiency of an ideal gas turbine operating with a pressure ratio of 20.

11.2. Determine the thermodynamic efficiency of an ideal Brayton cycle engine if the pressure increase across the compressor is 18 times the inlet pressure.

11.3. If the pressure ratio across the compressor of an ideal Brayton cycle engine is 22 to 1 and the working medium is air, determine the temperature ratio across the compressor.

11.4. If the inlet air temperature to the compressor of problem 11.2 (assuming air to be the working medium) is 100°F, determine the compressor work.

11.5. If the inlet air temperature to the compressor of problem 11.3 is 20°C, determine the work required of the compressor.

11.6. An ideal Brayton cycle engine operates with a pressure ratio of 8, the working medium is air, and the compressor inlet temperature is 85°F. If the entropy change across the combustor is 0.18 Btu/lbm-°R, determine the pressure, the specific volume, and the temperature at the four corners of the cycle. Assume the inlet pressure to the compressor is 15 psia.

11.7. Sketch the p-V and T-S diagrams for an ideal Brayton cycle engine. Then, if the heat added is 300 Btu/lbm air, $r_p = 10$, $p_1 = 14.7$ psia, $T_1 = 100°$F, and the working medium is air, determine the properties, p, ρ, and T at the four corners of the cycle.

11.8. A 4000-hp ideal gas turbine engine operates between 15 psia and 160 psia pressures. Determine the rate of heat addition to the cycle.

11.9. Starting from the steady flow energy equation, state the necessary assumptions for arriving at equations (11–1), (11–2), and (11–3).

11.10. Prove that the thermodynamic efficiency of an ideal Brayton cycle is given by

$$\eta_T = 1 - \frac{1}{(r_p)^{(k-1)/k}} \times 100$$

for a perfect gas working medium.

11.11. A turbine is designed with semicircular blades. If the blade velocity is expected to be 100 ft/s, determine the most desirable velocity of fluid impinging on the blades.

11.12. An ideal turbine receives 30 lbm/min of air at 2100°R and exhausts it at 700°F. Determine the power produced by the turbine if kinetic energy changes are neglected.

11.13. A turbine receives 100 lbm/min of air at 80 ft/s at 2500°R. The exhaust air is at 1200°R and 50 ft/s. Determine the work produced by the turbine per unit mass of air and the power produced.

11.14. A turbine nozzle shown in figure 11–37 receives air at 1600°F and 200 psia. Assuming the air impinging on the blades has a velocity of 800 ft/s, determine the blade velocity and the exhaust temperature if heat losses as well as kinetic energy changes are neglected in the turbine.

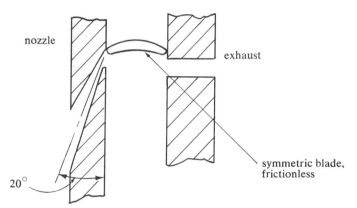

nozzle

exhaust

symmetric blade, frictionless

20°

Figure 11–37   Problems 11.14 and 11.15.

11.15. A gas turbine is designed to produce 3000 hp when using air at a maximum inlet temperature of 1400°F and assuming a heat loss of 70 Btu/s. If the exhaust temperature is desired to be no greater than 400°F, determine the mass flow rate of air required. Also determine the velocity of air leaving the nozzle and the blade velocity. If the blade is not to exceed sonic speeds, would you recommend using additional turbine stages? Yes

11.16. A turbine expands air from a region where the pressure and temperature are 300 psia and 1800°F to a region at 14.7 psia. Neglecting kinetic energy changes of the air and assuming the process through the turbine is polytropic with an exponent $n = 1.5$, determine
(a) wk_{turb}
(b) q
(c) Wk_{turb}, if mass flow of air is 12 lbm/s.

11.17. A turbine reversible adiabatically expands air from 280 psia to 15 psia and 800°F. Determine the inlet air temperature and the work gained from the turbine per lbm of air if kinetic and potential energy changes are neglected in the turbine.

11.18. If 120,000 lbm/hr of air are supplied to a combustor at 70 psia and 350°F and the leaving gases are to be at 1600°F, determine the rate of heat addition required by the combustion of fuel and air. Assume the kinetic energy changes and the mass of fuel to be negligible.

11.19. Air enters a combustor at 100 psia, 400°F, and 200 ft/s. Fuel having a heating value of 17,000 Btu/lbm is burned and the gases leaving the combustor (assuming they have similar properties to air) are at 1700°F and 400 ft/s. Determine the amount of fuel required per pound mass of air, $\dot{m}_f/\dot{m}_a$.

11.20. A compressor receives air at 14.7 psia, 90°F, and 20 ft/s. If the gases are compressed reversibly adiabatically to 100 psia and they leave the compressor at 280 ft/s velocity, determine the amount of work per lbm required to drive the compressor. Determine the pressure of the air leaving the compressor and the work per lbm of air required, if the process is reversible polytropic with $n = 1.34$ and if velocity changes of the air through the compressor are neglected. Assume that the temperature T_2 is 953°R.

11.21. Air having a density of 0.06 lbm/ft³ and at a pressure of 14.7 psia is compressed to 0.10 lbm/ft³. Determine the pressure of the air leaving the compressor and the work per lbm of air required if the process is reversible polytropic with $n = 1.34$ and the velocity changes of the air flow through the compressor are neglected.

Section 11.4

11.22. Air enters a reversible adiabatic converging nozzle at 20 ft/s, 1700°R, and 200 psia. Determine the velocity of air leaving the nozzle if the pressure there is 15 psia.

11.23. Determine the entrance and exit areas of the nozzle in problem 11.22 if the mass flow rate is to be 200 lbm/min.

11.24. A nozzle has 85% efficiency and operates between 180 psia and 12 inches of Hg pressure. Air flows in at 1600°R and leaves at 3200 ft/s. Determine the mass flow of air through the nozzle if the entrance area is 0.5 square inch.

11.25. For the nozzle of problem 11.24, determine the exit area of the nozzle.

11.26. Air at 1500 ft/s velocity enters a diffuser at 10.8 psia. If the diffuser is reversible and adiabatic and the exit pressure is 42 psia, determine the inlet air temperature. The air is quiescent when it leaves the diffuser.

11.27. A reversible adiabatic diffuser receives helium gas at 800°R and 1950 ft/s velocity. If the exhaust pressure must be 50 psia, determine the entrance pressure of the helium. Assume the exhaust velocity is zero.

11.28. A diffuser having 82% efficiency is operating between 15 psia and 35 psia pressures. Air enters at 1650°R and leaves at 2200°R. Determine the entering air velocity and the mass flow rate if the entrance area is 0.02 ft³.

11.29. An adiabatic diffuser is used to direct oxygen gas into a compressor. Oxygen is furnished at 650°K and 760 mm of Hg pressure. The diffuser is 88% efficient when the oxygen leaves at 3 meters of Hg pressure and 1000°K. Determine the entrance velocity and mass flow of oxygen if the entrance area is 10 cm².

11.30. Determine the mach number of the flow at the entrance in problems 11.24, 11.26, 11.28, and 11.29.

11.31. A supersonic deLaval nozzle has a throat area of 3 in² and a temperature of 850°R at the throat when air passes through. Determine the volume flow through the nozzle.

*11.32. The continuity equation (conservation of mass) in differential form is

$$\frac{dA}{A} + \frac{d\mathbb{V}}{\mathbb{V}} - \frac{dv}{v} = 0$$

and the steady flow energy equation can be written

$$\frac{\mathbb{V}d\mathbb{V}}{g} = -vdp$$

Using these two and the relation $pv^k = C$ prove that

$$\frac{dA}{A} = (M^2 - 1)\frac{d\mathbb{V}}{\mathbb{V}}$$

for reversible adiabatic nozzle flow. (M is the mach number.)

Section 11.5

11.33. An ideal Brayton cycle gas turbine operates with a pressure ratio of 7.6. The entering air is at a pressure of 14.6 psia and a temperature of 70°F. The air-fuel ratio is 35 to 1 and the fuel has a LHV of 15,000 Btu/lbm. Determine
  (a) p-v and T-s diagrams
  (b) q_{added}
  (c) $wk_{turbine}$
  (d) $wk_{compressor}$
  (e) wk_{cycle}
  (f) Thermodynamic efficiency.

11.34. An aircraft flying at 20,000 ft in a standard day atmospheric condition is traveling with an airspeed of 480 mph. The craft is propelled by an ideal gas turbine which operates on a pressure ratio of 9 to 1 and which burns kerosene. If the air-fuel ratio is 40 to 1, calculate the following:
  (a) wk_{turb}
  (b) wk_{comp}
  (c) wk_{cycle}
  (d) q_{rej}

11.35. An ideal closed gas turbine engine operates with air at 14.7 psia and 50°F. After compression the air is at 860°R. Then, entering the turbine the gases are at 1400°F. If the combustor is burning octane fuel, determine
  (a) Air-fuel ratio
  (b) wk_{turb}
  (c) wk_{comp}
  (d) Thermodynamic efficiency.

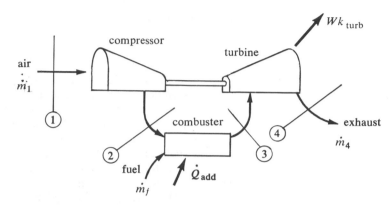

Figure 11–38   Problem 11.36.

11.36. For the gas turbine cycle shown in figure 11–38 the following data apply:

$$\eta_{comp} = 90\% \qquad p_2 = 100 \text{ psia}$$
$$\eta_{turb} = 90\% \qquad T_3 = 2500°R$$
$$\eta_{combust} = 65\% \qquad \text{fuel LHV} = 15,000 \text{ Btu/lbm}$$
$$p_1 = 14 \text{ psia} \qquad \dot{W}k_{turb} = 7,000 \text{ hp}$$
$$t_1 = 65°F$$

Determine
(a) $\dot{m}$ and $\dot{m}_f$
(b) Thermodynamic efficiency.

Section 11.6

11.37. An ideal Brayton cycle gas turbine using air operates with a pressure ratio of 7.5 to 1. The entrance conditions are 14.7 psia and 60°F. If 500 Btu/lbm of heat are added in the combustor, using the gas tables, determine
(a) p, v, t at the four cycle corners.
(b) Δs for compression and expansion.
(c) wk_{cycle}.
(d) Thermodynamic efficiency.

11.38. Using the gas tables, solve problem 11.34.

11.39. Using the gas tables, solve problem 11.35.

11.40. Using the gas tables, solve problem 11.36.

Section 11.7

11.41. Air enters an ideal stationary gas turbine engine at 20°C and $10^6 \backslash .986$ dynes/cm². The pressure ratio is 8 and the air leaves the combustor at $\frac{atm}{dyne}$ 1100°K. Assume the mass of fuel is negligible and the mass flow of air is 100 kg/s.
Determine
(a) Cycle thermodynamic efficiency.
(b) Heat added per unit time.
(c) Heat rejected per unit time.
(d) Power developed.

11.42. The effectiveness of a regenerator is defined by the ratio r_g given by

$$r_g = \left[\frac{(T_3 - T_2)\, \dot{m}_2}{(T_5 - T_6)\, (\dot{m}_2 + \dot{m}_{\text{fuel}})} \right] \times 100\%$$

where the subscripts refer to figure 11–29 and the specific heats are constant and equal for air and the exhaust. Determine the effectiveness of a regenerator which receives air at 100 psia and 500°F. The air going to the combustor is at 1000°F, the exhaust gases leaving the regenerator are at 900°F and the gases leaving the turbine are at 1600°F. Assume the air-fuel ratio is 30 to 1. Plot the *T-S* diagram of a cycle which could be using this regenerator, indicating the significant temperature data.

11.43. A regenerative gas turbine uses 200,000 lbm/hr of air and 4,000 lbm/hr of fuel having a LHV of 17,500 Btu/lbm. The air is taken in at 14.7 psia, 65°F, and compressed reversible adiabatically to 120 psia. A regenerator with an effectiveness of 60% (see problem 11.42) heats the air to 1100°R. Assume the turbine is reversible and adiabatic. Then determine
(a) $\dot{Q}_{\text{add}}$
(b) $\dot{W}k_{\text{cycle}}$
(c) Exhaust temperatures
(d) $\dot{Q}_{\text{rej}}$
(e) Cycle thermodynamic efficiency.

Section 11.8

11.44. Determine the thermodynamic efficiency of a ram jet engine operating between pressures of 12.8 psia and 80 psia.

11.45. An aircraft traveling at 4000 ft/s is powered by a ram jet engine. Given the data: $t_1 = 50°F$; $t_2 = 300°F$; $t_3 = 1800°F$; $t_4 = 600°F$; and $V_1 \approx V_2 \approx V_3 \approx V_4$, determine
(a) q_{add}
(b) q_{rej}
(c) $T*_1$
(d) $T*_4$
(e) Engine thermodynamic efficiency.

Section 11.10

11.46. Determine the specific impulse of a rocket which burns 3600 lbm fuel and air per second. The rocket nozzle has a cross-sectional area of 10 ft² allowing the exhaust gases to expand from the combustion chamber at 600 psia to a 10 psia atmospheric pressure surrounding. Assume the density of gases leaving the nozzle is 0.05 lbm/ft³.

11.47. A rocket engine burns 700 kg/s of fuel and oxidizer at 1300°C and 20 atmospheric pressure. The nozzle is frictionless and adiabatic and has a throat area equal to 0.6 square meter. Assume exhaust gases have a density of 0.004 g/cc³ and $c_p = 0.24$ cal/g°K. If the rocket is operating in an atmosphere at 760 mm Hg, determine
(a) Exhaust temperature.
(b) Impulse.
(c) Power produced by the rocket.

12

The Closed Steam Power Cycle and the Rankine Cycle

Electrical power represents the form in which the greatest amount of energy is directly used by society. However, the overwhelming majority of this power is generated by steam and the *steam turbine*. Fossil fuels (coal, oil, and gas), or nuclear fuels provide the chemical availability to produce the steam which in turn drives the steam turbine and electric generator. This cycle has demonstrated the highest thermodynamic efficiency for vast power production in the myriad of technical devices and it appears destined to be the major source of mechanical power for quite some time. A typical steam turbine power plant is shown in figure 12–1, where the raw material is coal, transported by river barge or railroads. When burned, the coal produces thermal energy which boils water to drive steam turbines and, in turn, electrical generators. The workings of this system constitute the typical *Rankine* or *steam turbine cycle*, which we will examine in this chapter.

The Rankine cycle, an idealization, is defined and compared to the operation of the actual steam turbine cycle. The major components of the closed steam cycle — boiler, condenser, feedwater pumps, and steam

Figure 12–1   The modern steam turbine, electric power generating station.
Reproduced with permission of Dayton Power and Light Company;
Dayton, Ohio.

turbine — are discussed to show the physical processes corresponding to
the Rankine cycle and to assist in the cycle analysis.

The *phase changes*, so important in steam power applications, are discussed along with the concept of· *superheated steam*. Property diagrams
(*T-s* and *p-v*) are used to help expand the details of the phase change
beyond the brief introduction from section 8.10.

Use of the *steam tables* and the *Mollier diagram* are explained (the perfect
gas assumption is never used with steam except at very high temperatures) to give the reader a storehouse of ready and useful data for problem solving. Some traditional steam turbine engines operating on the
Rankine cycle are then analyzed with the thermodynamic tools.

The modifications of the steam turbine cycle to increase power output
and/or efficiency are introduced with treatments of the *regenerative* and
reheat cycles. Finally, some general considerations are made regarding
the design of future steam turbine cycles.

**12.1
The Rankine
Cycle**

The thermodynamic cycle which most properly describes the workings of the ideal steam turbine is the Rankine cycle. This is defined by the four reversible processes:

(1–2) Adiabatic compression of liquid (water).
(2–3) Isobaric heat addition to convert liquid to a vapor (steam).
(3–4) Adiabatic expansion of vapor to low pressure.
(4–1) Isobaric heat rejection to condense vapor to liquid.

These four processes are the same combination which describes the gas turbine or Brayton cycle (see chapter 11), but here we are involved with phase changes in the working media which lend unique characteristics to the Rankine cycle not present in the Brayton cycle. In figure 12–2 are depicted the typical property diagrams of the Rankine cycle, and in figure 12–3 the schematic of the major components of the steam turbine cycle is described by these property diagrams. Notice that a saturation line is drawn in the property diagrams to give the reader an idea of the approximate relation of the process to the phase change of H_2O which occurs "under" the saturation line.

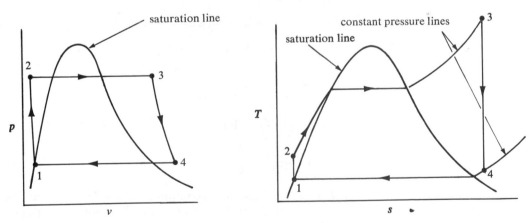

Figure 12–2   Property diagrams of the Rankine cycle.

For the ideal steam turbine cycle of figures 12–2 and 12–3, we can apply the steady flow energy equation to the individual components to arrive at the following pertinent equations:

$$q_{add} = h_3 - h_2 \tag{12–1}$$

$$wk_{turb} = h_3 - h_4 \tag{12–2}$$

$$q_{rej} = h_1 - h_4 \tag{12–3}$$

$$wk_{pump} = h_1 - h_2 \tag{12–4}$$

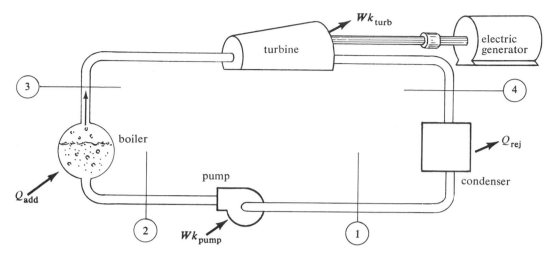

Figure 12-3   Typical closed steam turbine cycle.

For the full cycle then

$$wk_{\text{cycle}} = wk_{\text{turb}} + wk_{\text{pump}} \qquad (12\text{--}5)$$

and by using the results we get

$$wk_{\text{cycle}} = h_3 - h_4 + h_1 - h_2 \qquad (12\text{--}6)$$

Also we have, for the thermodynamic efficiency

$$\eta_T = \left(\frac{wk_{\text{cycle}}}{q_{\text{add}}}\right) \times 100 = \left(\frac{h_3 - h_4 + h_1 - h_2}{h_3 - h_2}\right) \times 100$$

$$= \left(1 - \frac{h_4 - h_1}{h_3 - h_2}\right) \times 100 \qquad (12\text{--}7)$$

We will now look at the individual components: the boiler, the condenser, the pump, and the steam turbine, to visualize equations (12–1) through (12–7). In this chapter, no perfect gas assumption will be made so that c_v or c_p and temperatures will generally not be used. Rather we will use steam tables to seek enthalpy values.

**12.2
Boilers**

Steam is generated in a boiler. Liquid H_2O is supplied to the boiler, and because of a heat addition, the water becomes vapor (or steam). A typical boiler unit is shown in figure 12–4, which indicates that our initial concept of a "teapot" is not quite correct. Tubes carry water from a lower tank past a combustion furnace and into an upper tank. From

Figure 12–4   Typical steam generating boiler. Field erected industrial boiler.
Reproduced with permission of Combustion Engineering, Inc.,
Windsor, Conn.

here (at which point the water is steam), superheater pipes pass the steam through the furnace again and elevate the steam temperature well beyond a saturated vapor state. In the construction of a boiler (a steam-generating unit), the combustion chamber is normally an integral part of the system. Whether coal, oil, gas, or nuclear fuel is used, the intimate relation between combustion and the water system prevents undue waste of heat. That is, the heat transfer to the water is maximized by design of the total unit. In addition, the furnace-type steady combustion provides the nearest to complete fuel combustion and consequently gives the best release of chemical energy from the fuel to be used in generating steam.

In the best combustion, we can approximate the heat added to the water in the boiler by the lower fuel heating valve (LHV). That is, in figure 12–3 which depicts a boiler

$$q_{add} \leq LHV \left(\frac{\dot{m}_{fuel}}{\dot{m}_{steam}} \right) \tag{12–8}$$

where LHV is based on a unit mass of fuel. We also can calculate heat added to the water from equation (12–1)

$$q_{add} = h_2 - h_1$$

With these two results we can define the boiler efficiency as

$$\eta_{boiler} = \left(\frac{h_2 - h_1}{LHV} \right) \left(\frac{\dot{m}_{steam}}{\dot{m}_{fuel}} \right) \tag{12–9}$$

The boiler is an important component in the steam cycle and with the search for new energy sources, it has been the center of attention.

12.3 Steam Turbines

Steam at a high pressure and temperature has a large amount of internal energy, but a device is needed to convert this energy into mechanical work or power. The steam turbine is the machine which does this converting and which operates on exactly the same principles as the historic waterwheel, or the gas turbine. The reader is advised to read section 11.2 at this time if he has not already done so, as the description of the gas turbine and its operation is completely analogous to that of the steam turbine. While the steam turbine is occasionally constructed as a reaction turbine, the majority of the machines are impulse-reaction turbines. In this device, steam is supplied from a boiler, and its internal energy is then converted into kinetic energy in a nozzle, which directs the steam against buckets attached to a turbine wheel. This steam is diverted by the buckets into an exhaust passage or another nozzle, and turbine wheel. The diverting of the steam causes the turbine wheel to rotate and thus produce a shaft work which can easily be utilized by mechanically connecting an electric generator or other device to the turbine shaft. For a frictionless turbine allowing reversible adiabatic expansion of the steam, we have found that by writing the steady flow energy equation, neglecting kinetic and potential energy changes, equation (12–2) resulted

$$wk_{\text{turb}} = h_3 - h_4$$

The work can also be obtained by using stream velocity values in equation (11–24)

$$wk = \frac{1}{g_c}(\bar{V}_{1x} - \bar{V}_{2x})V_b$$

The power is found by multiplying the work by the mass flow of steam through the turbine

$$\dot{W}k_{\text{turb}} = \dot{m}(h_3 - h_4) \qquad \textbf{(12–10)}$$

The expansion of steam through a turbine is most commonly conducted in a manner to achieve saturated vapor at the turbine exhaust. That is, at state (4) we wish to have extracted as much energy as possible from the steam in producing turbine work, but if any moisture or liquid particles are in the steam while it is anywhere in the turbine, these particles can act as abrasives and thus seriously damage parts of the turbine. Additionally, moisture in a turbine can induce corrosion of vital parts. For this reason we desire a steam which is a vapor that is ready to condense (saturated vapor) at the turbine exit.

For turbines which have irreversible effects (this includes *all* turbines), the work produced will be less than in the ideal case. Perhaps the best way to see what happens here is through a *T-s* diagram. In figure 12–5 is shown the ideal expansion from pressure p_3 to p_4. As is indicated, the steam has just achieved a saturated vapor state at the exhaust, at state (4). In the irreversible process, fluid friction expansion of steam between the same pressures p_3 and p_4 the entropy *increased* and thus the exhaust steam is still superheated at a temperature T_4, but at a pressure p_4. The work for the irreversible process is then gotten from

$$wk_{\text{turb}} = h_3 - h_{4'} \qquad \textbf{(12–11)}$$

with an entropy increase of $s_{4'} - s_3$ in the steam.

We use the reversible adiabatic expansion as the ideal process and define the turbine efficiency by

$$\eta_{\text{turb}} = \left(\frac{wk_{\text{actual}}}{wk_{\text{ideal}}}\right) \times 100 = \left(\frac{h_3 - h_{4'}}{h_3 - h_4}\right) \times 100 \qquad \textbf{(12–12)}$$

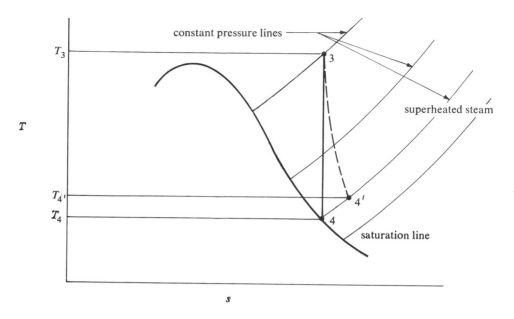

Figure 12–5   Expansion processes through a turbine.

12.4
Pumps

To complete a steam turbine cycle, that is, return the H_2O to the boiler from the turbine for generating more steam, mechanical means must be provided to force the fluid into the boiler. The steam is exhausted from the turbine at a low pressure (perhaps at atmospheric pressure) and the boiler is furnishing steam at a very high pressure. To raise the steam from a low to a high pressure and then force it into the boiler is the task of the pump. In some cycles we may call the device a *compressor*, but here we are handling water, and *pump* is a more common term. Pumps are continuous flow mechanisms which are essentially reversed turbines, or they are piston-cylinder devices which provide intermittent flow. In either case, we may consider the pump to be a steady flow open system when observed over a sufficient time span. By applying the steady flow energy equation, neglecting kinetic and potential energy changes, and assuming an adiabatic compression in the pump, we obtain equation (12–4)

$$wk_{pump} = h_1 - h_2$$

The power required to convey $\dot{m}$ lbm/s of fluid through the pump is calculated from the equation

$$\dot{W}k_{pump} = \dot{m}(h_1 - h_2) \tag{12–13}$$

In many cases the H_2O passing through the pump is saturated liquid. In this state, water is nearly incompressible (the density or specific volume remains constant with changes in pressure), so we can write the general equation for work in an open system

$$wk = -\int vdp$$

and since v = constant, we get

$$wk_{\text{pump}} = -v\int_{p1}^{p2} dp = -v(p_2 - p_1) \qquad \textbf{(12–14)}$$

A discrepancy may occur between the answers of equations (12–4) and (12–14) for pump work in which case the answer obtained from (12–4) should be used. The reason for this discrepancy is that in equation (12–14), the specific volume is assumed constant which is incorrect for all real fluids or gases. If we identified v as the average (v_{av}), then we would be correct to use equation (12–14) and obtain agreement with equation (12–4), but the average specific volume is difficult to predict in many problems.

Pumps are normally driven by electric motors or small steam turbines. In either case the pump work must be accountable in determining the

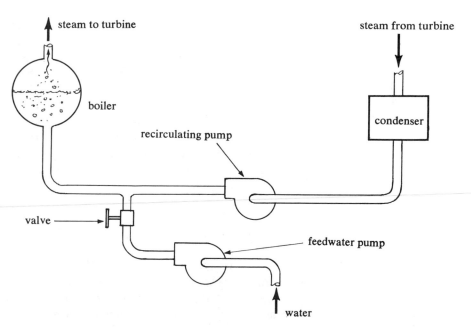

Figure 12–6   Pumps in a steam turbine cycle.

net cycle work. Also, if some of the H_2O is lost due to leakage in the cycle, fresh feed water can be added to the cycle by a pump. If this addition is accomplished by a separate pump, as shown in figure 12–6, then the pump we have been considering previously is called a *circulating pump* and the new one is the *feedwater pump.*

Occasionally, steam injectors are used instead of pumps to convey water to the boiler. A discussion of these devices is included in chapter 13.

12.5 Condensers

Steam exhausted from a turbine in a Rankine cycle is pumped back into the boiler to recycle the water. If, however, the steam exhausted by the turbine is directly furnished to the circulating pump, the ideal work required by the pump to deliver high pressure steam to the boiler would be exactly the work gotten out of the reversible adiabatic expansion in the turbine operating between the same pressures. In effect then, one steam turbine would be using all its output to drive the pump, and the cycle work would be zero. If there were irreversibilities in the cycle, the steam turbine work would not be enough to drive the pump and we would then need to add work from the surroundings. To prevent these undesirable results, a condenser receives the exhaust steam, condenses this steam to a liquid, and then feeds this liquid water to the pump. This device is shown in the cycle of figure 12–3 where it is indicated that heat is rejected. This is the precise role of the condenser — it transfers heat to the surroundings so that the heat engine can operate. We have seen that the second law of thermodynamics demands a transfer of heat between two regions for any cyclic heat engine. Of course, a mechanical device does not obey a law; a law rather obeys the operation of a mechanical device, but without the condenser, the steam turbine cycle is involved (ideally) in only one heat transfer, that at the boiler. In any case, if the condenser is used as an open steady flow system and if kinetic and potential energy changes in the system are neglected, the first law of thermodynamics gives us equation (12–3)

$$q_{rej} = h_1 - h_4$$

The rate of heat transfer $\dot{Q}_{rej}$ can be obtained from

$$\dot{Q}_{rej} = \dot{m}(h_1 - h_4) \qquad (12\text{–}15)$$

where $\dot{m}$ is the mass flow through the condenser. The mechanical reason that the condenser reduces the pump work, and thus makes the steam turbine practical, is that the steam vapor is converted (condensed) into a liquid. Consequently, the state of the medium leaving the condenser should ideally be a saturated liquid, and then enthalpy h_4 so determined.

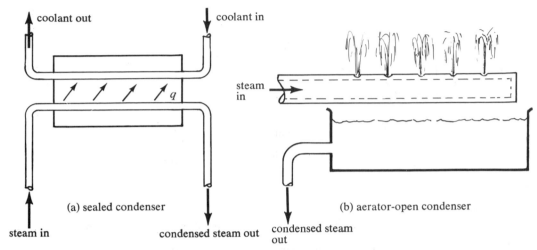

(a) sealed condenser   (b) aerator-open condenser

steam in   condensed steam out   condensed steam out

Figure 12–7   Types of condensers in their elemental form.

The condenser can be built in a number of different physical configurations. It may be a sealed condenser, as shown in figure 12–7(a), which transfers heat from the steam to a coolant across an intermediate boundary. The coolant may be air, river water, or some other fluid, but it is not in direct contact with the working medium (H_2O). Most condensers are constructed in this general manner. A variation of this device is the cooling tower, which relies on heat transfer to the surrounding air.

A second method of manifesting the phase change of steam vapor to liquid is the aerator or open condenser, illustrated in figure 12–7(b). This device literally exhausts the steam into the surroundings; due to this intimate mixing, the heat transfer is rapid and the condensing of the steam to vapor is equally rapid. The drawback to this type of condenser, while it is less expensive than the closed condenser, is that the exhaust pressure of the turbine cannot be less than the atmospheric pressure. The closed condenser can operate at pressures less than atmospheric and thus allow for exhaust vacuum pressures in the turbine.

**12.6
The Phase
Change and
Superheating**

The working medium used in the steam turbine cycle is water. Chemically, we symbolize it as H_2O and then call it a pure substance if no impurities are present. Water, of course, can take the three common phases with which we are familiar — solid (or ice), liquid, and vapor (or steam). In figure 12–8 is shown the phase diagram of water. Notice that at the triple point, all phases converge and that beyond the critical point (to

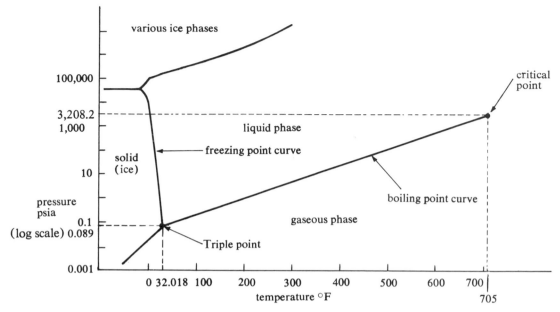

Figure 12–8   Phase diagram for water (H₂O) (not drawn to scale).

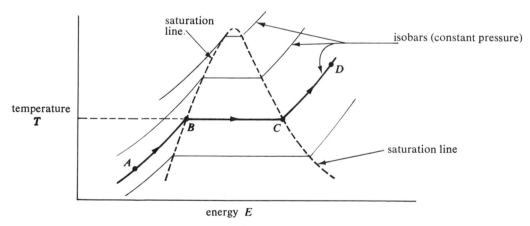

Figure 12–9   Temperature-energy relation of material going through liquid-gas phase change at various pressures.

the right of it) the liquid and vapor phases cannot be differentiated. For our purposes, the liquid-vapor phase change and the vapor or steam phase itself will be sufficient. Suppose, for instance, that we have a constant pressure container of water. If we add energy to the contents, monitoring the temperatures as we add energy, we notice that the water begins to have an increased temperature. In figure 12–9, we see the result of this plotted on a *T-E* diagram, indicated by the curve *A-B*. At point *B*, the water begins to boil; that is, it changes from a liquid to a vapor. If the boiler pressure is 14.7 psia, then the temperature at point *B* will be 212°F; but if the pressure is higher (or lower) than 14.7 psia, then the boiling temperature will be greater than (or less than) the 212°F. Various isobaric lines are shown in the graph of figure 12–9 to show this effect.

Once the water begins to boil, we may keep adding energy without changing the temperature until we reach a point when all the water is now converted to vapor or steam. This is the state *C* on our isobaric line, and if we continue adding energy, the temperature increases again for the steam. Any particular state beyond the point *C* is called a *superheated steam* (such as point *D*), and the unique points *B* and *C* are called the *saturated liquid* and *saturated vapor states* respectively. We could connect the various saturated liquid and vapor points for differing pressures, as indicated with the dashed line in figure 12–9, labeled the *saturation line*. In figures 12–10 and 12–11 are shown property diagrams in which this saturation line is clearly indicated. At the top of this dome-shaped line is shown the *critical point* which is the point where the phase change becomes undefined. The *T-s* diagram of figure 12–10 depicts isobaric lines as in figure 12–9. Also, the area under the *T-s* diagram curves should be visualized as "heat transfers," as we have always done.

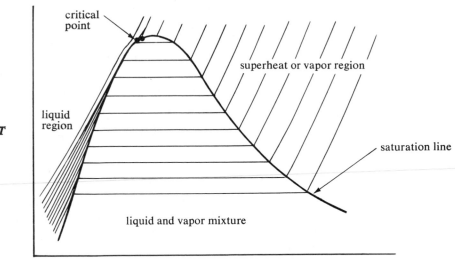

Figure 12–10   *T-s* diagram of steam-water phase change.

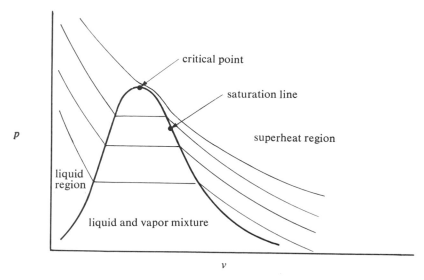

Figure 12–11   *p-v* diagram of steam-water phase change.

12.7 Use of Steam Tables

The appendix tables B.1, B.2, and B.3 will furnish the data for the properties of water at saturated conditions, superheated steam conditions, and high temperature liquid conditions. In table B.1 are listed the saturation properties as functions of temperature as well as of pressure. The properties specific volume, enthalpy, and entropy are tabulated. Notice that the subscript f denotes saturated liquid and g denotes saturated vapor. For enthalpy and entropy third and fourth terms are tabulated, h_{fg} and s_{fg}. These are the differences of the saturated states and are defined as

$$h_{fg} = h_g - h_f \qquad (12\text{–}16)$$

and

$$s_{fg} = s_g - s_f \qquad (12\text{–}17)$$

The term h_{fg} is obviously the measure of the energy added or lost during the vapor-liquid phase change. It is referred to as the *heat of vaporization, heat of condensation, latent heat, latent heat of vaporization,* or *latent heat of condensation.*

Through the above property data, we may determine the precise values of the properties of a state intermediate between saturated liquid and saturated vapor, provided we know the quality x of the state. We defined

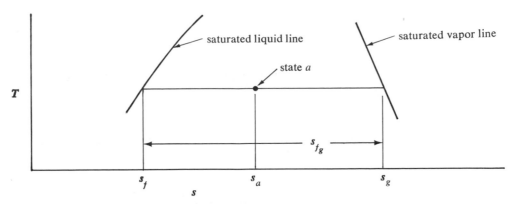

Figure 12–12   Relation between the entropy of a state intermediate between a saturated liquid and vapor and the entropy of the saturation points.

quality previously (see section 8.10) as the ratio of the vapor mass to the mixture mass. The state we wish to define, for example, state a, is depicted in figure 12–12 by use of the entropy property. Here we can write

$$100 \times \left(\frac{s_a - s_f}{s_g - s_f}\right) = x = \text{quality at state } a$$

and then

$$s_a = \frac{x}{100}(s_g - s_f) + s_f$$

or

$$s_a = \frac{x}{100} s_{fg} + s_f \tag{12–18}$$

Similarly, we could write

$$h_a = \frac{x}{100} h_{fg} + h_f \tag{12–19}$$

$$v_a = \frac{x}{100}(v_g - v_f) + v_f \tag{12–20}$$

and

$$u_a = \frac{x}{100}(u_g - u_f) + u_f \tag{12–21}$$

Internal energy can be easily calculated from the data in tables B.1 and B.2. The superheated steam properties are given in table B.2 where the pressure and temperature together determine the particular state. Specific volume, enthalpy and entropy are tabulated at each of these points.

Subcooled liquid properties are listed in an abbreviated table, B.3 which furnishes the data of differences between saturated liquid and the specified subcooled or compressed state properties.

The following example problems should help in seeing how data are extracted from these tables.

Example 12.1   A steam generating unit supplies 1,000 lbm/min of superheated steam at 400 psia and 800°F. Determine the entropy, enthalpy, and specific heat of the steam.

Solution   From table B.2 we obtain the answers

$$s = 1.6850 \text{ Btu/lbm°R}$$

$$h = 1417.0 \text{ Btu/lbm}$$

$$v = 1.8151 \text{ ft}^3/\text{lbm} \qquad \qquad \textit{Answer}$$

Example 12.2   Steam is condensing at a pressure of 10 psia. If the quality of the steam is 30% at a particular instant, determine the temperature, enthalpy, entropy and internal energy of the steam.

Solution   The steam is in the phase change, so we obtain the required data from the saturated steam tables, B.1. At a pressure of 10 psia, the temperature is given as 193.21°F.

We then use equations (12–9), (12–18), and (12–21)

$$h_a = \frac{x}{100} h_{fg} + h_f$$

$$s_a = \frac{x}{100} s_{fg} + s_f$$

$$u_a = \frac{x}{100} (u_g - u_f) + u_f$$

and from table B.1 find

$$h_{fg} = 982.1 \text{ Btu/lbm} \qquad \qquad h_f = 161.2 \text{ Btu/lbm}$$

$$s_{fg} = 1.5043 \text{ Btu/lbm°R} \qquad \qquad s_f = 0.2836 \text{ Btu/lbm°R}$$

$$u_f = h_f - v_f p = 161.2 - (0.0166)(10)(144)/778$$
$$\cong 161.17 \text{ Btu/lbm}$$

$$u_g = h_g - v_g p = 1143.3 - (38.42)(10)(144)/778$$
$$= 1072.2 \text{ Btu/lbm}$$

and proceed to calculate the answers

$$h_a = 0.3(982.1) + 161.2 = 455.8 \text{ Btu/lbm} \qquad \textit{Answer}$$

$$s_a = 0.3(1.5043) + 0.2836 = 0.7349 \text{ Btu/lbm}°\text{R} \qquad \textit{Answer}$$

$$u_a = 0.3(1072.2 - 161.17) + 161.17 = 434.48 \text{ Btu/lbm} \quad \textit{Answer}$$

Example 12.3   A circulating pump compresses saturated liquid water at 200°F to 400 psia and 200°F. Determine the enthalpy of the water leaving the pump.

Solution   The enthalpy of the entering fluid is easily found in table B.1:

$$h_f = 168.1$$

Note that the pressure is 11.53 psia at this point. The difference between the enthalpy of the existing water and that at the inlet is obtained from table B.3:

$$h - h_f = +0.88 \text{ Btu/lbm}$$

Then

$$h = +0.88 + h_f = 0.88 + 168.1$$
$$= 168.98 \text{ Btu/lbm} \qquad \textit{Answer}$$

Example 12.4   Determine the work done on each pound of steam for a pump isentropically compressing steam from a saturated liquid at 200°F to 400 psia.

Solution   From equation (12–4)

$$wk_{\text{pump}} = h_1 - h_2$$

We may be tempted to use the values for enthalpy listed in the solution of example 12.3. That is, we can write

$$h_f = h_1 = 168.1 \text{ Btu/lbm}$$

and then say

$$h = h_2 = 168.98 \text{ Btu/lbm}$$

This is not precisely correct since the relation $h - h_f$ tabulated in B.3 refers to constant temperature changes, but in the pump we normally assume reversible adiabatic or isentropic conditions. This would require a value $\Delta s = s - s_f = 0$, whereas table B.3 gives us a value of -0.5×10^{-3} Btu/lbm°R for $s - s_f$ at 200°F. By judicious interpolation of the data of table B.3, we may determine the final temperature of an isentropic process beginning with saturated liquid and ending at 400 psia pressure.

The error involved in the above simpler method is not appreciable, however, so that we can say

$$h = 168.98 \text{ Btu/lbm} \approx h_2$$

and then

$$wk_{\text{pump}} \approx 168.1 - 168.98 = -0.88 \text{ Btu/lbm} \qquad \textit{Answer}$$

We could also obtain the pump work from equation (12–14)

$$wk_{\text{pump}} \approx -v_{\text{av}}(p_2 - p_1)$$

where specific volume at state (1)

$$v_1 = 0.016634$$

from table B.1. If we then assume $v_{\text{av}} \approx v_1$ (not an appreciable error), then

$$wk_{\text{pump}} \approx -0.016634(400 - 11.526) \times \frac{144}{778} \text{ Btu/lbm}$$

$$= -1.196 \text{ Btu/lbm} \qquad \textit{Answer}$$

As can be seen, there is a discrepancy between our answers which points up the nature of approximate solutions. Either answer can be used for an analysis of a steam turbine cycle.

Example 12.5

The steam generating unit of example 12.1 burns coal having a heating value of 12,000 Btu/lbm. If the effectiveness of the boiler is 90% and the pump of example 12.4 furnishes the water to the unit, determine the rate of coal consumption.

Solution

Let us first define the generating unit and the flux involved in it (see figure 12–13), and we can quickly obtain $\dot{Q}_{\text{added}}$ from

$$Q_{\text{added}} = \dot{m}_{\text{steam}}(h_3 - h_2)$$

Then

$$\dot{Q}_{\text{added}} = 1000 \text{ lbm/min } (1417.0 - 168.98) \text{ Btu/lbm}$$
$$= 1,248,020 \text{ Btu/min}$$

The boiler efficiency η_{boiler} is identified as

$$\eta_{\text{boiler}} = \frac{\dot{Q}_{\text{add}}}{\dot{m}_{\text{fuel}} (\text{HV})} \times 100 \qquad \textbf{(12–22)}$$

so that, for our problem

$$90\% = \frac{1,248,020 \text{ Btu/min}}{\dot{m}_{\text{fuel}} (12,000 \text{ Btu/lbm})} \times 100$$

or

$$\dot{m}_{\text{fuel}} = 115.6 \text{ lbm/min} \qquad \textit{Answer}$$

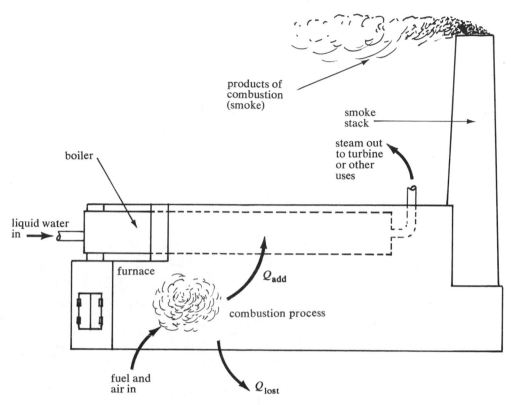

products of
combustion
(smoke)

smoke
stack

steam out
to turbine
or other
uses

boiler

liquid water
in

furnace

Q_{add}

combustion process

fuel and
air in

Q_{lost}

Figure 12–13   Elements of a typical steam generating unit.

12.8
The Mollier
Diagram

Many people prefer to extract data from graphs or charts rather than from tables. While the steam tables contain the required data to properly analyze the behavior of steam in the vapor-liquid, superheated, and subcooled regions, the Mollier diagram duplicates this information in chart form. This diagram is included in appendix B in a reduced form. Mollier diagrams are quite popular in sizes fourfold to tenfold larger than that given here, but the basic method for use of this chart can be grasped from this small version. Notice that the Mollier diagram is essentially an *h-s* diagram with various iso-lines, as illustrated in figure 12–14. Notice that the saturation line and critical point are identified on the sketch as well as on the detailed Mollier diagram. The region below the saturation line, the phase change, is crisscrossed by isobars and constant-moisture lines. Using these lines, we may easily find enthalpy and entropy from a known pressure and quality since the moisture is related to quality by the equation

$$\text{moisture percentage} = (1 - x) \times 100 \qquad \textbf{(12–23)}$$

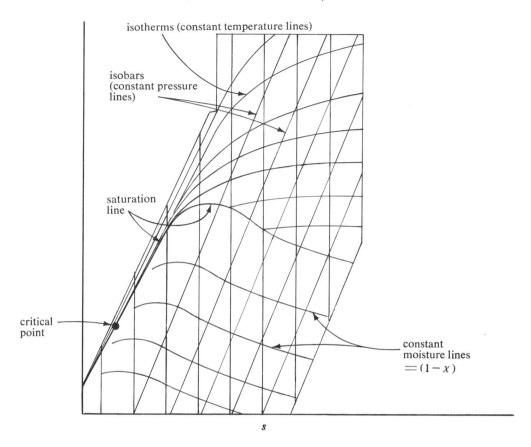

Figure 12–14   Sketch of Mollier diagram (not drawn to scale).

The isobars extend into the superheated steam region above the saturation line. This region is also mapped by isotherms and between a given
temperature and pressure, the enthalpy and entropy may be found from
the Mollier diagram.

One of the most useful characteristics of the Mollier diagram is the
tracing of reversible adiabatic processes. These follow vertical lines
(constant entropy), so that from given initial and final pressures,
we may easily extract enthalpy from the chart. There are other con-
veniences to be discovered from this useful graph and those who have
trouble relating tabulated data with actual physical processes may find
the Mollier diagram much more convenient and more descriptive of the
steam processes. We will use both the steam table data and Mollier
diagram data in the following example problems.

**12.9
Analysis of
Conventional
Rankine Cycle**

The following example problems should tie together the various components in the simple steam turbine cycle and serve to show the use of the steam data in analyzing the complete cycle.

Example 12.6

An ideal steam turbine uses 10,000 lbm/hr of steam. The steam is superheated to 800°F and 250 psia when supplied to the turbine. The exhaust pressure is 15 psia and the fluid leaving the condenser is assumed to be a saturated liquid. Determine the power generated by the cycle, the rate of heat rejected, the amount of coal consumed if the boiler unit is assumed to be 100% effective while the LHV of coal is 13,500 Btu/lbm, and the steam rate. The steam rate $\dot{m}_{sr}$ is defined by the equation

$$\dot{m}_{sr} = \frac{\dot{m}_{\text{steam}}}{\dot{W}k_{\text{cycle}}} \qquad (12\text{--}24)$$

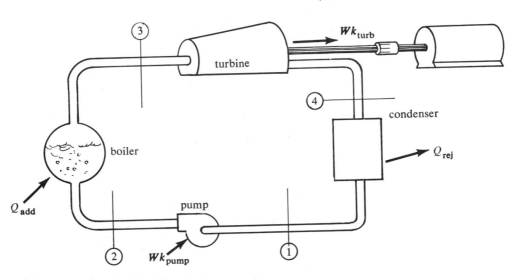

Figure 12–15

Simple steam turbine cycle.

Solution

This heat engine is sketched in figure 12–15 and the property diagrams are depicted in figures 12–16 and 12–17. The property data labeled in the graphs are either given in the problem statement or easily obtained from the steam tables. The power generated by the cycle is obtained from

$$\dot{W}k_{\text{cycle}} = \dot{m}wk_{\text{cycle}}$$

where

$$wk_{\text{cycle}} = wk_{\text{turb}} + wk_{\text{pump}} \qquad (12\text{--}25)$$

The ideal cycle involves adiabatic processes in the pump and turbine, so we use equations (12–2) and (12–4). We obtain from table B.1 the

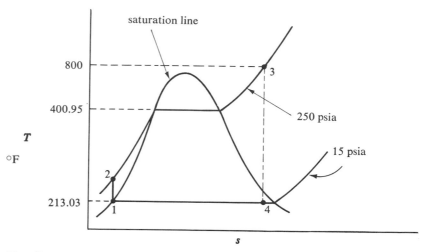

Figure 12–16   *T-s* diagram of steam turbine cycle of example 12.6.

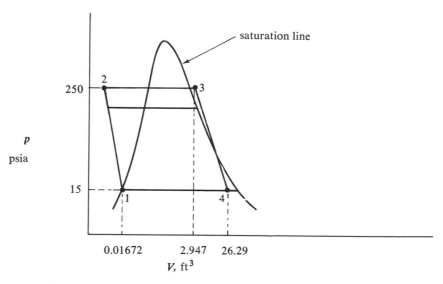

Figure 12–17   *p-V* diagram of steam turbine cycle of example 12.6.

value

$$h_1 = h_{f_1} = 181.2 \text{ Btu/lbm}$$

From table B.2 we get

$$h_3 = 1423.4 \text{ Btu/lbm}$$

and from table B.3 we interpolate for a value of 0.75 for $h - h_f$. Then

$$h_2 = h_1 + 0.75 = 181.95 \text{ Btu/lbm}$$

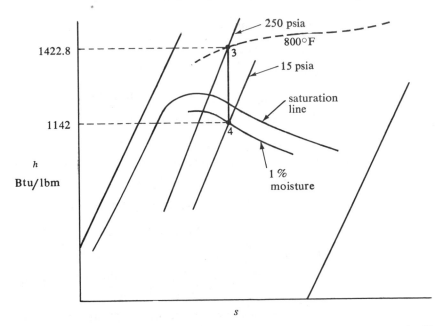

Figure 12-18   Mollier diagram of reversible adiabatic expansion in example 12.6.

From the Mollier diagram, we find h_4 by first locating the position of state (3) in the superheated region. Then we drop vertically down the isentropic line to the pressure line (isobar) of 15 psia. This technique is sketched in figure 12–18 where we also see that the moisture can be determined. We read*

$$h_4 = 1142 \text{ Btu/lbm}$$

and moisture = 1%, so that $x = 0.99$. Quickly then

$$
\begin{aligned}
wk_{\text{cycle}} &= h_3 - h_4 + h_1 - h_2 \\
&= 1423.4 - 1142 + 181.2 - 181.95 \\
&= 280.65 \text{ Btu/lbm}
\end{aligned}
$$

This gives us for the power generated, that

$$
\begin{aligned}
\dot{W}k_{\text{cycle}} &= \dot{m}(280.65) = 10{,}000 \text{ lbm/hr} \times 280.65 \text{ Btu/lbm} \\
&= 2{,}806{,}500 \text{ Btu/hr} \hspace{3cm} \textit{Answer}
\end{aligned}
$$

or

$$\dot{W}k_{\text{cycle}} = 2{,}806{,}500 \text{ Btu/hr} \times 1 \text{ hp-hr}/2545 \text{ Btu} = 1103 \text{ hp}$$
Answer

---

*We could get h_4 from table B.1 and equation (12–18). Here we would equate $s_a = s_4 = s_3$ and solve for x, then using equation (12–19), obtain h_4.

From equation (12–3) we obtain

$$q_{rej} = h_1 - h_4 = 181.2 - 1142 = -960.8 \text{ Btu/lbm}$$

and then

$$\dot{Q}_{rej} = \dot{m}q_{rej} = -9.608{,}000 \text{ Btu/hr} \qquad \textit{Answer}$$

The rate of heat added we get from

$$\begin{aligned}\dot{Q}_{added} &= \dot{m}q_{added} = \dot{m}(h_3 - h_2) \\ &= (10000 \text{ lbm/hr})(1423.4 \text{ Btu/lbm} - 181.95 \text{ Btu/lbm} \\ &= 12{,}414{,}500 \text{ Btu/hr}\end{aligned}$$

Since the boiler is perfect (100%) efficiency we write

$$\dot{m}_{fuel}(\text{LHV}) = \dot{Q}_{added} = 12{,}414{,}500 \text{ Btu/hr}$$

Hence

$$\dot{m}_{fuel} = \frac{12{,}414{,}500 \text{ Btu/hr}}{13{,}500 \text{ Btu/lbm}} = 919.6 \text{ lbm/hr} \qquad \textit{Answer}$$

The thermodynamic efficiency can easily be obtained here

$$\eta_T = \frac{wk_{cycle}}{q_{add}} \times 100 = \frac{\dot{W}k_{cycle}}{\dot{Q}_{add}} \times 100$$

$$= \frac{2{,}806{,}500}{12{,}414{,}500} \times 100 = 22.6\%$$

The steam rate we calculate from equation (12–24)

$$\dot{m}_{sr} = \frac{\dot{m}_{steam}}{\dot{W}k_{cycle}} = \frac{10{,}000 \text{ lbm/hr}}{1103 \text{ hp}} = 9.07 \text{ lbm/hp-hr} \qquad \textit{Answer}$$

Let us now see what irreversibilities do to the cycle.

Example 12.7

A steam turbine cycle uses 10,000 lbm/hr of steam. The steam generator furnishes superheated steam to the turbine at 800°F and 250 psia. The exhaust pressure is 15 psia and the temperature of the steam here is 250°F. Assume the water leaving the condenser is a saturated liquid and assume the boiler has an effectiveness of 91%. Determine power produced, turbine efficiency, and steam rate.

Solution

We can define our system as that sketched in figure 12–15; however, our property diagrams given in figures 12–19 and 12–20 deviate slightly at state (4) from those of example 12.6. The enthalpy values correspond to the data of example 12.6 except at state (4)

$$h_1 = 181.2 \text{ Btu/lbm}$$

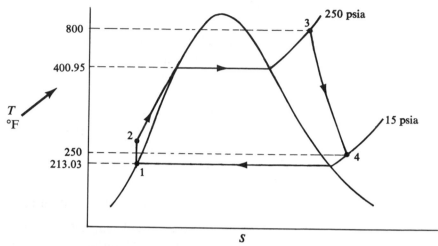

Figure 12–19   Example 12.7.

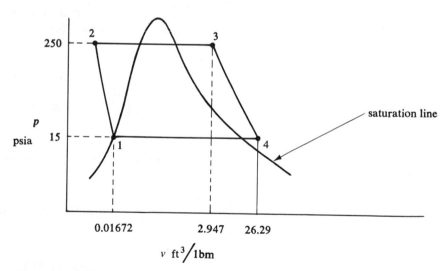

Figure 12–20   Example 12.7.

$$h_2 = 181.95 \text{ Btu/lbm}$$
$$h_3 = 1423.4 \text{ Btu/lbm}$$

At state (4) we find the enthalpy from the Mollier diagram: $t_4 = 250°F$ and $p_4 = 15$ psia, so that

$$h_4 = 1169 \text{ Btu/lbm}$$

Then the power is calculated

$$\dot{W}k_{cycle} = \dot{m}(h_3 - h_4 + h_1 - h_2)$$
$$= 10000 \text{ lbm/hr} (1423.4 - 1169 + 181.2 - 181.95) \text{ Btu/lbm}$$
$$= 2,536,500 \text{ Btu/hr}$$

or

$$\dot{W}k_{cycle} = 997 \text{ hp} \qquad\qquad\qquad Answer$$

In this irreversible process, notice that the steam is still superheated at the exhaust of the turbine, while in the reversible adiabatic process the steam was a mixture of vapor and liquid. The entropy increase during the irreversible process here is

$$s_4 - s_3 = 1.78 - 1.74 = 0.04 \text{ Btu/lbm}^\circ\text{R}$$

Various theories have been postulated, and research continues in methods to predict the increase in entropy during irreversible processes, yet there exists at this time no way to proceed from a predicted entropy change to the defined state. That is, if we had a rational, analytical method for predicting the change in entropy for an irreversible process, we could easily predict the outcome of processes and the efficiency of unbuilt engines. But we cannot.

The efficiency of the turbine we calculate from equation (12–12)

$$\eta_{turb} = \left(\frac{h_3 - h_{4'}}{h_3 - h_4}\right) \times 100$$

where

$$h_3 - h_4 = 1423.4 - 1142 = 281.4 \text{ Btu/lbm}$$

and

$$h_3 - h_{4'} = 1423.4 - 1169 = 254.4 \text{ Btu/lbm}$$

Then

$$\eta_{turb} = \frac{254.4}{281.4} \times 100 = 90.4\% \qquad\qquad Answer$$

The rate of heat rejected can be found from the equation

$$Q_{rej} = \dot{m}(h_4 - h_1)$$
$$= (10000 \text{ lbm/hr}) (1169 - 181.2) \text{ Btu/lbm}$$
$$= 9,878,000 \text{ Btu/hr} \qquad\qquad Answer$$

Then the thermodynamic efficiency is

$$\eta_T = \left(\frac{2,536,500}{2,536,500 + 9,878,000}\right) \times 100$$

$$\eta_T = 20.4\%$$ *Answer*

The steam rate we calculate from equation (12–24)

$$\dot{m}_{sr} = \frac{\dot{m}_{steam}}{\dot{W}k_{cycle}}$$

$$= \frac{10000 \text{ lbm/hr}}{997 \text{ hp}} = 10.03 \text{ lbm/hp-hr}$$

Obviously, the efficiency of the cycle decreased and the steam rate increased for the irreversible steam turbine when compressed to the ideal cycle. The question of how we may improve the efficiency of the heat engine will now be taken up.

12.10 The Reheat Cycle

The most obvious method to increase the efficiency of the steam turbine cycle, apparent from the *T-s* diagram of figure 12–19, is to increase the steam pressure and temperature — thereby increasing the turbine work the same amount as the additional heat added (ideally). There are, unfortunately, upper limits to operating pressures and temperatures, generally determined by the boiler, turbine, pump, pipe, and condenser materials. Consequently, to achieve higher thermodynamic efficiencies than the simple Rankine cycle, the reheat device is used. This involves the addition of energy to steam after it has expanded through a high pressure turbine and is subsequently expanding through a low pressure turbine. This cycle is sketched in figure 12–21, and the property diagrams, in figure 12–22. The cycle is still closed, and the system is all the water contained in the various components. Notice in the property diagrams that the shaded areas represent the additional heat and work of the reheat cycle over the simple Rankine cycle. Of course, these diagrams are descriptive of the reversible processes only, and with this restriction we can determine the following relationships for the ideal reheat cycle:

$$q_{add} = h_3 - h_2 + h_5 - h_4 \tag{12–26}$$

$$q_{rej} = h_1 - h_6 \tag{12–27}$$

$$wk_{pump} = h_1 - h_2 \tag{12–28}$$

$$wk_{turb} = h_3 - h_4 + h_5 - h_6 \tag{12–29}$$

The reheat cycle is best adapted to systems where very high pressure and temperature superheated steam are used so that the initial expansion through the high pressure turbine will leave a sufficiently high steam pressure for reheating. Let us now look at an example of this engine.

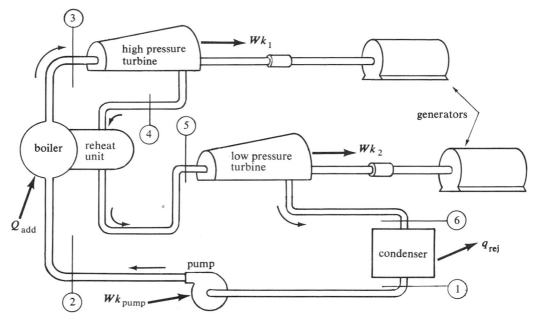

Figure 12–21   Reheat cycle.

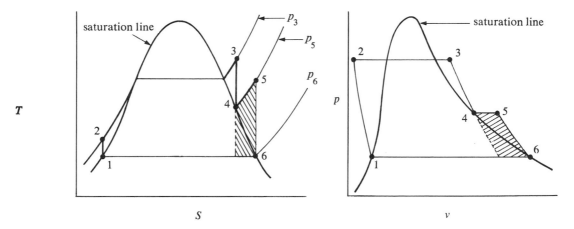

Figure 12–22   Reheat cycle property diagram.

Example 12.8   Steam at 2000 psia and 1600°F enters the high pressure turbine stages of
a reheat cycle and expands reversibly and adiabatically to 200 psia. This
steam is then directed through the reheater from which it emerges at a
temperature of 1200°F. The low pressure turbine expands the steam reversibly and adiabatically to 2 psia. If 7 lbm/s of steam are used, determine the power produced, the rate of heat added, the cycle efficiency, and

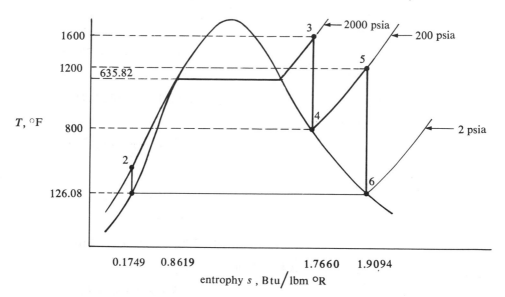

Figure 12–23   *T*-*s* diagram of reheat cycle (not drawn to scale).

Solution

the steam rate. Assume the circulating pump operates ideally and the fluid leaving the condenser is saturated liquid.

We will obtain the power from the relation $\dot{W}k_{\text{cycle}} = \dot{m}wk_{\text{cycle}}$ and where we have for the ideal cycle

$$wk_{\text{cycle}} = h_3 - h_4 + h_5 - h_6 + h_1 - h_2$$

The subscripts refer to the physical points of the reheat cycle of figure 12–21. We shall assume the engine here fits the same description. In figure 12–23 is shown the T-s diagram of the cycle with certain states defined by given temperatures. From table B.2 we extrapolate the following:

$$h_3 = 1832.5$$

We directly read

$$h_5 = 1633.7$$

Also, since $s_4 = s_3$, we can see that at state (4) the steam is still superheated (from table B.2 or the Mollier diagram), we obtain

$$h_4 \approx 1424.8 \ @ \ T_4 \approx 800°F$$

We have that $s_5 = s_6$ and since $s_5 = 1.9109$ we get that

$$1.9109 = \frac{x}{100} s_{fg_6} + s_{f_6}$$

where

$$s_{f_6} = 0.1749 \quad \text{at 2 psia (table B.1)}$$

and

$$s_{fg_6} = 1.7451 \quad \text{(table B.1)}$$

Then

$$\frac{x}{100} = \frac{1.9109 - s_{f_6}}{s_{fg_6}} = \frac{1.9109 - 0.1749}{1.7451} = 0.99$$

and then

$$h_6 = \frac{x}{100} h_{fg_6} + h_{f_6}$$

$$= 0.99(1022.2) + 93.99 = 1105.97$$

From table B.1 we have

$$h_1 = h_{f1} = 93.99 = h_{f6}$$

The pump work is obtained from equation (12–28)

$$wk_{\text{pump}} = h_1 - h_2 = h_{f_6}$$

or from equation (12–14)

$$wk_{\text{pump}} = -\int_{p1}^{p2} vdp \approx -v_{\text{av}}(p_2 - p_1)$$

We will here use equation (12–14). We first approximate the average specific volume by v_1. Then

$$v_1 \approx v_{f1} \approx 0.01623 \text{ ft}^3/\text{lbm}$$

and the work is calculated

$$wk_{\text{pump}} = -(0.01623 \text{ ft}^3/\text{lbm})(2000 - 2 \text{ lbf/in}^2)\left(\frac{144 \text{ in}^2/\text{ft}^2}{778 \text{ ft-lbf/Btu}}\right)$$

$$= 6.0 \text{ Btu/lbm}$$

We then have

$$h_2 = h_1 - wk_{\text{pump}}$$
$$= 93.99 - (-6.0) = 99.99 \text{ Btu/lbm}$$

and we can then calculate the net work of the cycle

$$wk_{\text{cycle}} = h_3 - h_4 + h_5 - h_6 + h_1 - h_2$$
$$= 1832.5 - 1424.8 + 1633.7 - 1105.97 + 93.99 - 99.99$$
$$= 929.43 \text{ Btu/lbm}$$

The power is quickly obtained by the computation

$$\dot{W}k_{cycle} = \dot{m}wk_{cycle} = 7 \text{ lbm/s} \times 929.43 \text{ Btu/lbm} \qquad \textit{Answer}$$

or

$$\dot{W}k_{cycle} = 1.41 \frac{\text{hp}}{\text{Btu/s}} \times 6506 \text{ Btu/s} = 9173 \text{ hp} \qquad \textit{Answer}$$

The rate of heat addition to the cycle is obtained from

$$\dot{Q}_{add} = \dot{m}q_{add}$$

and q_{add} is calculated from equation (12–26)

$$q_{add} = h_3 - h_2 + h_5 - h_4$$
$$= 1832.5 - 99.99 + 1633.7 - 1424.8 = 1941.4 \text{ Btu/lbm}$$

Then

$$\dot{Q}_{add} = 7 \text{ lbm/s} \times 1941.4 \text{ Btu/lbm} = 13,590 \text{ Btu/s} \qquad \textit{Answer}$$

The cycle efficiency or thermodynamic efficiency is gotten from the computation

$$\eta_T = \frac{\dot{W}k_{cycle}}{\dot{Q}_{add}} \times 100 = \frac{6506}{13590} \times 100 = 47.9\% \qquad \textit{Answer}$$

and the steam rate is obtained from equation (12–24)

$$\dot{m}_{sr} = \frac{\dot{m}_{steam}}{\dot{W}k_{cycle}} = \frac{7 \text{ lbm/s}}{9173 \text{ hp}} \times 3600 \text{ s/hr}$$
$$= 2.75 \text{ lbm/hp-hr} \qquad \textit{Answer}$$

12.11
The Regenerative Cycle

A method which can increase steam cycle efficiency without increasing the superheated steam pressure and temperature is the regenerative heating process which is, essentially, a method of adding heat, like a reversible process at a higher temperature. Regenerative heating, in practice, is the process of expansion of steam in the turbine well into the phase change region. Moisture is withdrawn mechanically from the turbine to reduce the effects of wear and corrosion. In figure 12–24 are shown the physical features of the regenerative steam turbine cycle and the property diagrams corresponding to this cycle are depicted in figure 12–25. The saturated liquid from the condenser is fed to a mixing chamber by a pump (c). The chamber allows the liquid to be heated by mixing with steam bled from the turbine; then this mixture is fed to the next chamber and ultimately back to the boiler for recirculation. In the

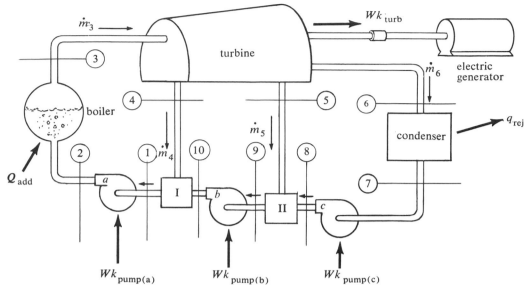

Figure 12–24   Regenerative cycle.

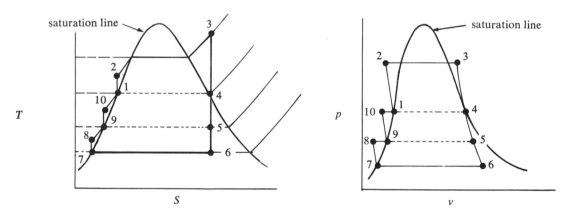

Figure 12–25   Property diagrams of regenerative cycle.

cycle of figure 12–24, two mixing chambers are utilized, but more could easily be added.

The computational techniques for the regenerative cycle are identical to those of the simple Rankine cycle with the following major deviation: mass must be accounted for throughout the cycle in a more detailed manner than in the simple Rankine cycle analysis. If we look at the conservation of mass through the turbine we get

$$\dot{m}_3 = \dot{m}_4 + \dot{m}_5 + \dot{m}_6 \tag{12–30}$$

and from this relationship we can obtain the total power produced by the turbine from the equation

$$\dot{W}k_{turb} = \dot{m}_3[wk_{turb(3-4)}] + (\dot{m}_5 + \dot{m}_6)[wk_{turb(4-5)}] + (\dot{m}_6)[wk_{turb(5-6)}] \tag{12-31}$$

where the $wk_{turb(3-4)}$ term is the work produced per unit mass of steam expanding from state (3) to state (4). Disregarding kinetic and potential energy changes, we have

$$wk_{turb(3-4)} = h_3 - h_4 \tag{12-32}$$

and, in a like manner

$$wk_{turb(4-5)} = h_4 - h_5 \tag{12-33}$$

and

$$wk_{turb(5-6)} = h_5 - h_6 \tag{12-34}$$

Consequently we can write for an ideal turbine with regeneration

$$\dot{W}k_{turb} = \dot{m}_3(h_3 - h_4) + (\dot{m}_5 + \dot{m}_6)(h_4 - h_5) + \dot{m}_6(h_5 - h_6) \tag{12-35}$$

If we look at the mixing chambers we can write the steady flow energy equation for them. Neglecting kinetic or potential energy changes we have for chamber (I)

$$\dot{m}_1 h_1 = \dot{m}_4 h_4 + \dot{m}_{10} h_{10}$$

but

$$\dot{m}_{10} = \dot{m}_1 - \dot{m}_4$$

so that

$$\dot{m}_1 h_1 = \dot{m}_4 h_4 + (\dot{m}_1 - \dot{m}_4) h_{10} \tag{12-36}$$

For mixing chamber (II) we get

$$\dot{m}_9 h_9 = \dot{m}_8 h_8 + \dot{m}_5 h_5$$

or, since $\dot{m}_9 = \dot{m}_{10}$ and $\dot{m}_8 = \dot{m}_6$

$$(\dot{m}_1 - \dot{m}_4) h_9 = \dot{m}_6 h_8 + \dot{m}_5 h_5$$

Also we can write from conservation of mass in the last turbine stages that

$$\dot{m}_6 = \dot{m}_3 - \dot{m}_4 - \dot{m}_5 = \dot{m}_1 - \dot{m}_4 - \dot{m}_5$$

so that

$$(\dot{m}_1 - \dot{m}_4) h_9 = (\dot{m}_1 - \dot{m}_4 - \dot{m}_5) h_8 + \dot{m}_5 h_5 \tag{12-37}$$

Using these relations we can analyze quite precisely the expected flow rates from a knowledge of the enthalpy values. The following example problem will show some of the numerical computation involved in a regenerative heating cycle.

Example 12.9

An ideal regenerative cycle operates with 400,000 lbm/hr of steam. The steam is at 2000 psia and 1000°F furnished to the turbine, and the exhaust is at 2 psia. Assume that steam is extracted at two stages for regeneration. The first stage corresponds to saturated vapor pressure and the second when the moisture content is 14% in the turbine steam. Determine the amounts of steam extracted for regenerating in the two steps, the power produced, and the thermodynamic efficiency.

Solution

The system is shown in figure 12–24 since we are considering two stage regeneration. The *T-s* diagram is sketched in figure 12–26 to show the given or easily identified property values. From table B.2 we get $h_3 =$ 1474.1. We can then proceed to determine the enthalpy at states (4), (5), and (6) from table B.1 or from the Mollier diagram. Using the Mollier diagram we obtain $h_4 = 1197$, $h_5 = 1022$, and $h_6 = 906$.

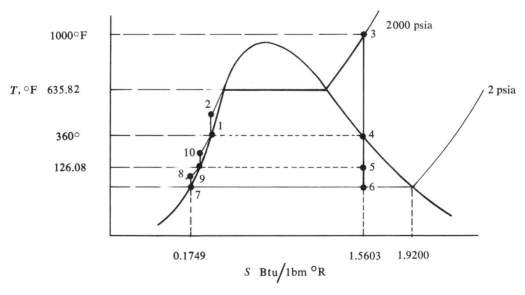

Figure 12–26   *T-s* diagram for two-stage regenerative cycle (not drawn to scale).

From table B.1 we find $h_7 = 93.99$, $h_9 = 175$ (corresponding to a pressure at states (5) and (9) of 13.6 psia), and $h_1 = 339$. Using equation (12–28)

$$h_2 = h_1 - wk_{\text{pump } (a)}$$
$$h_{10} = h_9 - wk_{\text{pump } (b)}$$
$$h_8 = h_7 - wk_{\text{pump } (c)}$$

and from equation (12–14) we get

$$wk_{\text{pump } (a)} \cong -v_{f1}(p_2 - p_1)$$

$$= -0.01817(20000 - 165) \times \frac{144}{778} \text{ Btu/lbm}$$

$$= -6.17 \text{ Btu/lbm}$$

yielding, for the enthalpy

$$h_2 = 339 - (-6.17) = 345.17$$

Similarly

$$wk_{\text{pump } (b)} \cong v_{fg}(p_{10} - p_9)$$

$$= -0.01667(165 - 13.6)\frac{144}{778} = -0.47 \text{ Btu/lbm}$$

and

$$wk_{\text{pump } (c)} \cong v_{f7}(p_8 - p_7)$$

$$= -0.01623(13.6 - 2)\frac{144}{778} = -0.035 \text{ Btu/lbm}$$

so that

$$h_{10} = 175 - (-0.47) = 175.47$$

and

$$h_8 = 93.99 - (-0.035) = 94.03$$

From these data we can then proceed with our analysis. Using equation (12–36), we can obtain the flow rate of steam at state (4)

$$\dot{m}_1 h_1 = \dot{m}_4(h_4 - h_{10}) + m_1 h_{10}$$

or

$$\dot{m}_4 = \frac{\dot{m}_1(h_1 - h_{10})}{h_4 - h_{10}} = \frac{400{,}000 \text{ lbm/hr } (339 - 175.47)}{1197 - 175.47}$$

$$= 64{,}033 \text{ lbm/hr} \qquad \qquad \textit{Answer}$$

Then, from equation (12–37)

$$(\dot{m}_1 - \dot{m}_4)h_9 = (\dot{m}_1 - \dot{m}_4 - \dot{m}_5)h_8 + \dot{m}_5 h_5$$

we get

$$(\dot{m}_1 - \dot{m}_4)(h_9 - h_8) = \dot{m}_5(h_5 - h_8)$$

or

$$\dot{m}_5 = \frac{(\dot{m}_1 - \dot{m}_4)(h_9 - h_8)}{(h_5 - h_8)}$$

$$= \frac{(400,000 - 64,033)(175 - 94.03)}{1022 - 94.03} = 29,315 \text{ lbm/hr } Answer$$

To obtain the power we use equation (12–35)

$$\dot{W}k_{turb} = \dot{m}_3(h_3 - h_4) + (\dot{m}_5 + \dot{m}_6)(h_4 - h_5) + \dot{m}_6(h_5 - h_6)$$

We have that $\dot{m}_3 = \dot{m}_1 = 400,000$ lbm/hr

$$\dot{m}_6 = \dot{m}_1 - \dot{m}_4 - \dot{m}_5 = 306,652 \text{ lbm/hr}$$

giving us

$$\dot{w}k_{turb} = [400,000 \text{ lbm/hr } (1474.1 - 1197) \text{ Btu/lbm}$$
$$+ (335,967 \text{ lbm/hr})(1197 - 1022)] \text{ Btu/lbm}$$
$$+ (306,652) \text{ lbm/hr } (1022 - 906) \text{ Btu/lbm}$$
$$= 205.2 \times 10^6 \text{ Btu/hr} = 80,631 \text{ hp} \qquad\qquad Answer$$

The net power produced we obtain by subtracting the power of the pumps; i.e.

$$\dot{W}k_{cycle} = \dot{W}k_{turb} + \dot{W}k_{pump \ (a)} + \dot{W}k_{pump \ (b)} + \dot{W}k_{pump \ (c)}$$

we have

$$\dot{W}k_{pump \ (a)} = \dot{m}\dot{W}k_{pump \ (a)} = 400,000 \times (-6.17)$$
$$= -2,468,000 \text{ Btu/hr} = -970 \text{ hp}$$

$$\dot{W}k_{pump \ (b)} = (\dot{m}_1 - \dot{m}_4)wk_{pump \ (b)} = (335,967)(-0.47)$$
$$= -157,904 \text{ Btu/hr} = -62.0 \text{ hp}$$

and

$$\dot{W}k_{pump \ (c)} = (\dot{m}_1 - \dot{m}_4 - \dot{m}_5)wk_{pump \ (c)} = (306,652)(-0.035)$$
$$= -10,733 \text{ Btu/hr} = -4.2 \text{ hp}$$

Then

$$\dot{W}k_{cycle} = 80,000 - 970 - 62.0 - 4.2$$
$$= 79,595 \text{ hp} \qquad\qquad Answer$$

The rate of heat addition is accounted for in the equation

$$\dot{Q}_{add} = \dot{m}_1(h_3 - h_2)$$

Then

$$\dot{Q}_{add} = (400,000 \text{ lbm/hr}) (1474.1 - 345.17) \text{ Btu/lbm}$$
$$= 4.52 \times 10^9 \text{ Btu/hr}$$

or

$$\dot{Q}_{\text{add}} = 1.254 \times 10^6 \text{ Btu/s} = 1.77 \times 10^6 \text{ hp}$$

Then the thermodynamic efficiency is found from the computation

$$\eta_T = \left(\frac{\dot{W}k_{\text{cycle}}}{\dot{Q}_{\text{add}}}\right) \times 100 = \left(\frac{0.79595 \times 10^6 \text{ hp}}{1.77 \times 10^6 \text{ hp}}\right) \times 100$$

$$= 45.0\% \qquad\qquad\qquad\qquad\qquad\qquad \textit{Answer}$$

12.12 Other Considerations of the Rankine Cycle

We have seen that the Rankine cycle in operation is a closed steam turbine cycle and that we can improve its efficiency by using reheat or regenerative heating devices. Steam turbines operating with both reheaters and regenerators are the most efficient power-producing devices available for large power demands. Increased improvements in thermodynamic efficiencies can also be realized with higher superheated steam pressures and temperatures; but, of course, an upper limit is set here by material properties.

One of the most attractive aspects of the steam turbine cycle is its independence of the source of thermal energy. While internal combustion engines are critically dependent on their working media (requiring a precise refined fuel), the steam turbine merely requires energy to boil water. This energy can come from coal, petroleum, or natural gas combustion, from high temperature exhaust gases of some other engine, from solar energy, or from nuclear reactors. Beyond the steam generating unit there are few, if any, design alterations needed to accommodate these different types of heat sources. We can see then that the steam turbine cycle is well adapted to convert thermal energy to mechanical and electrical energy, and it allows us to hope that future power requirements may be met.

The combustion of fuel (if this type of heat source is utilized for generating steam) can be accomplished under well-controlled conditions and thus effect more efficient combustion and less obnoxious exhaust or products of combustion.

An attractive use of the steam turbine cycle is "behind" or tandem with another engine operating at a much higher temperature. This leading engine is commonly proposed to be a magneto-hydrodynamic (MHD) converter which operates on very high temperature; ionized gases and exhaust are also at a high temperature. The steam generator could then produce superheated steam from this exhaust and get additional power that might be lost. The MHD converter produces electrical power, and the two separate units then can operate as an efficient generating station.

The steam turbine has been characterized as a constant speed device, probably because its greatest development has come in large power units. In small units, however, the steam turbine may be applicable to variable speed applications although this utilization has not been fully attempted.

Practice Problems

Section 12.1

12.1. A closed steam power plant delivers 10 megawatts (10 million watts) of electrical power. If 10,000 watts are required to drive water-recirculating pumps and 1.2×10^6 Btu/min of heat are added to the cycle, determine the amount of heat rejected and the cycle efficiency.

12.2. Sketch the p-v and T-s diagrams for a Rankine cycle and indicate which processes have work and heat present.

12.3. What are the source of power and the working medium in the common steam turbine cycle?

Sections 12.2 and 12.7

12.4. A boiler must supply steam at 250 psia and 650°F. If the water supplied to the boiler is saturated liquid at 3 psia, determine the amount of heat added per unit mass steam.

12.5. There are 20 lbm/s of water furnished to the boiler at 12 psia and saturated liquid condition. If the steam is at 180 psia and 600 Btu of heat are added per pound mass of steam determine
(a) Rate of heat addition.
(b) Steam temperature.

12.6. A steam generator operates with 85% efficiency when using coal as a fuel. If the incoming water is saturated liquid at 14.7 psia and if there are 2 tons of steam produced per hour at 200 psia and 650°F, determine the amount of coal needed per hour. Assume the coal has a lower heating value of 13,000 Btu/lbm.

12.7. A gas-fired steam generator uses a fuel with a LHV of 16,000 Btu/lbm. A burning rate of 6 lbm/min of fuel is expected and the boiler is assumed to have an efficiency of 90%. If water is supplied to the boiler at conditions comparable to a saturated liquid at 130°F and if the steam produced is at 400 psia and 600°F, determine the mass flow of water through this unit.

Sections 12.3 and 12.7

12.8. A turbine reversibly and adiabatically expands steam from 400 psia and 650°F to a saturated vapor. Determine the final steam temperature and the work produced per pound-mass of steam.

12.9. In a turbine 700 lbm/min of steam are isentropically expanded from 350 psia and 700°F to 10 psia. Estimate the power produced by the turbine.

12.10. A turbine receives steam at 600 psia and 800°F. It expands the steam to 10 psia and 200°F. Determine the work done per pound-mass and the turbine efficiency.

12.11. A turbine has an efficiency of 90% in expanding steam from 300 psia and 650°F to 3 psia. Determine the work done and the final steam temperature.

12.12. A steam turbine is comprised of a single wheel with a two-foot diameter. If steam is impressed against the blades of the wheel at 800 ft/s and at an angle of 10 degrees as shown in figure 12–27, determine the work gotten out of the turbine per lbm of steam, if the blade is frictionless and symmetrical.

Figure 12–27   Problem 12.12.

12.13. A turbine produces 20 Btu of work per lbm of steam. Determine the blade speed V_b if the blade is frictionless and symmetrical. The steam is directed against the blades at an angle of 18°.

Sections 12.4 and 12.7

12.14. A pump moves 50,000 lbm/min of 180°F water from 2 psia to a boiler at 380 psia. Determine the size of motor you would recommend to drive the pump.

12.15. Saturated liquid water at 5 psia is pumped into a steam generator at 600 psia. Assuming the water has a constant density through the pump of 60 lbm/ft³, determine the enthalpy of the steam entering the boiler.

12.16. A pump reversibly and adiabatically pumps saturated liquid water from a pipe at 10 psia to a boiler at 600 psia. Using the steam tables in appendix B, determine the work done by the pump per lbm of water.

12.17. What is the general equation to calculate pump work, starting from the steady flow energy and assuming an adiabatic process as the only restriction.

Sections 12.5 and 12.7

12.18. A condenser receives steam at 10 psia and 95% quality. Determine the heat rejected per unit mass of H_2O if the exit condition is 10 psia and saturated liquid.

12.19. If a condenser is to reject 2000 Btu/s and the mass flow of H_2O is 150 lbm/min, determine the steam quality entering the condenser at 5 psia if the leaving water is saturated liquid.

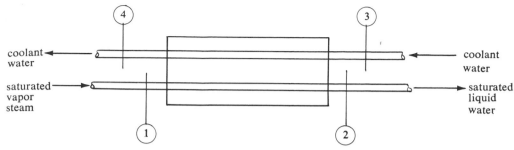

Figure 12–28   Problem 12.20.

12.20. A sealed condenser is used to reject heat from a steam turbine cycle in which 10 lbm/s of steam are converted from a saturated vapor to a saturated liquid at 10 psia (figure 12–28). The coolant is water at 15 psia and has a mass flow rate of 200 lbm/s. If the specific heat of the coolant is assumed to be 1 Btu/lbm°R, determine the temperature increase of the coolant through the condenser.

12.21. A condenser rejects heat from a steam turbine cycle and thereby converts the working medium (steam) from a saturated vapor to a saturated liquid at 120°F. Determine the entropy decrease per lbm of steam in the condenser and the entropy increase of the atmosphere. What is the total increase in entropy of the universe for the condenser process?

Section 12.8   Use the Mollier diagram to determine the following:

12.22. Enthalpy of steam at 1000°F and 150 psia.

12.23. Entropy and enthalpy of steam at 30 psia and 6% moisture (94% quality).

12.24. Temperature and enthalpy of steam at 1.5 psia and with an entropy of 1.9 Btu/lbm°R.

12.25. Enthalpy and pressure of steam with a temperature of 500°R and an entropy of 1.65 Btu/lbm°R.

Section 12.9   **12.26.** A steam turbine produces 700 hp through a reversible adiabatic expansion of from 200 psia and 800°F to 10 psia. Neglecting pump work, determine the steam rate of the cycle.

12.27. An ideal closed steam turbine cycle uses 5500 lbm/hr of steam. The steam leaves the boiler at 250 psia and is expanded reversibly and adiabatically to 8 psia and 95% quality in the turbine. The water leaving the condenser is a saturated liquid at 8 psia when it is then recirculated to the boiler by a pump. Calculate

(a) Turbine power produced.
(b) Heat rejected.
(c) Pump work.
(d) Heat added.
(e) Cycle efficiency and steam rate.

12.28. A steam generator unit supplies 12,000 lbm/hr of steam at 400 psia and 1000°F. Assume that the generator burns pulverized coal having an

LHV of 12,500 Btu/lbm and that the heat transfer efficiency is 88%. A turbine expands this steam at 90% efficiency to 10 psia and a condenser allows the steam to revert to a saturated liquid. If we neglect the inefficiencies in the pump, determine

(a) Heat added to cycle.
(b) Heat added to water in boiler.
(c) Work of turbine.
(d) Heat rejected.
(e) Net cycle work.
(f) Overall cycle efficiency.
(g) Steam rate.

12.29. A 200-megawatt power station operates on a closed steam turbine cycle, as shown in figure 12–29. The mechanical efficiency of the turbine-electric generator unit is 95% and the turbine efficiency is 90% in extracting energy from steam. The steam expands from 300 psia and 1050°F to 12 inches Hg in the turbine and is condensed to a saturated liquid in the condenser. The condenser is assumed to lose 5% of the steam to the atmosphere. A feedwater pump inserts 65°F water into the system after the circulating pump. Determine for the system

(a) Mass flow rate of steam through turbine.
(b) Heat added to water in boiler.
(c) Heat rejected in condenser.
(d) Work of both pumps. (Assume the water to be incompressible at all points.)
(e) Cycle efficiency.

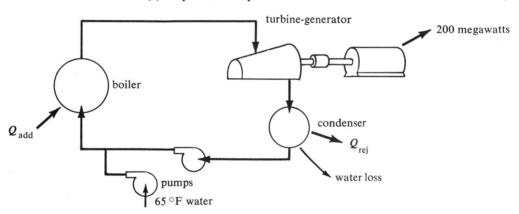

Figure 12–29   Problem 12.29.

12.30. A steam turbine cycle produces a net power of 10 megawatts. The turbine generator unit has a mechanical efficiency of 97% while steam is expanded reversibly and adiabatically through the turbine from 190 psia and 900°F to 5 psia. The water leaving the condenser is saturated liquid and is incompressible. If the boiler unit has a heat transfer efficiency of 90% and burns coal having an LHV of 11,000 Btu/lbm determine

(a) Pump work per unit mass of steam.

(b) Power produced by turbine.
(c) Rate of coal consumption (lbm/hour).
(d) Rate of heat rejection.

Section 12.10   **12.31.** In an ideal steam turbine reheat cycle, the turbine expands reversibly adiabatically 60 lbm/s of steam from 450 psia, 850°F, to 50 psia. This steam is then reheated to 800°F and expanded further to 1.5 psia. Determine
(a) Steam rate.
(b) Cycle efficiency.

12.32. For the reheat cycle shown in figure 12–30, determine
(a) Turbine output (horsepower).
(b) Rate of heat added.
(c) Steam rate.
(d) Cycle efficiency.

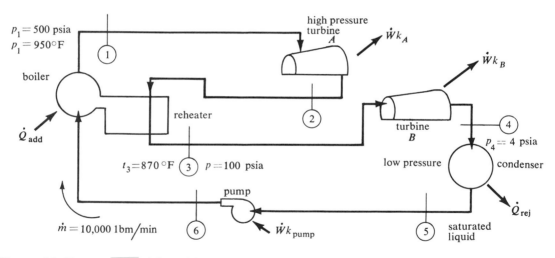

Figure 12–30   Problem 12.32.

Section 12.11   **12.33.** Steam at 800 psia and 1100°F is supplied to a regenerative steam turbine with a single extraction. If the cycle is ideal, the steam is expanded in the turbine to 1 psia, and steam is extracted at 80 psia, determine, per lbm of incoming steam
(a) Amount of steam extracted.
(b) wk_{turb}
(c) Efficiency of cycle.

12.34. For the ideal regenerative cycle shown in figure 12–31, determine
(a) $\dot{m}_4$ and $\dot{m}_5$
(b) $\dot{W}k_c$, $\dot{W}k_b$, and $\dot{W}k_a$
(c) $\dot{W}k_{turb}$
(d) Cycle efficiency
(e) $\dot{Q}_{rej}$

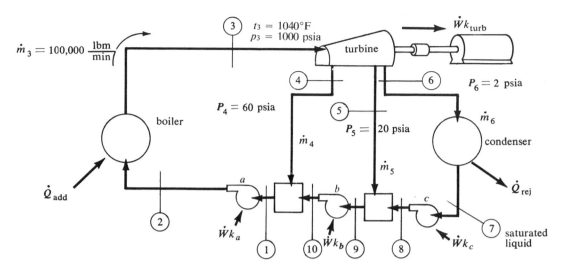

Figure 12–31   Problem 12.34.

13

The Steam Engine

There is no other device that has contributed as much to the development of modern society as the *steam engine* has. While it is now regarded almost as a museum piece, there are subtle features about the steam engine which could very easily return it to the wide use it enjoyed during the early years of the industrial revolution. From the early 1800s to the mid 1900s, the steam engine represented the major power-producing heat engine of society, and its popularity was the single most important reason for the development of thermodynamics. Early engineers and scientists, in seeking to explain and predict the steam engine's operation, refined and expanded the concepts and tools we now consider classical thermodynamics. We will discuss the steam engine here, not only to provide historical information, but also to see the promised capabilities of this most ingenious device.

The *ideal thermodynamic cycle of the steam expansion cycle* (which approximates the working of the actual steam engine) is defined. The work of the steam engine is then derived and calculated for typical engines, followed by a discussion of the efficiency. The mechanics which make up the typical steam engine are described in general terms to provide the reader with some idea of the components involved. Two of the most critical aspects of this engine are *valve arrangement* and *kinematics*. We will point out some of the important variations of these details and finally summarize the inherent advantages and problems associated with the steam engine.

13.1 The Steam Expansion Cycle

The steam engine is a heat engine utilizing the piston-cylinder device to provide mechanical work. It typically operates on a Rankine cycle which we define, ideally, by the following reversible processes:

(1–2) Adiabatic compression (liquid).
(2–3) Isobaric heat addition (vaporization).
(3–4) Adiabatic expansion (steam).
(4–1) Isobaric heat rejection (condensation).

These processes are sketched in the property diagrams in figures 13–1 and 13–2. Notice that these diagrams and the processes they describe are identical to the steam turbine cycle of section 12.1; the only difference is in the components which execute the various processes. The components of the closed cycle steam engine are shown in figure 13–3, The historically more common steam engine is the open cycle, shown in figure 13–4, which eliminates the condenser and circulating pump but which must continuously be supplied feedwater by an injector or a pump. The open cycle is generally simpler to operate in practice, but it is not as efficient

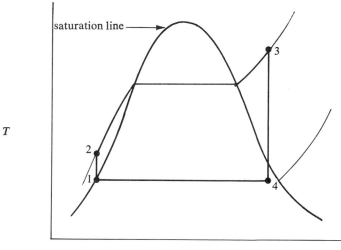

Figure 13–1   *T-s* diagram of steam engine.

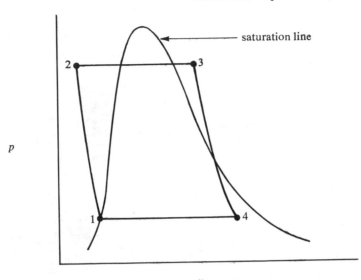

Figure 13-2   *p-v* diagram of steam engine.

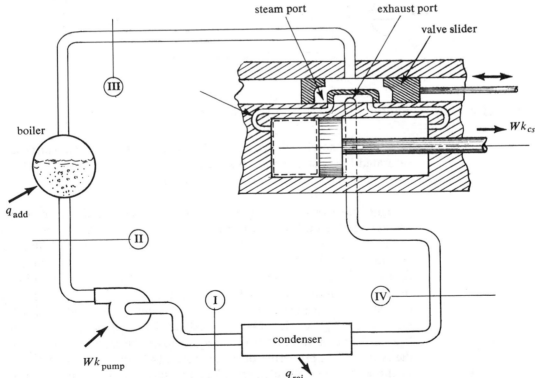

Figure 13-3   Sketch of typical closed cycle steam engine.

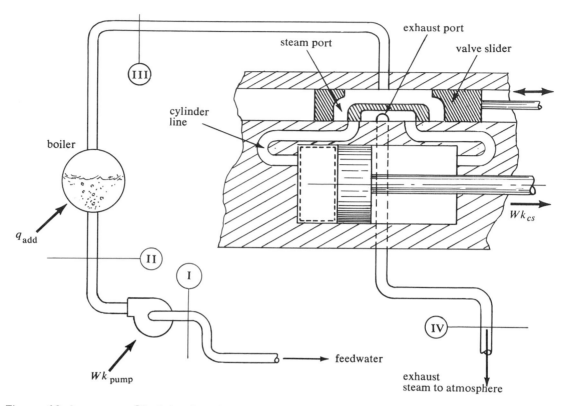

Figure 13–4   Sketch of typical open system steam engine.

(ideally) as the closed cycle unless a free supply of water exists at a saturated liquid state equal to state (1), as indicated in figures 13–1 and 13–2. Also, the open cycle uses large amounts of water.

In figure 13–3 we define our system for analysis as the chamber volume enclosed by the dashed line. In the piston and valve position shown, the engine is proceeding through process (3–4), the reversible adiabatic expansion. In figure 13-5(d) the piston is shown in the bottom-dead-center (BDC) position, state (4). Notice that the steam is enclosed inside the chamber during process (3–4); but at the beginning of process (4–1), the valve exhaust port is aligned with the cylinder line and retains this position through process (4–1), as indicated in figures 13-5(a) and (d). We see then that process (4–1) merely removes the expanded steam from the cylinder. The properties of steam at state (4) should ideally correspond to those at state (IV) in figure 13–3. Pressure losses and heat transfer, however, will reduce the pressure and temperature of the steam entering the condenser at state (IV) below those at state (4) for actual engines.

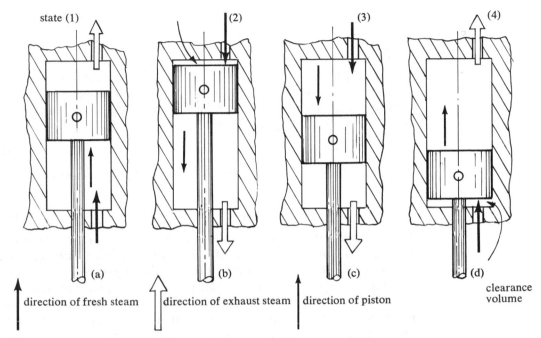

Figure 13-5   Piston positions during one cycle of a steam engine.

When the piston reaches a position such that the remaining steam in the cylinder condenses to state (1), the engine furnishes fresh, superheated steam at state (3) to the cylinders by repositioning the valve, as indicated in figure 13–5(b). As the engine then proceeds through process (1–2), the steam mixture (fresh and some expanded steam) is reversibly and adiabatically compressed to pressure p_2. State (2) represents the extreme innermost position of the piston inside the cylinder, called the *top-dead-center* (TDC), from which the piston is forced back out at constant steam pressure. At state (3), the steam port is closed to the cylinder and expansion process (3–4) then proceeds again.

To repeat, process (3–4) is ideally a reversible adiabatic expansion of steam carried out in a closed system. (The slide valve does not permit steam to enter or leave the cylinder.) Process (4–1) is an exhaust process which eliminates much of the expanded steam from the cylinder. Process (1–2) is a reversible adiabatic compression of the remaining steam in a closed system to a pressure and state equal to the state (2), indicated in the diagrams of figures 13–1 and 13–2. Process (2–3) is a constant pressure process, carried out with the steam port positioned over the cylinder line in the mechanism of figure 13–3. This is an open system process, since steam is added to the system or cylinder. The valve slide is then repositioned to that position shown in figure 13-3 and the reversible adiabatic process (3-4) proceeds again.

There are some apparent inconsistencies which should be explained. First of all, we have stated that ideally states (IV) and (4) are identical (in figure 13-3 and 13-1 respectively). Writing the steady flow energy equation, with kinetic and potential energy changes ignored, for the condenser as an open system we get

$$q_{\text{rej}} = h_1 - h_{\text{IV}} \qquad (13\text{--}1)$$

For ideal or reversible cycles the heat rejected by the piston cylinder is just

$$q_{\text{rej}} = h_1 - h_4 \qquad (13\text{--}2)$$

and since we have already stated that, ideally, $h_4 = h_{\text{IV}}$ we can imply $h_1 = h_{\text{I}}$ and surmise that states (I) and (1) are identical although they are physically removed from each other in the engine. Also, the steam entering the engine, described as state (III), can be equated to state (3) if pressure and thermal losses are neglected in the lines and more importantly, if we neglect the loss of specific enthalpy and temperature due to the initial mixing of the steam at state (III) with the remaining steam from process (4-1). This allows us to write $h_3 = h_{\text{III}}$ and since

$$q_{\text{add}} = h_{\text{III}} - h_{\text{II}} = h_3 - h_2 \qquad (13\text{--}3)$$

we can then say $h_{\text{II}} = h_2$ and infer that states (II) and (2) are identical.

Quite generally we can write

$$q_{\text{cycle}} = wk_{\text{cycle}}$$

from which we obtain

$$q_{\text{add}} - q_{\text{rej}} = wk_{\text{cycle}} \qquad (13\text{--}4)$$

or

$$q_{\text{add}} - q_{\text{rej}} = wk_{\text{engine}} - wk_{\text{pump}} \qquad (13\text{--}5)$$

Easily then, from equations (13-2), (13-3), and (13-4), we get

$$wk_{\text{cycle}} = h_3 - h_2 + h_1 - h_4 \qquad (13\text{--}6)$$

and from a standard definition of thermodynamic efficiency

$$\eta_T = \frac{wk_{\text{cycle}}}{q_{\text{add}}} \times 100$$

we obtain

$$\eta_T = \frac{h_3 - h_2 + h_1 - h_4}{h_3 - h_2} \times 100 = \left(1 - \frac{h_4 - h_1}{h_3 - h_2}\right) \times 100 \quad (13\text{--}7)$$

13.2
Work of the
Steam Engine

We may be tempted to further reduce the equation for calculating net work of a steam engine, equation (13–6), by recalling the caloric relation $h = c_p T$. This would be correct, but over the temperature ranges of the steam engine, the specific heat at constant pressure c_p is not a constant value and in fact has sharp changes during a phase change. For this reason we will use equation (13–6) for obtaining net work of the ideal steam engine. However, where there are small temperature changes and where no phase changes exist, $h = cT$ can be used. It might also be mentioned that net work should be equivalent to the area enclosed on the *p-v* diagram of figure 13–2. In this case, then, geometric means may be used to calculate the cyclic work.

The net work can also be equated to the average pressure of the steam cylinder during the full piston travel. If we call the average pressure the *mean effective pressure*, mep, the piston area A, and the piston travel L, then we have that

$$(\text{mep})(L)(A) = Wk_{\text{cycle}} \qquad (13\text{–}8)$$

and the rate of work produced, the power, is obtained from

$$\dot{W}k_{\text{cycle}} = Wk_{\text{cycle}} \times N = (\text{mep})(LAN) \qquad (13\text{–}9)$$

Equation (13–9) is frequently written in the mnemonic form

$$\dot{W}k_{\text{cycle}} = PLAN \qquad (13\text{–}10)$$

where

$$P = \text{mep}$$

In the actual steam engine test procedure, an engine indicator is used to generate an engine diagram. The engine diagram is the actual *p-V* diagram of the steam piston-cylinder device. (A typical one is shown in figure 13–6.) When measured with a planimeter, the enclosed area will approximate the work produced by the piston. The actual work which can be gotten from the engine is, however, bound to be slightly less than the area, due to mechanical friction. Of course, the engine diagram is considerably altered for increasing or decreasing speeds, and for this reason the work of a cycle will be a function of the engine speed. In figure 13–7 is shown a typical efficiency curve of a steam engine which reflects this variation in cyclic work due to engine speed.

In figure 13–6, the ideal cycle is superimposed to provide a visual comparison between the actual and ideal cases. Notice that the actual engine cycle is truncated at the right side; that is, the volume never expands to state (4), but rather is "cut off" at state (4'). This is a common procedure in actual steam engines, as the final portion of the expansion of steam is

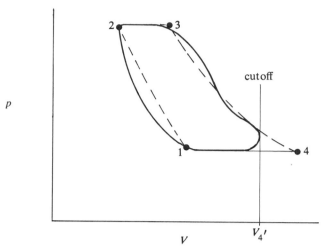

Figure 13–6   Typical engine diagram.

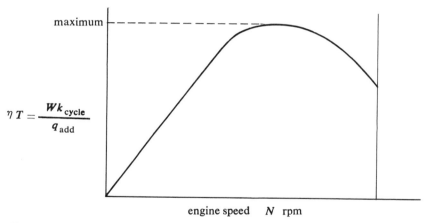

Figure 13–7   Typical relationship between engine efficiency and speed.

carried on at very slow speeds. In order to accommodate more acceptable engine speeds, therefore, the piston is reversed at a point corresponding to state (4′) rather than (4). Some potential work is lost due to this variation, but the increase in operating speeds is generally considered a good replacement for the loss.

In actual engine analysis, equation (13–10) is frequently utilized to obtain the indicated power data, but a more accurate approach would be to calculate work from

$$wk_{cycle} = h_{III} - h_{IV} + h_I - h_{II} \qquad (13\text{–}11)$$

and then use the relation

$$\dot{W}k_{cycle} = Nwk_{cycle} \qquad \text{(13–12)}$$

Here the subscripts I, II, III, and IV correspond to the points indicated in figure 13–3. Using this approach, we see that line energy losses may be more realistically included in the engine analysis, although we still assume adiabatic processes in the cylinder and circulating pump to arrive at equation (13–11).

Example 13.1   A steam engine diagram is shown in figure 13–8. The area enclosed by this diagram is 9000 ft-lbf which was recorded at an engine speed of 250 rpm. Determine the indicated power produced, the mep, and the cylinder dimensions if the bore and stroke of the cylinder are equal.

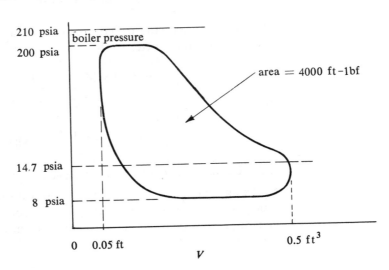

Figure 13-8   Engine diagram.

Solution   We can obtain the power produced by using equation (13–12)

$$\dot{W}k_{cycle} = Nwk_{cycle}$$

where

$$N = 250 \text{ rpm}$$

and

$$wk_{cycle} = 9000 \text{ ft-lbf/min}$$

or

$$wk_{cycle} = \frac{250 \times 9000}{33000} \text{ hp} = 68.2 \text{ hp} \qquad \textit{Answer}$$

The mean effective pressure we can calculate from equation (13–9)

$$\text{mep} = \frac{\dot{W}k_{\text{cycle}}}{LAN}$$

Since here

$$LA = \text{volume swept by piston} = 0.5 - 0.05$$
$$= 0.45 \text{ ft}^3$$

we get

$$\text{mep} = \frac{9000 \text{ ft-lbf} \times 250 \text{ rpm}}{0.45 \text{ ft}^3 \times 250 \text{ rpm}} = 20,000 \text{ lbf/ft}^2$$

$$= \frac{20,000}{144} \text{ psi} = 139 \text{ psi} \qquad\qquad \textit{Answer}$$

The dimensions of the cylinder can be obtained since we know $L =$ stroke = piston diameter so that

$$LA = L \times \pi \times \frac{L^2}{4} = \frac{\pi}{4} L^3 = 0.45 \text{ ft}^3$$

Then

$$L = \sqrt[3]{\frac{4 \times 0.45}{\pi}} = 0.83 \text{ ft} \qquad\qquad \textit{Answer}$$

and the cylinder diameter and piston stroke are both 0.83 ft.

Notice in figure 13–8 that the boiler pressure is indicated as 210 psia. Can you justify why the boiler pressure does not correspond to the highest pressure realized in the steam cylinder as indicated by the closed curve?

13.3 Cycle Components

The major parts of a common steam engine have been indicated in figure 13–3: the boiler, the piston-cylinder, the condenser, and the circulating pump or injector.

The boiler we recognize as the component which generates steam and we could quite properly consider it as a steam generator. Physically, it may function as a simple pressure cooker, but in most applications, it is a series of tubes to allow for faster, more efficient heat transfer between the water and the heat source. The combustion of fuel such as coal, oil, wood, or gas has commonly been the heat source to generate steam; but nuclear or thermoelectric heat sources could also be used. The steam generator for turbines is identical to those used for steam engines.

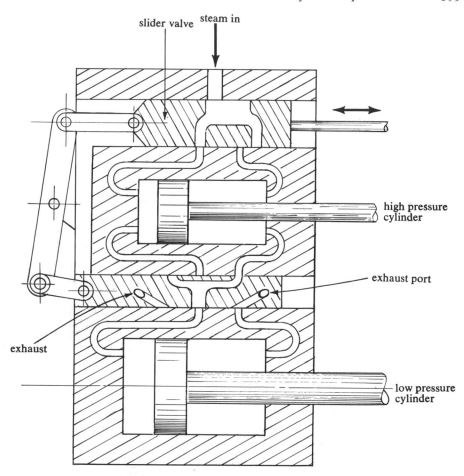

Figure 13-9   Compound engine or double expansion cylinder.

The piston-cylinder which transforms thermal or internal energy of steam into mechanical work represents the heart of the cycle. We have already considered ways in which the work produced by the operating cylinder may be determined, so let us here look only at some configurations which have been used. The double-acting cylinder, shown in figure 11–3, operates with two power strokes per revolution. That is, there are two separate systems operating in opposition, with the result that for every revolution or cycle of the piston, both systems traverse a cycle and produce a power stroke. The advantage to the double-acting cylinder is, of course, its compactness.

Another piston-cylinder arrangement which has had frequent use is the double-expansion cylinder, or compound engine. This is sketched in figure 13–9 along with a typical *p-V* diagram, shown in figure 13–10, to

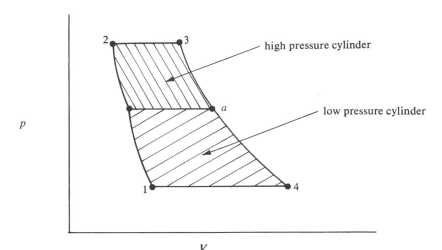

high pressure cylinder

low pressure cylinder

p

V

Figure 13-10   Double expansion cylinder.

illustrate that the double expansion is carried out in two separate cylinders, a high pressure cylinder and a low pressure one. The operation is, basically, one in which high pressure steam is expanded in the high pressure cylinder to state (*a*), as depicted on the *p-V* curve of figure 13-10. At this point, the steam is then shuttled to the low pressure cylinder where it completes the expansion to state (4). The primary reasons for the double-expansion configuration have been the increased power and the smoother running characteristics. The practical reason for gaining power over the simple steam engine is that a more complete expansion can be achieved without a premature cutoff as is necessitated in the simple engine. (See figure 13-6.)

The condenser is required on all steam engines which recirculate the H_2O. This device is demanded from the implications of the second law of thermodynamics; that is, we must reject heat to a sink to have an operable heat engine. There are numerous configurations to the condenser, but here we will merely state that moist steam (mixture of liquid and vapor) enters the condenser and will leave, ideally at a saturated vapor state — we say the steam has been condensed.

The water or circulating pump is a necessary part of any continuously operating steam engine. The function of this device is to recirculate the water in closed cycles and feed fresh water into the boiler in open cycles. In either case, the pump must, figuratively, push the water up a pressure gradient from a low pressure to a high pressure in the boiler or steam generator.

In the open cycle, illustrated in figure 13–4, the injector is commonly used in place of the pump. This device will be exposed in the following example problem to show the application of thermodynamics to an almost mystical process. This problem should also serve to show that the injector may be used in closed steam cycles, as well as in the open cycle.

Example 13.2

A nonlifting injector operates with superheated steam at 200 psia and 500°F. The injector receives water at 10 psia and 180°F and discharges it at 200 psia and 280°F into a boiler feed pipe. Determine the amount of steam required to supply 10 lbm of water to the boiler.

Solution

A nonlifting injector, shown in figure 13–11, receives water and steam through separate ports. The water is free to flow into a chamber and steam is forced into this same chamber through a nozzle. Irreversible mixing of the steam and water takes place in the region designated as *a* in the chamber and this mixture then reaches an equilibrium at a sufficiently high dynamic pressure p_D $[p_D = p + (\rho V^2 / 2g_c)]$ that it can be introduced into the boiler or other high pressure region. The check value shown in the figure merely prevents backward flow when the injector is not operating — it does not retard forward fluid flow.

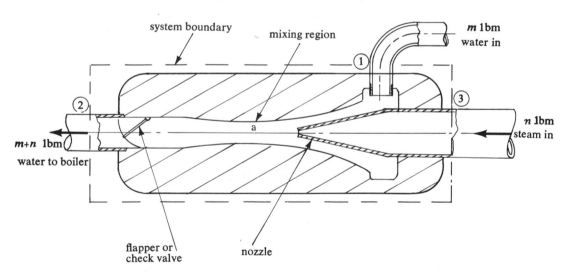

Figure 13–11   Nonlifting injector.

We now proceed to write the steady flow energy equation for the defined system enclosed by the boundary in figure 13–11. First we assume that the process is adiabatic (Q = 0), and that no changes in potential energy exist. Also, we observe that no work is being done. Then using subscripts which are compatible with the figure, we have

$$(m + n)h_2 - mh_1 - nh_3 + \frac{m + n}{2g_c} V_2^2 - \frac{m}{2g_c} V_1^2 - \frac{n}{2g_c} V_3^2 = 0$$

Further, it can be shown that the velocity terms are negligible with respect to the enthalpy terms so that we can then reduce our energy equation to

$$(m + n)h_2 - mh_1 - nh_3 = 0 \qquad \text{(13–13)}$$

If we desire, we may now write a general equation for the amount of steam required to transport 1 lbm of water through the injector. This term n/m is found by rearranging equation (13–13) to read

$$\frac{n}{m} = \frac{h_1 - h_2}{h_2 - h_3} \qquad \text{(13–14)}$$

For our problem, since the water flowing through the injector is well below the boiling point and since we do not know the value of h_1 nor can we extract it from the steam tables, we shall use the approximation $h = cT$ where c is near 1 Btu/lbm°R. Using this we can write that

$$h_f - h_1 = c(T_f - T_1) = c(t_f - t_1) \qquad \text{(13–15)}$$

where $h_f = 161.2$ Btu/lbm corresponding to the saturation temperature t_f of 193.21°F as given in table B.1 for steam at 10 psia. Then we can directly calculate h_1; that is

$$\begin{aligned} h_1 &= h_f - c(t_f - t_1) \\ &= 161.2 - 1.0(193.21 - 180) = 147.99 \text{ Btu/lbm} \end{aligned}$$

Using this result, we calculate h_2 from the equation

$$h_2 - h_1 = 1 \text{ Btu/lbm°R } (t_2 - t_1)$$

since we assume that the water has not boiled. Then

$$h_2 = h_1 + 1(t_2 - t_1)$$

where

$$t_2 = 280°F \quad \text{and} \quad t_1 = 180°F$$

and

$$\begin{aligned} h_2 &= 147.99 + 1(280 - 180) \\ &= 247.99 \text{ Btu/lbm} \end{aligned}$$

The steam supplied to the injector is at 500°F and 200 psia so $h_3 = 1269.0$ Btu/lbm from table B.2. Then, using equation (13–14)

$$\frac{n}{m} = \frac{h_1 - h_2}{h_2 - h_3}$$

we get

$$\frac{n}{m} = \frac{147.99 - 247.99}{247.99 - 1269.0} = 0.098 \frac{\text{lbm steam}}{\text{lbm water}}$$

and for 10 lbm of water supplied to the boiler, we say $n + m = 10$ lbm. But we now can use the result from above, $m = n/0.098$, to get

$$n + \frac{n}{0.098} = 10$$

or

$$n = \frac{0.980}{1.098} = 0.892 \text{ lbm of steam} \qquad \qquad \textit{Answer}$$

for every 10 lbm of water supplied.

This device, the steam injector, is irreversible in its operation, yet the internal energy of the steam which is degraded to a lower temperature cannot be counted as lost — it merely recirculates through the boiler or steam generator to be used again. Also, in this example problem we see that the water received by the injector could very likely be coming from the condenser and this operation we would associate with a closed cycle. As we have earlier stated, injectors have commonly been used in open systems to supply feedwater to a boiler.

The nonlifting injector, which we have discussed here, operates with a slight vacuum pressure in the mixing chamber *a*. (See figure 13–11.) With a slightly different physical configuration, the steam and water mixture may have a higher velocity in this chamber with a corresponding decrease in pressure. This arrangement is depicted in figure 13–12 and is called a *lifting injector* since the decreased pressure in the mixing chamber allows water to be drawn up from a well or reservoir below the injector. Of course, this "lifting" of water can only occur if the pressure in the mixing chamber is less than the atmospheric pressure and consequently there is no conceptual difference between lifting and nonlifting injectors.

**13.4
The Typical
Steam Engine**

Now that we have investigated the parts of the steam engine and the thermodynamic cycle which suitably describes the engine, let us look at two examples of the steam engine: an open cycle engine (which fits the description of the museum pieces) and a closed cycle (which has an attractive operation for modern applications).

Example 13.3

A simple double-acting steam engine is required to develop 200 hp at 180 rpm. The boiler can generate superheated steam at 600°F and 300

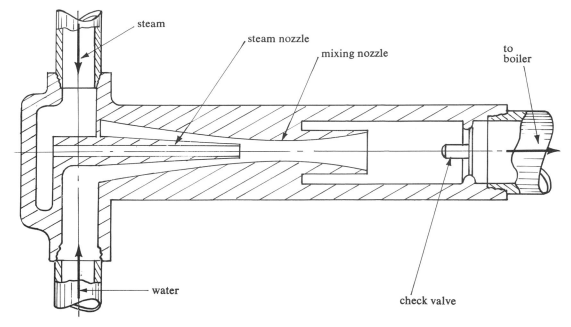

Figure 13–12   Lifting injector.

psia and the piston-cylinder is assumed to expand the steam reversibly and adiabatically to 14.7 psia. Assuming the injection receives water at 60°F and delivers it to the boiler at 185°F and 300 psia, determine the mass flow rates through the cylinder and the boiler, the thermodynamic efficiency, and the recommended bore and stroke for the cylinder. The engine is diagrammed in figure 13–13.

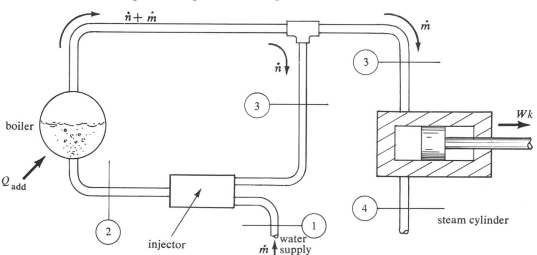

Figure 13–13   The steam engine.

Solution

We obtain the mass flow rate of the steam through the cylinder by first identifying the cylinder as an open system and writing the steady flow energy equation

$$h_4 - h_3 = -wk_{os}$$

This equation results from the assumption

$$\Delta KE = \Delta PE = 0$$

and the statement that the cylinder is adiabatic, $Q = 0$. The power generated is then found from

$$\dot{m}(h_4 - h_3) = -\dot{W}k_{os} = -200 \text{ hp}$$

Notice that we have used as the cycle work the term $h_4 - h_3$ rather than equation (13–6) because the additional enthalpy values h_1 and h_2 do not represent work. That is, the injector does not require work to operate.

We have from table B.2 that $h_3 = 1315.2$ and from table B.1 or the Mollier diagram, $h_4 = 1061$ since $s_3 = s_4 = 1.6274$. Then

$$\dot{m} = \frac{200 \text{ hp}}{(1315.2 - 1061) \text{ Btu/lbm}} \times 42.4 \text{ Btu/min-hp}$$

$$= 33.4 \text{ lbm/min} \qquad\qquad\qquad Answer$$

Using equation (13–15), we obtain the enthalpy of the feedwater h_1. From table B.1 we get $h_f = 180.2$ at a pressure of 14.7 psia and 212°F. Then

$$h_1 = h_f - c(T_f - T_1)$$
$$= 180.2 - 1(212 - 60) = 28.2 \text{ Btu/lbm}$$

Also, we approximate the enthalpy of the mixture entering the boiler h_2 from the relation

$$h_2 - h_1 = c(T_2 - T_1)$$

where $C \approx 1$ Btu/lbm°R. This gives us

$$h_2 = h_1 + c(T_2 - T_1)$$
$$= 28.2 + 1(185 - 60) = 153.2 \text{ Btu/lbm}$$

Using equation (13–14)

$$\frac{n}{m} = \frac{h_1 - h_2}{h_2 - h_3}$$

and the observation that $\dot{n}/\dot{m} = n/m$ we have

$$n = \dot{m}\left(\frac{h_1 - h_2}{h_2 - h_3}\right) = 33.4\left(\frac{28.2 - 153.2}{153.2 - 1315.2}\right)$$

$$n = 3.6 \text{ lbm/min} \qquad \qquad \textit{Answer}$$

The mass flow through the boiler is

$$\dot{n} + \dot{m} = 33.4 + 3.6 = 37.0 \text{ lbm/min}$$

The thermodynamic efficiency is

$$\eta_T = \frac{\dot{W}k_{\text{cycle}}}{\dot{Q}_{\text{add}}} \times 100 \qquad \qquad \textbf{(13–16)}$$

and in our problem, $\dot{W}k_{\text{cycle}} = \dot{W}k_{os} = 200$ hp.
Also

$$\begin{aligned}
\dot{Q}_{\text{add}} &= (\dot{n} + \dot{m})(h_3 - h_2) \\
&= (37.0)(1315.2 - 153.2) = 42{,}994 \text{ Btu/min} \\
&= 1{,}014 \text{ hp}
\end{aligned}$$

From equation (13–16) we then get

$$\eta_T = \frac{200}{1014} \times 100 = 19.7\% \qquad \qquad \textit{Answer}$$

We may now recommend cylinder dimensions by using equation (13–9)

$$\dot{W}k_{\text{cycle}} = (\text{mep})(LAN)$$

and suggest a mean effective pressure of 180 psia. This mep value was reached by our assuming it to be near 60% of the boiler pressure. Then, since we have a double acting cylinder we may assign a value of 180 rpm $\times$ 2 = 360 cycles/min to N. We then get from equation (13–9) that

$$LA = \frac{200 \text{ hp/cycle} \times 33{,}000 \text{ ft-lbf/min-hp}}{360 \text{ cycles/min} \times 180 \text{ lbf/in}^2 \times 144 \text{ in}^2/\text{ft}^2}$$

$$= 0.708 \text{ ft}^3$$

If we use a bore B and stroke L, which are equal, we get

$$L\frac{B^2}{4} = 0.708 \text{ ft}^3 = \frac{L^3}{4}$$

and

$$L = \sqrt[3]{4 \times 0.708} = 1.415 \text{ ft} \qquad \qquad \textit{Answer}$$

If we used more than one cylinder we may reduce the bore and/or stroke, but for our problem the bore and stroke are both equal to 1.415 ft.

Example 13.4   Consider the engine of example 13.3 and assume that the cycle is closed; that is, the steam exhausted from the engine is condensed to a saturated liquid and returned to the boiler by a circulating pump. If the pump is reversible, determine the mass flow rate through the boiler and the thermodynamic efficiency.

Solution

In figure 13–14 is shown the configuration of the cycle. We have replaced the injector by a pump, but this was done primarily to show the diversity of arrangements possible. Of course, the injector would require an additional flow path for the steam. We now can readily identify the work and heat terms from equations (13–2), (13–3), and (13–6).

$$q_{rej} = h_1 - h_4$$
$$q_{add} = h_3 - h_2$$

and

$$Wk_{cycle} = h_3 - h_2 + h_1 - h_4$$

From problem 13.3 we have

$$h_3 = 1315.2 \quad \text{and} \quad h_4 = 1061$$

From table B.1 we obtain $h_1 = 180.2$ since the steam is a saturated liquid at state (1). Since the pump is reversible, we write

$$h_2 - h_1 = -wk_{pump} = \int_{p2}^{p1} v\,dp$$

and, assuming the steam is incompressible with a specific volume v_1 of 0.0167 (table B.1), we get

$$h_2 = h_1 + v\int_{p1}^{p2} dp = h_1 + v(p_2 - p_1) = 180.2 + 0.0167 \text{ ft}^3/\text{lbm}$$
$$(300 - 14.7) \text{ psia}$$

$$= 180.2 + \frac{0.0167 \times 285.3 \times 144}{778} = 181.08 \text{ Btu/lbm}$$

Then, we quickly obtain

$$wk_{cycle} = 1315.2 - 181.08 + 180.2 - 1061$$
$$= 253.32 \text{ Btu/lbm}$$

The power Wk_{cycle} is 200 hp so we write

$$200 \text{ hp} = \dot{m}wk_{cycle} = \dot{m}(253.32 \text{ Btu/lbm})$$

or

$$\dot{m} = \frac{200 \text{ hp}}{253.32 \text{ Btu/lbm}} \times 42.4 \text{ Btu/min-hp} = 33.7 \text{ lbm/min}$$

Answer

This is the mass flow rate which passes through each of the cycle components.

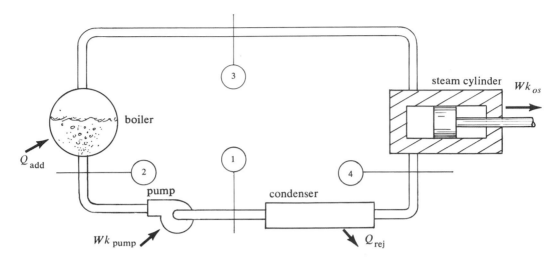

Figure 13–14   Closed cycle steam engine.

Now we may use equation (13–7) to obtain the efficiency;

$$\eta_T = \left(1 - \frac{h_4 - h_1}{h_3 - h_2}\right) \times 100 = \left(1 - \frac{1061 - 180.2}{1315.2 - 181.08}\right) \times 100$$

$$= 22.3\% \qquad\qquad Answer$$

Notice that we have increased the efficiency of the engine by using the closed instead of the open cycle. This increase can be attributed mainly to the energy differences between the condenser water (in the closed cycle) and the feedwater (in the open cycle).

**13.5
Valving**

As we have just seen, we may improve the efficiency of a steam engine by using a closed cycle rather than an open cycle. Other obvious methods for improving the efficiency of the engine are (1) increasing the superheated steam temperature and pressure, (2) increasing the vacuum (or decreasing the absolute pressure) of the region to which the steam is exhausted from the cylinder, and (3) reducing the irreversibilities in operation. Of these three methods, the first two are open-ended and limited only by the practical ability to handle high pressures and temperatures as well as high vacuums. The third method is one which concerned engineers almost from the time the steam engine was conceived. While we found efficiencies of 19% to 22% in examples 13.3 and 13.4, the efficiencies realized by actual steam engines have been closer to 4% to 15%. This pronounced discrepancy between theoretical and actual efficiencies can generally be attributed to the fact that a piston-cylinder does not expand steam reversibly and adiabatically (although higher superheats would make this more nearly realized) and to the fact that the valves are inefficient. We will here consider the common types of valves used in steam engines to emphasize that valve design improvements constituted major developments in the steam engine and would be one of the prime concerns in future designs.

The cylinder shown in figure 13–3 utilizes a valve which oscillates back and forth, shuttling steam through either of two paths. This valve is called a *slide valve* and represents one of the earliest methods of controlling or switching steam flow. It is reliable, simple to operate, and inexpensive, but it generally is limited to low speeds (due to inertia problems) and low steam temperatures. At superheated conditions, the steam will create a considerable lubrication problem in the slide valve.

In order to provide for higher steam temperatures and pressures, the *piston valve*, shown in figure 13–15, was developed. This device accomplished its purpose, but the friction between the valve piston and body was still high.

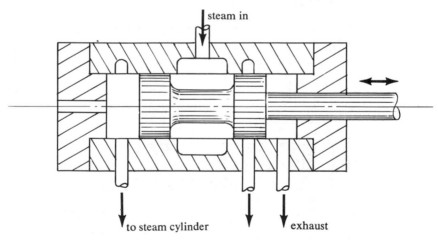

Figure 13-15   Piston valve.

The *Corliss valve* arrangement, shown in figure 13–16, represents an attempt to further reduce inefficiencies in valves. The valves *A*1, *A*2, *B*1, and *B*2 are synchronized, and with a slight rotation they can each be opened or closed. In the position shown, valves *A*1 and *B*2 are open, while the remaining two are closed, and the piston will move to the right.

The devious path taken by the steam, however, in reaching the cylinder still allows for thermal inefficiencies.

The most advanced valve to be used on steam engines has been the *poppet valve* or mushroom valve, shown in figure 13–17. This valve, when used with cams and springs, as in internal combustion engines, can be very efficient and the flow path of the steam may be simplified to reduce thermal losses.

In the future, other valve designs or concepts may very well provide better control and conveyance of steam for more efficient engine operations.

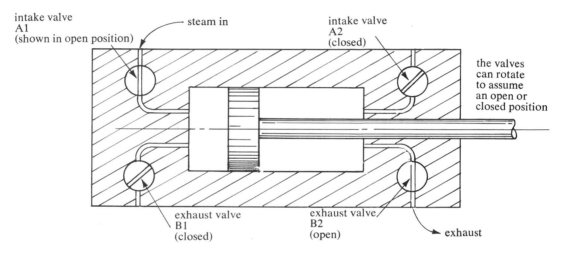

Figure 13–16   Corliss valve.

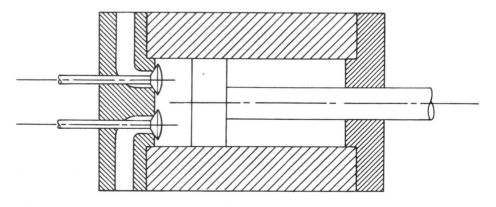

Figure 13–17   Poppet valves.

**13.6
Applications**

The steam engine enjoyed a long period of extensive use. It challenged and replaced the animals in power production and was the single most important heat engine for social needs for almost 100 years. It has been used in powering ground and water transportation vehicles and various stationary power devices and in producing electric power. The reasons it was replaced by other heat engines can be listed as follows:

(1) Reciprocating motion which tends to create vibrations, non-smooth motion, and maintenance problems.
(2) Low thermodynamic efficiency.
(3) Inconvenience due to a slow start-up time.

The inherent advantages of the steam engine which made it one of the first developments of the industrial revolution and which could return it to the prominence are as follows:

(1) Simple and inexpensive construction.
(2) Reliable operation.
(3) Potential of high power outputs at slow speeds.
(4) Relatively low engine weight to output power ratio.
(5) Efficient fuel combustion.
(6) Adaptability to any heat source.

Further, with imaginative designing, the start-up time can be reduced to a matter of a few seconds. This can be achieved by reducing the mass of water in the steam generator at any time, and concentrating the area of heat transfer to small regions. Therefore, it seems that while the steam engine may not replace the large steam turbine cycles or other large engines, it may very well be the engine for small and medium power demands in vehicles of the future.

Practice Problems

Sections 13.1 and 13.2

13.1. Steam at 200 psia and 400°F is supplied to a steam engine cylinder. If the exhaust steam is at 14.7 psia and 90% quality, determine the maximum work obtainable from the engine per unit mass of steam.

13.2. Water at 60°F and 200 psia is supplied to a boiler. Determine the heat added per lbm water to produce steam at 200 psia and 440°F if the specific heat of water is assumed to be 1 Btu/lbm°F between 60°F and 381°F at 200 psia.

13.3. A boiler is supplied 2 lbm/s of saturated liquid water at 300 psia. Determine the rate of heat added to produce steam at 300 psia and 500°F.

13.4. A steam engine is furnished 600°F steam at 160 psia. If the engine expands the steam reversibly and adiabatically to 50 psia, determine the work produced per lbm of steam.

13.5. Determine the work per lbm steam produced by a steam engine in expanding steam from 150 psia and 500°F to saturated vapor at 20 psia.

13.6. A steam engine runs at 3000 rpm, has a 2-inch bore, a 2.2-inch stroke, and a mean effective pressure of 110 psia. Calculate the power produced.

13.7. Determine the speed required of a steam engine to produce 20 hp with a bore of 6 inches and a stroke of 5 inches. Assume the average cylinder pressure cannot exceed 120 psia.

13.8. An engine diagram encloses an area of 6500 ft-lbf for a steam engine running at 1000 rpm. Determine the power produced, the mep, and the dimensions of the bore and stroke (assuming they are equal) if the volume swept by the piston is 4 ft³.

Section 13.3

13.9. List the major components of a steam engine cycle and their functions.

13.10. Define *cutoff*.

13.11. Sketch the *p-V* and *T-s* diagrams of a steam engine.

13.12. A nonlifting injector supplies 30 lbm of 190°F water at 100 psia to a boiler. If the water supplied to the injector is at 14.7 psia and 170°F, determine the rate of steam required if the steam is at 190 psia and 450°F.

Section 13.4

13.13. A simple, double-acting steam engine operates at 150 rpm and uses 1 lbm/s of steam. The water furnished to the boiler by an injector is at 180°F and 150 psia. The steam delivered to the engine is at 500°F and the steam is expanded reversibly and adiabatically to 14.7 psia pressure. Assume the feedwater is at 65°F and 14.7 psia.
Determine
(a) Power produced.
(b) Rate of heat addition.
(c) Thermodynamic efficiency.

13.14. A closed cycle, ideal steam engine having a bore of 6 cm and a stroke of 7 cm is required to deliver 95 hp at 1500 rpm. If steam is furnished at 200 psia and 500°F and is expanded reversibly and adiabatically to 30 psia, and the steam leaving the condenser is saturated liquid, determine
(a) Mass flow rate of steam.
(b) Thermodynamic efficiency.
(c) mep of piston.
(d) Sketch of the *p-V* and *Ts* diagrams.

14

Refrigeration Cycles

In this chapter we will consider the operation of the devices which provide the refrigeration and cooling for society: the freezer, the refrigerator, and the air conditioner. These devices can all be classified as heat pumps, which we have defined as reversed heat engines.

We will first consider the *reversed Carnot cycle* in order to introduce the basic concept of refrigeration and some of the terminology associated with it. Next we will look at the *reversed Brayton cycle* to illustrate the mechanics of how a practical cycle might be designed. The *vapor compression cycle* will then be introduced and analyzed with primary emphasis placed on the *dry compression cycle*, rather than on the wet compression cycle. Since the vapor compression is the most common of the actual refrigeration cycles, example problems of a realistic and practical nature will be presented. Also, various types of cycle-working media or refrigerants will be listed and briefly discussed. Another cycle which is used in actual refrigerating devices, the *ammonia absorption cycle*, is described. Finally, a brief description is given for the manner of gas *liquefaction* — one of the more recent applications of the reversed heat engine.

14.1
The Reversed
Carnot Cycle

We have discussed the Carnot cycle as a heat engine operating between two temperature regions and producing work. In reverse, the Carnot cycle transfers energy from a lower temperature region to a higher temperature one; we have already been introduced to this device, called the *heat pump*. With this arrangement, the Carnot cycle is defined by the following four reversible processes:

(1–2) Isothermal heat addition.
(2–3) Adiabatic compression.
(3–4) Isothermal heat rejection.
(4–1) Adiabatic expansion to initial state (1).

The property diagrams are given in figure 14–1 and the system diagram in figure 14–2. Notice that the enclosed areas in the property diagrams must be equal for the reversible cycle so that

$$Wk_{cycle} = Q_{net} = \Sigma Q = Q_{add} + Q_{rej} \qquad (14\text{–}1)$$

For the Carnot engine we have that

$$Q_{add} = T_L \Delta S \qquad Q_{rej} = -T_H \Delta S \qquad (14\text{–}2)$$

Then,

$$Q_{net} = (T_L - T_H)\Delta S \qquad (14\text{–}3)$$

and the coefficient of performance COP defined by equation (8–22)

$$\text{COP} = \frac{\text{heat rejected}}{\text{net cycle work}} = \frac{Q_{rej}}{Wk_{cycle}}$$

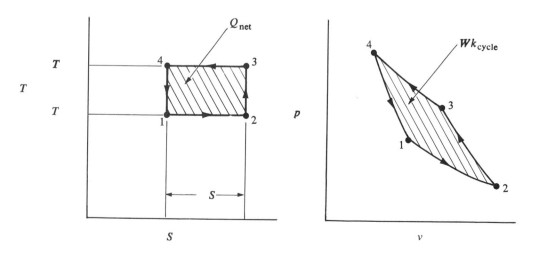

Figure 14–1   Property diagrams for Carnot heat pump.

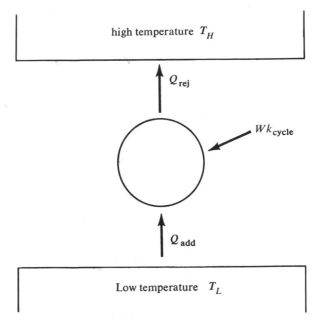

Figure 14–2   System diagram of Carnot heat pump.

reduces to

$$COP = \frac{T_H}{T_H - T_L} \qquad (14\text{–}4)$$

We earlier defined the coefficient of refrigeration COR by equation (8–23)

$$COR = \frac{-Q_{add}}{Wk_{cycle}}$$

which is, for the Carnot heat pump, given by

$$COR = \frac{T_L}{T_H - T_L} \qquad (14\text{–}5)$$

One of the common units for describing the refrigerating capacity of a heat pump device (such as an air conditioner or freezer) is the *ton*. The definition is as follows:

1 Ton of Refrigeration: *Amount of energy removed from 1 ton of water at 32°F and 14.7 psia, in converting from a pure liquid to a solid (ice) over a period of 24 hours.*

It can be seen then that the unit *ton* is a description of the rate of heat transfer and we can give it further meaning by noting that 144 Btu are required to convert 1 lbm of saturated liquid H_2O at 32°F and 14.7 psia into ice at 32°F. Then

$$1 \text{ ton refrigeration} = 144 \text{ Btu/lbm} \times 2000 \text{ lbm/ton} \times 1 \text{ day/24 hours}$$
$$= 12,000 \text{ Btu/hr}$$
$$= 200 \text{ Btu/min}$$
$$= 3.33 \text{ Btu/s}$$

Example 14.1

A room needs to be kept at 72°F when the atmospheric temperature is 102°F. If the room requires a $2\frac{1}{2}$-ton air conditioner, determine the minimum amount of power required and the minimum amount of heat put in the atmosphere.

Solution

We note that the best device conceivable for air conditioning would be a reversible Carnot heat pump. For this device the coefficient of refrigeration is, from equation (14–5)

$$\text{COR} = \frac{72 + 460}{(120 + 460) - (72 + 460)}$$
$$= 11.1$$

Then, since the heat added $\dot{Q}_{\text{add}}$ is given as $2\frac{1}{2}$ tons we can say

$$\dot{Q}_{\text{add}} = 2.5 \text{ tons} \times 200 \text{ Btu/min-ton} = 500 \text{ Btu/min}$$

and, from equation (8–23)

$$\text{COR} = 11.1 = \frac{-\dot{Q}_{\text{add}}}{\dot{W}k_{\text{cycle}}}$$

we get

$$\dot{W}k_{\text{cycle}} = \frac{-\dot{Q}_{\text{add}}}{11.1} = \frac{-500}{11.1} = -45.1 \text{ Btu/min}$$

We can recall that 42.4 Btu/min = 1 hp, to obtain

$$\dot{W}k_{\text{cycle}} = \frac{-45.1 \text{ Btu/min}}{42.4 \text{ Btu/min-hp}} = -1.06 \text{ hp} \qquad \textit{Answer}$$

The amount of heat dumped into the atmosphere is found from equation (14–1)

$$Q_{\text{rej}} = Wk_{\text{oycle}} - Q_{\text{add}}$$

or

$$\dot{Q}_{\text{rej}} = \dot{W}k_{\text{cycle}} - \dot{Q}_{\text{add}}$$

Substituting values in this equation, we get

$$\dot{Q}_{\text{rej}} = -45.1 \text{ Btu/min} - 500 \text{ Btu/min} = -545.1 \text{ Btu/min}$$

Interestingly, although the room is cooled for a time, the net effect of an air conditioner or any other heat pump is to increase the temperature of the universe, which includes the room itself.

14.2
The Reversed
Brayton Cycle

While the Carnot heat pump theoretically represents a most attractive device, a practical Carnot cycle is generally not convenient. As a result, a reversed Brayton cycle (see chapter 11 for the Brayton power cycle) has often been proposed as a practical heat pump. The Brayton cycle normally functions with air or some perfect gas as a working medium so that the reversed cycle is sometimes called a *gas refrigerating* cycle. The property diagrams and the schematic of the physical components are shown in figures 14–3 and 14–4 respectively. As in the case for a Carnot heat pump, the enclosed areas on the *T-s* and *p-V* diagrams are equal if the processes are all reversible and we then have

$$Wk_{\text{cycle}} = Q_{\text{net}} = Q_{\text{rej}} + Q_{\text{add}} \qquad (14\text{–}6)$$

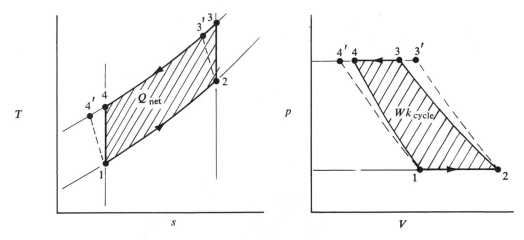

Figure 14–3   Property diagrams of ideal gas refrigeration cycles.

For the ideal gas refrigeration cycle, the processes (2–3) and (4–1) are reversible adiabatic, but in actual applications they are polytropic processes, slightly different from the adiabatic case. For this deviation we may visualize the processes to be (1–2), (2–3'), (3'–4'), and (4'–1) as shown in figure 14–3. Quite generally,

$$Wk_{\text{cycle}} = Wk_{\text{expander}} + Wk_{\text{compressor}} \qquad (14\text{–}7)$$

and if the gas circulating through the device is a perfect gas we may use the polytropic relations from chapter 6

$$Wk_{\text{comp}} = \frac{-n}{n-1}(p_3V_3 - p_2V_2) \qquad (14\text{–}8)$$

$$Wk_{expand} = \frac{-n}{n-1}(p_1V_1 - p_4V_4) \qquad (14\text{--}9)$$

to calculate the cyclic work. Also, the net heat of the cycle will be given by

$$Q_{net} = Q_{add} + Q_{rej} + Q_{expand} + Q_{comp} \qquad (14\text{--}10)$$

where the last two terms represent the heat transferred during polytropic processes in the expander and compressor.

For the ideal cycle, with reversible adiabatic processes through the compressor and expander, we get for any perfect gas that

$$Wk_{comp} = \frac{-k}{k-1}(p_3V_3 - p_2V_2) \qquad (14\text{--}11)$$

and

$$Wk_{expand} = \frac{-k}{k-1}(p_1V_1 - p_4V_4) \qquad (14\text{--}12)$$

which reduce to

$$Wk_{comp} = mc_p(T_3 - T_2) \qquad (14\text{--}13)$$

and

$$Wk_{expand} = mc_p(T_1 - T_4) \qquad (14\text{--}14)$$

for perfect gases with constant specific heats.

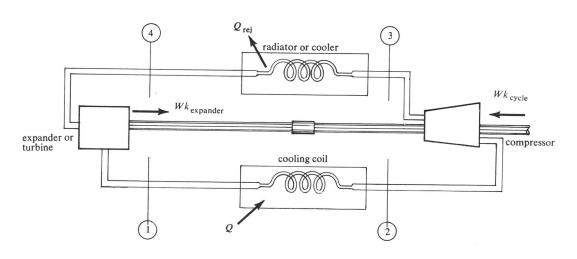

Figure 14–4   Schematic of gas refrigeration cycle.

Example 14.2

An ideal gas refrigeration cycle describes the operation of a 3-ton air conditioner which keeps a room at 68°F while the surrounding atmosphere is at 105°F. Assume the radiator and cooling coil operate with perfect heat transfer ($t_2 = 68°F$ and $t_4 = 105°F$) and the pressure ratio across the compressor is 15 to 1. Determine the rate of heat addition from the room, the rate of heat rejection to the atmosphere, the power required, the mass flow rate of air, the COP, and the COR.

Solution

The rate of heat addition from the room $\dot{Q}_{add}$ is identified as the capacity of the air conditioner

$$\dot{Q}_{add} = 3 \text{ tons} = 3 \text{ tons} \times 200 \text{ Btu/min-ton}$$
$$= 600 \text{ Btu/min} \qquad \qquad \text{\textit{Answer}}$$

From this result and the equation

$$\dot{Q}_{add} = \dot{m}q_{add} = \dot{m}(h_2 - h_1) = \dot{m}c_p(T_2 - T_1) \qquad \textbf{(14–15)}$$

we can obtain the mass flow rate. First we note that, for the reversible adiabatic expansion

$$\frac{T_4}{T_1} = \left(\frac{p_4}{p_1}\right)^{(k-1)/k}$$

and then

$$T_1 = T_4 \left(\frac{p_1}{p_4}\right)^{(k-1)/k} = (105 + 460)\left(\frac{1}{15}\right)^{0.4/1.4}$$
$$= 260°R$$

In this calculation, k was assumed to have the value 1.4. From this result and assuming a value of 0.24 Btu/lbm°R for c_p, we obtain from equation (14–15) that

$$600 \text{ Btu/min} = \dot{m}(0.24 \text{ Btu/lbm-°R})(68 + 460 - 260°R)$$

or

$$\dot{m} = \frac{600 \text{ Btu/min}}{(0.24)(268) \text{ Btu/lbm}}$$
$$= 9.62 \text{ lbm/min} \qquad \qquad \text{\textit{Answer}}$$

The rate of heat rejection to the surroundings $\dot{Q}_{rej}$ is obtained from

$$\dot{Q}_{rej} = \dot{m}q_{rej} = \dot{m}c_p(T_4 - T_3) \qquad \textbf{(14–16)}$$

First we calculate the temperature T_3 from

$$\frac{T_3}{T_2} = \left(\frac{P_3}{P_2}\right)^{(k-1)/k} \quad \text{or} \quad T_3 = T_2 \left(\frac{P_3}{P_2}\right)^{(k-1)/k}$$

so that

$$T_3 = (528°R)(15)^{0.4/1.4} = 1146°R$$

Then,

$$\dot{Q}_{rej} = (9.62 \text{ lbm/min}) \times (0.24 \text{ Btu/lbm°R}) \times (565 - 1146)°R$$
$$= -1341 \text{ Btu/min} \qquad \qquad \textit{Answer}$$

From equation (14–6) we calculate the net power of the cycle,

$$\dot{W}k_{cycle} = \dot{Q}_{rej} + \dot{Q}_{add}$$
$$= -1341 + 600 = -741 \text{ Btu/min} \qquad \textit{Answer}$$

The coefficient of performance is given by equation (8–22)

$$COP = \frac{-1341}{-741} = 1.81 \qquad \qquad \textit{Answer}$$

and the coefficient of refrigeration by equation (8–23)

$$COR = \frac{-600}{-741} = 0.810 \qquad \qquad \textit{Answer}$$

Notice that the Carnot heat pump operating between 68° and 105°F would have COP and COR values obtained from equations (14–4) and (14–5)

$$COP = \frac{105 + 460}{105 - 68} = 15.27$$

and

$$COR = \frac{68 + 460}{105 - 68} = 14.27$$

A quick comparison of the gas refrigeration cycle and the Carnot heat pump reveals that the COP and COR values are much lower for the gas refrigeration cycle. This is the primary reason that the reversed Brayton or gas refrigeration cycle is not more attractive. It can be used to refrigerate at very low temperatures and it is flexible enough to be used with various working media, but the low efficiency, as reflected in the COP and COR values, detracts from its popularity.

**14.3
The Vapor
Compression
Cycle**

The Carnot heat pump represents the ultimate in refrigeration cycles. A practical method of approaching this pump is with the vapor compression cycle, defined by the following processes:

(1–2) Constant pressure heat addition during a phase change of the working medium, or refrigerant.

(2–3)   Adiabatic compression.

(3–4)   Constant pressure heat rejection.

(4–1)   Throttling expansion at constant enthalpy.

If the cycle is considered reversible, then all these processes must also be reversible. The typical pressure-volume diagram for the ideal reversible vapor compression cycle is depicted in figure 14–5. The corresponding temperature-entropy diagram is shown in figure 14–6 (a) and (b). Notice that a vapor compression can be performed with dry vapor (superheated)

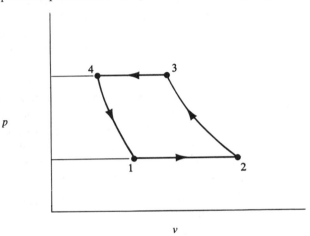

Figure 14–5   *p-V* diagram for ideal vapor compression refrigeration cycle.

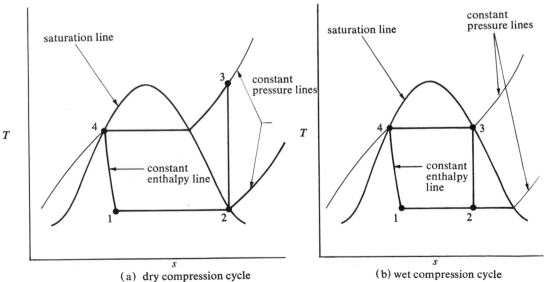

(a) dry compression cycle   (b) wet compression cycle

Figure 14–6   *T-s* diagram for ideal refrigeration cycles.

or a mixture of saturated vapor and liquid. As the descriptions of the cycles imply, the dry compression cycle involves a compression process (2–3) with a dry vapor, while the wet vapor compression cycle involves a mixture of vapor and liquid through the compression.

The dry compression cycle seems to be more popular in application to actual refrigeration devices even though the wet compression more closely approximates the reversed Carnot cycle; that is, the COP of the wet compression cycle would be expected to exceed the COP of the dry compression, both operating between the same pressures. The reason for the success of the dry compression cycle is that compressors typically perform better with a pure vapor than with a mixture of vapor and liquid.

Another tool often used to evaluate and analyze vapor compression cycles is the *pressure-enthalpy diagram*. In figure 14–7 are shown these diagrams for normal dry and wet vapor compression refrigeration cycles with the saturation line again determining the limits of the cycle as it does in the *T-s* diagram. From this observation, we may expect that a critical decision in the design of a refrigeration unit is in the selection of the working medium or *refrigerant*.

Refrigerants are normally selected for vapor compression cycles by the following criteria:

(1) Economical.
(2) Nontoxic or harmless to the surroundings.
(3) Nonflammable.

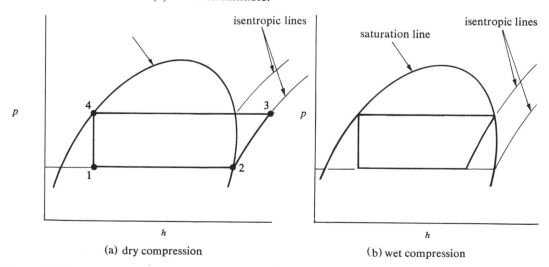

(a) dry compression   (b) wet compression

Figure 14–7   Pressure-enthalpy diagram for vapor compression refrigeration cycle.

(4) High latent heat of condensation (h_{fg}) at the refrigerating temperature.

(5) Low saturation pressure at operating temperature.

Thermodynamically, criterion (4) represents the most significant aspect of working media. A high latent heat of condensation reflects a capability for a high amount of heat addition to the cycle refrigerant per unit of refrigerant mass. In table 14–1 are listed the heats of condensation h_{fg} for some common refrigerants at typical saturation pressures, in order that a quick comparison can be made. Remember, however, that the heat of vaporization alone should not determine the most desirable refrigerants. Even some properties other than those listed above, such as viscosity, solubility, or thermal conductivity, could enter into the requirements for selecting a working medium for refrigeration cycles.

Table 14–1

Heat of Vaporization of Some Common Refrigerants

Refrigerant	Formula	Saturation Pressure, Psia	Saturation Temperature °F	h_{fg} Btu/lbm
Ammonia	$\overline{N}H_3$	48.21	20	533.10
Sulfur Dioxide	SO_2	17.18	20	165.30
Carbon Dioxide	CO_2	45.87	−40	136.50
Methyl Chloride	CH_3Cl	29.30	20	178.40
Freon-12	CCl_2F_2	35.75	20	67.94
Freon-22	$CHClF_2$	34.75	−5	94.89
Freon-22	$CHClF_2$	57.72	20	90.54

The most common substances used for refrigeration processes are as follows:

> *ammonia*
> *sulfur dioxide*
> *carbon dioxide*
> *methyl chloride*
> *dichlorodifluoromethane (Freon-12)*
> *chlorodifluoromethane (Freon-22)*
> *propane*
> *butane*
> *dichloromethane (carrene)*

In the appendix are tables of thermodynamics properties of ammonia,

Freon-12, and Freon-22 for quantitation analysis of the vapor compression cycles.

The typical physical components which comprise a vapor compression refrigeration cycle are shown in figure 14–8. Here we see that an evaporator or cooling coil absorbs energy or accepts heat addition from an external region by utilizing a working medium during its phase change. (See figure 14–6 or 14–7.) The refrigerant collects energy and is a saturated vapor (or near to that state) as it enters the compressor. The compressor increases the pressure and, more importantly, the temperature of the refrigerant as it leaves. Here the working medium is probably a superheated vapor or, in the case of a wet vapor compression, a saturated vapor. The condenser allows the refrigerant to release much of its energy in a heat rejection. Then the refrigerant returns through an expander of restriction to the evaporator.

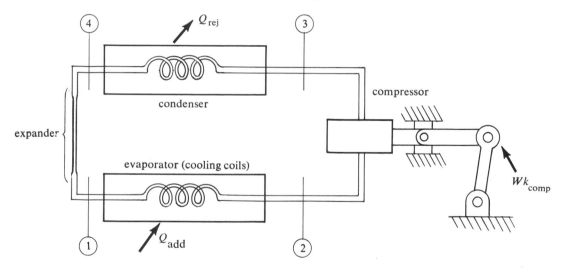

Figure 14–8   Sketch of typical vapor compression refrigerator.

By applying the steady flow energy equation to the various components, we get (making appropriate assumptions on the individual items)

(a)   for the evaporator:

$$h_2 - h_1 = q_{add} \qquad (14\text{--}17)$$

(b)   for the compressor:

$$h_3 - h_2 = -wk_{comp} \qquad (14\text{--}18)$$

(c)   for the condenser:

$$h_4 - h_3 = q_{rej} \qquad (14\text{--}19)$$

(d)   and for the expander:

$$h_1 - h_4 = 0 \qquad (14\text{--}20)$$

Then, for the complete cycle we apply equation (8–22)

$$\text{COP} = \frac{Q_{\text{rej}}}{Wk_{\text{cycle}}} = \frac{q_{\text{rej}}}{wk_{\text{cycle}}}$$

which can be written as

$$\text{COP} = \frac{h_4 - h_3}{h_2 - h_3} \qquad (14\text{--}21)$$

for the vapor compression cycle. Also, the coefficient of refrigeration is

$$\text{COR} = -\frac{h_2 - h_1}{h_2 - h_3} \qquad (14\text{--}22)$$

here.

**14.4
Applications of
the Vapor
Compression
Cycle**

Attention is given here to the details of the application of the thermodynamic concepts to actual devices. First we will look at a refrigeration device which can be described by a wet vapor compression cycle, followed by a similar arrangement which is best described by a dry compression cycle.

Example 14.3

A refrigerator operates with ammonia as the working medium and the system removes 1000 Btu per hour from a freezer compartment at 26°F. If the heat is rejected to a room which is at 85°F and the refrigerator can be described by an ideal wet vapor compression cycle with saturated vapor ammonia leaving the compressor, determine the following: (a) COP and COR, (b) flow rate of the ammonia through the cycle, and (c) the power required.

Solution

This refrigerator can be described by the sketch given in figure 14–8 and the property diagrams for the cycle as those given in figures 14–5, 14–6(b), and 14–7(b).

(a)   From the ammonia table, table B.11 in the appendix, we obtain directly, or through interpolation, the values

$$h_3 = h_g @ 85°F = 631.4 \text{ Btu/lbm}$$
$$h_4 = h_f @ 85°F = 137.8 \text{ Btu/lbm}$$
$$h_1 = h_4 \text{ and } s_3 = 1.1919 \text{ Btu/lbm°R}$$

Since $s_2 = s_3$, we can write

$$s_2 = \chi(s_{g2} - s_{f2}) + s_{f2} = 1.1919$$

Also, from table B.11 we have, since $t_2 = 26°F$

$$s_{g2} = 1.2861$$
$$s_{f2} = 0.1573$$

and then we can calculate the quality χ

$$\chi = \frac{1.1919 - 0.1573}{1.2761 - 0.1573} = 0.925$$

Hence,

$$h_2 = \chi(h_{fg2}) + h_{f2}$$
$$= (0.925)(548.1) + 71.3 = 578.2$$

then, from equation (14–17)

$$q_{add} = h_2 - h_1 = 578.2 - 137.8 = 440.4 \text{ Btu/lbm}$$

and from equation (14–19)

$$q_{rej} = h_4 - h_3 = 137.8 - 631.4 = -493.6 \text{ Btu/lbm}$$

From equations (14–21) and (14–22) we get

$$\text{COP} = \frac{-493.6}{578.2 - 631.4} = 8.64 \qquad \textit{Answer}$$

and

$$\text{COR} = -\frac{440.4}{578.2 - 631.4} = 7.71 \qquad \textit{Answer}$$

(b) From the given condition that

$$\dot{Q}_{add} = 1000 \text{ Btu/lbm}$$

we can then write

$$\dot{Q}_{add} = \dot{m}q_{add} = \dot{m}(440.4 \text{ Btu/lbm})$$

Then

$$\dot{m} = \frac{1000 \text{ Btu/hr}}{440.4 \text{ Btu/lbm}} = 2.27 \text{ lbm/hr} \qquad \textit{Answer}$$

(c) The power required can be determined from the relationship

$$\dot{W}k_{cycle} = \dot{m}wk_{cycle}$$
$$= (2.27 \text{ lbm/hr})(578.2 - 631.4 \text{ Btu/lbm})$$
$$= 120.8 \text{ Btu/hr}$$

or

$$\dot{W}k_{cycle} = (120.8 \text{ Btu/lbm})(1/2545 \text{ Btu/hp-hr})$$
$$= 0.047 \text{ hp} \qquad\qquad\qquad \textit{Answer}$$

Example 14.4 An air conditioner operates in 110°F weather to keep a room at 70°F by withdrawing 12,000 Btu/hour of heat. Assuming that the evaporator and condenser have perfect heat conduction, that the cycle is an ideal dry vapor compression cycle, and that the working medium is Freon-22, determine the following: (a) rate of heat rejected to the atmosphere, (b) power required, and (c) the COP and COR.

Solution The air conditioner is physically describable by the sketch in figure 14–8 and the operating cycle is depicted by the property diagrams in figures 14–7(a), 14–6(a), and 14–5. To determine the heat rejection, the power, and the coefficient of performance, we must determine the enthalpy values. From table B.9 in the appendix we obtain the following values:

$$h_2 = h_g @ 70°F = 110.4$$
$$s_2 = 0.21456 = s_3$$

and the pressure at state (3) must be the saturation pressure at 110°F namely 241 psia. Then $h_3 = 116$ (approximately, from interpolating in the superheated tables). Also,

$$h_4 = h_{f4} @ 110°F = 42.45 = h_1$$

We can now proceed to calculate the requested values.

(a) The heat rejected is obtained from equation (14–19)

$$q_{rej} = h_4 - h_3 = 42.45 - 116 = -73.55$$

and the rate of rejection from

$$\dot{Q}_{rej} = \dot{m}q_{rej}$$

The mass flow is calculated from,

$$\dot{Q}_{add} = \dot{m}q_{add} = \dot{m}(h_2 - h_1) = 12,000 \text{ Btu/hr}$$

or

$$\dot{m} = \frac{12,000 \text{ Btu/hr}}{110.4 - 42.45} = 176.6 \text{ lbm/hr}$$

so that

$$\dot{Q}_{rej} = (176.6)(-73.55) = -12,989 \text{ Btu/hr} \qquad \textit{Answer}$$

(b) The power is obtained from

$$\dot{W}k_{cycle} = \dot{m}wk_{cycle} = \dot{m}(h_2 - h_3)$$

$$\dot{W}k_{\text{cycle}} = (176.6)(110.4 - 116) = -989 \text{ Btu/hr}$$

or

$$\dot{W}k_{\text{cycle}} = 989 \times \frac{1}{2545} = 0.389 \text{ hp} \qquad \textit{Answer}$$

(c)   From equations (14–21) and (14–22) we calculate

$$\text{COP} = \frac{-12,989}{-989} = +13.13 \qquad \textit{Answer}$$

and

$$\text{COR} = \frac{-12,000}{-989} = 12.13 \qquad \textit{Answer}$$

14.5
Ammonia
Absorption
Refrigeration

The ammonia absorption cycle represents an attempt to reduce the mechanical work required to compress refrigerants and replace this energy demand by a heat transfer process. The typical cycle, shown in figure 14–9, indicates that the condenser, the expansion value, and the evaporator are components having identical functions in the ammonia absorption and vapor compression cycles. That is, the condenser serves to remove heat from the system; the expansion valve allows isenthalpic (constant enthalpy) expansion of the ammonia to a low pressure; and the evaporator allows for heat addition to the low pressure ammonia from a region being refrigerated or cooled. In the ammonia absorption cycle, the compressor is replaced by an auxiliary water system. The absorber receives a low pressure mixture predominately of water with some ammonia (weak liquor) and low pressure ammonia vapor. These two fluids are mixed together in the absorber so that the ammonia is "absorbed" or dissolved in the weak liquor. The fluid flowing out of the absorber is then a strong solution of water and ammonia (strong liquor) and is pumped to a generator under a high pressure. Heat is supplied to the generator by steam, electric coils, or other appropriate means, which increases the temperature of the water-ammonia liquor. A consequence of this is that ammonia vapor boils off and recirculates to the condenser, while the remaining liquor returns to the absorber. All of this operates because ammonia is more soluble in cold water than in hot water, and the absorber and the generator provide the means of taking advantage of this phenomenon.

For a complete analysis of the ammonia absorption refrigeration cycle, the reader is referred to publications specifically concerned with refrigeration.

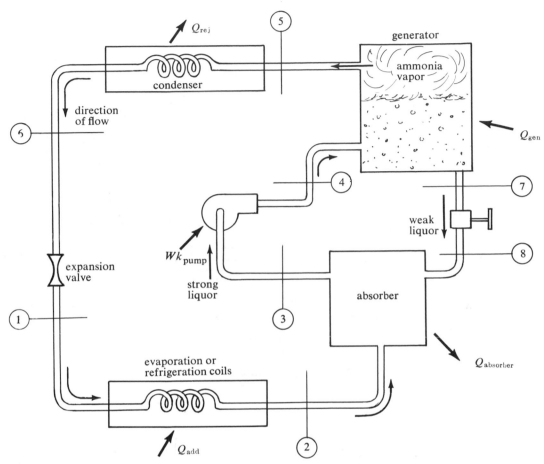

Figure 14-9   Ammonia absorption refrigeration cycle.

**14.6
Gas
Liquefaction**

One of the most recent applications of the heat pump is to the *liquefaction* or condensation of gases. This requires that the gases be brought to very low temperatures so that the phase change from gas to liquid may begin. Table 14–2 lists the boiling temperature t_B or temperature at which the listed gas exhibits the liquid-gaseous phase change at 1 atmosphere of pressure (14.7 psia) and it can easily be seen that very low temperatures are required.

The most common and typical method of producing liquid gases (such as oxygen, hydrogen, and air) is depicted in figure 14–10. Here the gas itself is a working medium (refrigerant) and is converted to a saturated liquid through an expansion valve after losing much of its energy through

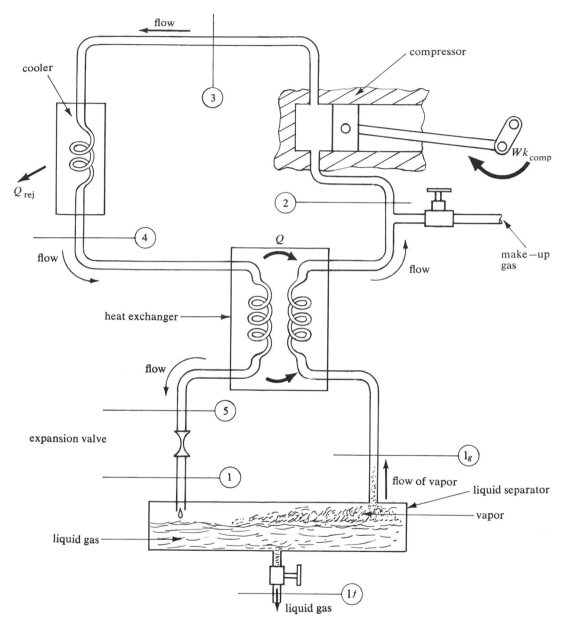

Figure 14–10   Schematic of gas-liquefaction process.

heat transfer at a high pressure and a relatively high temperature. The heat exchanger has the sole function of transferring energy from the high pressure gas to the low pressure gas through heat transfer, but frequently more than one heat exchanger is used in the actual cycle.

Table 14–2

Saturation Temperature of Gases at 14.7 psia

Substance	Saturation Temperature t_B-°F
Air	—317.6
Helium	—352.0
Hydrogen	—422.9
Nitrogen	—320.4
Oxygen	—297.4
Chlorine	—30.3
H_2O (water)	212.0

Note: Condensed from the International Critical Tables with permission of the National Academy of Sciences

In figure 14–11 is presented the temperature-entropy diagram of the gas liquefaction process, assuming reversible and ideal situations for all the components. In this diagram the temperature at state (4) (entering the high pressure side of the heat exchanger) is indicated as being equal to that at state (2), but in actual cases T_4 must be greater than T_2 if heat transfer is to be effected over a finite period of time. Also, at state (5) the gas may not be a saturated liquid but may be a mixture of liquid and vapor, a saturated vapor, or even somewhat superheated. However, if the whole cycle is to have meaning, the expansion (5–4) through the

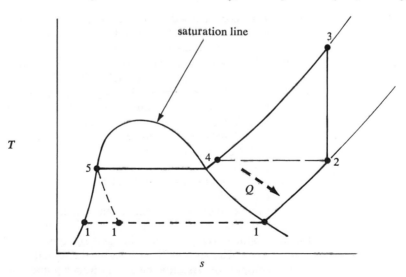

Figure 14–11   *T-s* diagram of gas liquefaction.

expansion valve must result in a mixture of liquid and vapor. This allows for the separation of the mixture and appearance of liquid gas at low pressure.

Practice Problems

14.1. A Carnot heat pump operates between a high temperature region of 80°C and a low temperature region of −10°C. Determine the COP and COR.

14.2. A reversed Carnot cycle operates between 100°F and 10°F. Determine the COR and COP.

14.3. A reversible Carnot refrigerator receives heat at 25°F and requires 3 horsepower for each ton of refrigerating. Determine the COP.

14.4. For the refrigerator of problem 14.3 determine
(a) Temperature of rejection of heat.
(b) COR.

14.5. A Carnot air conditioner has a COP of 4 and a capacity of 5 tons. Determine the power requirement of this device.

14.6. A 60-ton Carnot freezer operates between −10°C and 30°C. Determine the rate of heat rejected to the atmosphere and the power required.

14.7. A Carnot freezer is used to freeze water at 14.7 psia. What is the maximum COR achieved by this device if the minimum atmospheric temperature is 28°C?

14.8. Sketch the p-V, T-s, and h-s diagrams of the ideal Carnot cycle.

14.9. A reversed ideal Brayton cycle refrigerator uses 5000 lbm/hr of air with a pressure ratio of 10 to 1. If the refrigerator is rated at 2 tons, determine the power required by the unit.

14.10. A reversed ideal Brayton cycle refrigerator operates with a pressure ratio of 18 to 1. Determine the COP and COR.

14.11. An ideal gas refrigerator keeps a freezer at −10°C when operating and rejects heat to a surrounding at 35°C. The freezer unit requires a $\frac{1}{2}$-hp electric motor to run a compressor and the pressure ratio of the compressor is 16 to 1 assuming air to be the working medium. Determine
(a) Capacity rating of the freezer in tons.
(b) COP.

14.12. For the unit of problem 14.11 determine
(a) COR.
(b) Rate of heat rejected to surrounding.
(c) Mass flow rate of air.

14.13. An air refrigerating machine uses 100 lbm/min of air to cool a storage room. The air is 75°F as it leaves the radiator and 30°F as it enters the compressor. Assume the air is reversibly and adiabatically compressed from 15 psia to 55 psia in the compressor and expanded reversibly and adiabatically through the turbine. Determine

(a) Temperature of air entering the radiator.
(b) Temperature of air entering cooling coils.
(c) $\dot{Q}_{net}$
(d) COP and COR.
(e) Net work rate to cycle.

14.14. Use the same cycle as in problem 14.13 except that the compression of the air is reversible and polytropically done with $n = 1.34$ and the expansion is reversible and polytropic with $n = 1.45$. Determine
(a) Temperature of air entering the radiator.
(b) Temperature of air entering the cooling coil.
(c) $\dot{Q}_{net}$
(d) COP.
(e) Refrigeration rating of device.

Section 14.3

14.15. Saturated liquid ammonia is throttled through an expansion valve from 90°F to 30.42 psia. Determine the enthalpy and the quality of the ammonia leaving the expansion valve.

14.16. Determine the heat added per pound-mass of ammonia if the ammonia has 20% quality and is at -20°F entering the evaporator and saturated vapor upon leaving.

14.17. At a rate of 6 lbm/sec, Freon-22 enters an evaporator as 12% liquid and at 20 psia. If the freon leaves with a quality of 88%, determine the rate of heat addition.

14.18. An ideal dry compression refrigerator operates on a cycle diagrammed in figure 14–6(a). If the working medium is Freon-12 and the pressures are 15 psia and 50 psia, determine
(a) wk_{comp}
(b) q_{add}
(c) q_{rej}

14.19. An ideal wet compression air conditioner using Freon-22 uses a cycle as diagrammed in figure 14–6(b). If the Freon is at 30°F in the evaporator and 110°F in the condenser, determine
(a) wk_{comp}
(b) q_{add}
(c) q_{rej}

14.20. Calculate the COP for
(a) The cycle of problem 14.18.
(b) The cycle of problem 14.19.

Section 14.4

14.21. A wet compression cycle refrigerator using ammonia as a working medium operates such that the ammonia is 15°F in the evaporator and 90°F in the condenser. Determine the power required per ton of refrigeration.

14.22. Saturated liquid Freon-12 at 75°F is expanded to 40 psia through an expansion valve. It leaves the evaporator at 100% quality and 40 psia. If 12 lbm/min of freon flows through the device, what is its rating? Is this a dry or wet compression cycle?

14.23. An air conditioner rejects heat to air at 150°F while cooling a room at 68°F. If the cycle is an ideal dry vapor compression with Freon-12 as the working medium, and if heat transfer is assumed to be reversible in the condenser and the evaporator, determine

(a) q_{rej}

(b) q_{add}

(c) COP and COR.

(d) Power required per ton of refrigeration.

14.24. A freezer using Freon-22 removes 1500 Btu/hr from a compartment at −8°C. The Freon is at −5°C in the evaporator and the quality is 85% leaving the evaporator. If the freezer operates on an ideal wet compression cycle determine

(a) $\dot{m}$ of freon.

(b) $\dot{Q}_{rej}$

(c) $\dot{W}k_{cycle}$

(d) COP and COR.

15

Mixtures

Various concepts and descriptions for the combining of separate systems will be presented here. We will first consider a mixture of perfect gases and define *partial pressure*, *mass fractions*, and other terms necessary for an accurate, useful description of the system. Then the common mixture of air and water vapor will be considered in terms of the thermodynamics of meteorology. Chemical potential will be defined in an intuitive manner and related to the water-air interaction. To give the reader the barest concept of the mechanism of mixing, *diffusion* will be briefly introduced through *Fick's law* and *mass balance*.

The concluding topic to be considered will be a mixture of chemically active systems. Since the reader is not expected to be well versed in chemical systems, only an outline of the approach in handling chemical reactions can be given. This topic is introduced mainly to show how to determine heats of combustion in the fast chemical reaction of fuels and air or oxygen.

**15.1
The Perfect
Gas Mixture**

Suppose we take 1 ft³ of oxygen at 14.7 psia and 70°F and mix it with 1 ft³ of nitrogen at 14.7 psia and 70°F in a 2-ft³ container. After these two gases have been sufficiently mixed, we notice that each gas is occupying the full volume of 2 ft³ and that the temperature is 70°F. For this final condition then, if we assume both gases are perfect, we can write

$$p_O V = m_O R T$$

for the oxygen, and

$$p_N V = m_N R T$$

for the nitrogen.

Then if we assume an "average" gas constant, R_{av}, for the mixture we have

$$p_O V = m_O R_{av} T \qquad (15\text{--}1)$$

and

$$p_N V = m_N R_{av} T \qquad (15\text{--}2)$$

If we add these two results we get

$$(p_O + p_N)V = (m_N + m_O)R_{av}T$$

but $m_N + m_O$ = mass of the mixture = m. As a consequence we must have that

$$p_O + p_N = \text{pressure of mixture} = p$$

which is known as *Dalton's law of partial pressures*. Using this result,

$$pV = m R_{av} T \qquad (15\text{--}3)$$

and equations (15–1) and (15–2) we get that

$$\frac{p_O}{p} = \frac{m_O}{m} \quad \text{and} \quad \frac{p_N}{p} = \frac{m_N}{m} \qquad (15\text{--}4)$$

so that we call the pressures p_O and p_N the partial pressures of the oxygen and nitrogen, respectively. If we have a multi-component mixture (three or more substances), it is convenient to number each substance or gas for bookkeeping purposes. For example, in air we have oxygen (call it gas 1), argon (call it gas 2), and nitrogen (call it gas 3), so the partial pressure of oxygen would be denoted as p_1. Similarly for argon, the partial pressure would be p_2 and for nitrogen, p_3. Thus we could abbreviate the notation as p_i where i could be 1, 2, or 3, depending on which gas is being considered. Of course, a mixture of four or more substances then requires that i be able to be assigned any of these other numbers. If we consider an "i-th" component or gas, we mean a typical substance in the mixture. The partial pressure of a gas 1 in a certain mixture is given by

$$p_1 = \frac{m_1}{m} p$$

For a gas 2

$$p_2 = \frac{m_2}{m} p$$

and in general, the partial pressure of the *i*-th gas would be

$$p_i = \frac{m_i}{m} p \qquad (15\text{–}5)$$

Also, Dalton's law is then written for the general case as

$$p = \sum_{i=1}^{n} p_i = \sum_{i=1}^{n} \frac{m_i}{m} p = p \sum_{i=1}^{n} \frac{m_i}{m} \qquad (15\text{–}6)$$

where *n* is the number of components. It can be easily seen from the above analysis that the mass fraction m_i/m of the *i*-th component of the mixture determines the partial pressure of that particular component.

Example 15.1

There are 2 lbm of air at 80°F and 15 psia found to be composed of 75.5% (by mass) nitrogen, 23.2% oxygen, and 1.3% argon. Calculate the partial pressures of each of the components.

Solution

Using equation (15–5)

$$p_i = \frac{m_i}{m} p$$

we have that *p* is the air pressure, 15 psia. The partial pressure of nitrogen p_N we then get from

$$p_N = (0.755)(15 \text{ psia}) = 11.325 \text{ psia} \qquad \textit{Answer}$$

Similarly, for oxygen,

$$p_O = (0.232)(15) = 3.480 \text{ psia} \qquad \textit{Answer}$$

and for argon

$$p_{Ar} = (0.013)(15) = 0.195 \text{ psia.} \qquad \textit{Answer}$$

Mixing is an irreversible process and accordingly the total entropy of components increases after they have been mixed. To see this we write the change in entropy of a mixture during the mixing process as

$$\Delta S = \sum_{i=1}^{n} m_i \Delta s_i \qquad (15\text{–}7)$$

where we can recall, from chapter 7

$$\Delta s_i = \int \frac{c_v dT}{T} + R_i \ln \frac{V}{V_i}$$

or, if c_V is constant

$$\Delta s_i = c_V \ln \frac{T}{T_i} + R \ln \frac{V}{V_i} \qquad (15\text{--}8)$$

where T_i and V_i are the temperature and volume of the i-th component before mixing. From equation (15–8) it can be deduced that the sum of the various entropy changes must be positive since generally V is greater than V_i and since the sum of the ratios T/T_i will be near zero. An example should serve to show the details.

Example 15.2   At 70°F and 14.7 psia 1 ft³ of oxygen is mixed with 1 ft³ of nitrogen at 70°F and 14.7 psia. The mixing process occurs in a 2-ft³ container at 70°F. Determine the entropy change due to this process.

Solution   From equation (15–7) we have

$$\Delta S = m_O \Delta s_O + m_N \Delta s_N$$

where "O" and "N" denote the properties of oxygen and nitrogen respectively. From the perfect gas laws we have

$$m_O = \frac{p_O V_O}{R_O T_O} = \frac{14.7 \times 1 \times 144}{48.291 \times 530} = 0.0827 \text{ lbm}$$

and

$$m_N = \frac{p_N V_N}{R_N T_N} = \frac{14.7 \times 1 \times 144}{55.158 \times 530} = 0.0724 \text{ lbm}$$

Using equation (15–8) we have

$$\Delta s_O = c_{VO} \ln \frac{T}{T_O} + R_O \ln \frac{V}{V_O}$$

and

$$\Delta s_N = c_{VN} \ln \frac{T}{T_N} + R_N \ln \frac{V}{V_N}$$

But $T = T_N = T_O$ and $V_O = V_N = V/2$ in this problem so that we obtain from equation (15–7)

$$\Delta S = m_O R_O \ln \frac{V}{V/2} + m_N R_N \ln \frac{V}{V/2}$$

$$= (0.0827)(48.291) \ln 2 + (0.0725)(55.158) \ln 2$$

$$= 5.54 \text{ Btu/°R} \qquad \qquad \textit{Answer}$$

This problem and its solution tells us that the entropy is indeed increased without reversible heat transfer — in fact, the above entropy increase is entirely due to irreversibilities in mixing.

15.2 Water and Air Mixtures

Although air is a mixture of nearly perfect gases, it is commonly considered to be a homogeneous, single component having distinct properties associated with it. These properties, such as the gas constant R, would be the "average" property, as mentioned in the previous section, or the "total" property in the case of pressure. In the cases where we consider air a uniform, homogenous substance, we are assuming it to be dry; that is, no water is mixed with the air. Dry air is generally assumed to be composed of 78% nitrogen, 21% oxygen, and 1% argon by volume. In reality, however, air seeks an equilibrium with its surroundings which are commonly liquid water. (The sea and sky are common neighbors.) When equilibrium is achieved, the rate of evaporation of the liquid water equals the rate of condensation of water vapor (see figure 15–1) — the water vapor being a gas mixed with the air. To describe properly the air-water mixture, we may define, for some precise volume of this mixture, the mass ratio of water vapor to air. This ratio ω is written

$$\omega = \frac{m_V}{m_a} \qquad (15\text{–}9)$$

and is called *specific humidity*, where m_V is the mass of water vapor and

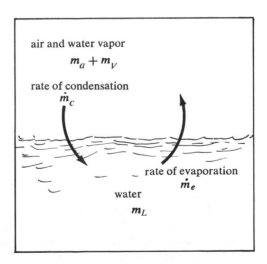

Figure 15–1   Equilibrium of liquid water and air in a closed container.

m_a is the mass of air. If we assume the water vapor acts as a perfect gas then the specific humidity can be written

$$\omega = \frac{p_V V / R_V T}{p_a V / R_a T} = \frac{R_a}{R_V} \frac{p_V}{p_a} = 0.622 \frac{p_V}{p_a} \qquad (15\text{–}10)$$

Now, when water and its vapor are in equilibrium we say the water is in a phase change (the liquid-vapor phase change, in fact). This liquid and vapor mixture is called *saturated* and depending on the quality (see section 12.6 for a definition of *quality*), it is somewhere between a saturated liquid and a saturated vapor. If the water vapor is in a mixture with air and is in equilibrium with liquid water, then the water vapor is a saturated vapor, and we call the mixture *saturated air*.

If the air-water vapor mixture is saturated air, then the partial pressure of the water vapor p_V is equal to the saturation pressure p_g corresponding to the temperature of the air. Use of the saturated steam tables can be made (table B.1) to obtain the corresponding saturation pressure at a given temperature. Air which is not saturated is either dry air or some intermediate condition. To describe the relative condition we define relative humidity β as

$$\beta = \left(\frac{p_V}{p_g}\right) \times 100\% \qquad (15\text{–}11)$$

which gives us the conditions

$$\beta = 100\% \text{ for saturated air } (p_V = p_g)$$

and

$$\beta = 0\% \text{ for dry air } (p_V = 0)$$

Air which has a relative humidity between 100% and 0% is called *unsaturated air* and is a mixture of dry air and superheated water vapor. By using equations (15–10) and (15–11) we see that relative and specific humidities are related by

$$\omega = 0.622 \, \beta \left(\frac{p_g}{p_r}\right) \frac{1}{100} \qquad (15\text{–}12)$$

Unsaturated air could become saturated through a temperature change. The temperature at which an air-water vapor mixture is saturated air is called the *dew point temperature*. In figure 15–2 is shown on a *T-s* diagram an example state *a* which is an unsaturated air. By reducing the temperature at constant partial vapor pressure p_V, we reach the dew point temperature at state *b*.

Another manner of reaching a saturation condition from an unsaturated

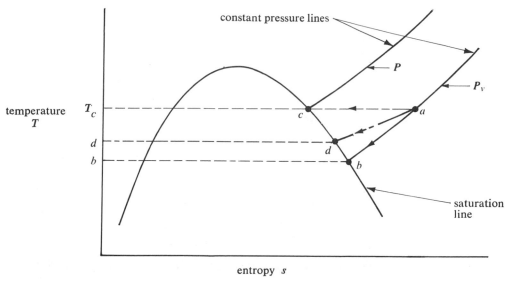

Figure 15–2

Temperature-entropy diagram for water vapor.

state is to increase the partial vapor pressure p_v at constant temperature until the pressure p_g is reached. This process is indicated by the curve *a-c* in figure 15–2. Notice that an increase in vapor pressure p_v is normally associated with a corresponding reduction in partial air pressure p_a since by Dalton's law we must have $p = p_a + p_v$ where p is the atmospheric pressure.

For further descriptions of the air-water mixture, the dry-bulb and wet-bulb temperatures are commonly given. The dry-bulb temperature is that which is recorded in normal atmospheric conditions. The wet-bulb temperature, however, is measured in an atmosphere of saturated air and ideally is measured in a saturated condition. This process, the adiabatic saturation one, is depicted by curve *a-d* in figure 15–2 so that T_d would represent the wet-bulb temperature of air having a dry-bulb temperature T_a. In this process, no heat is transferred out of the air-water vapor mixture (thus the term *adiabatic*), but saturated vapor is added to the mixture to increase the vapor pressure.

The wet- and dry-bulb temperatures are recorded by a psychrometer shown in figure 15–3. The air flow is induced either by moving the psychrometer in the air (which is then called a *sling psychrometer* since it is normally rotated by hand) or by using a fan to force air through a stationary psychrometer. When measured in either of these two manners, however, the wet-bulb temperature is not precisely that which would be achieved

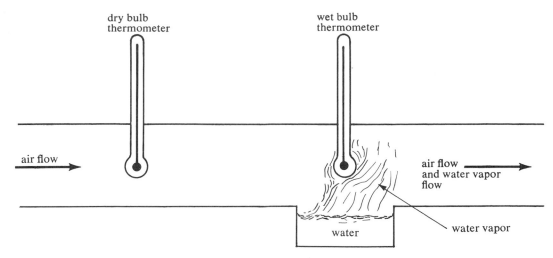

Figure 15-3　　Wet- and dry-bulb psychrometer or sling psychrometer.

through an adiabatic saturation process, but is rather achieved through a saturation process with minor heat transfer.

The various properties of air-water vapor mixtures, which we must consider in describing the normal atmosphere, are correlated on a psychrometric chart given in the appendix. In figure 15–4 the basic elements of this chart are shown in order to assist in understanding the manner of data extraction. The chart given here and in the appendix is based on an atmospheric pressure of 14.7 psia. If the atmospheric pressure

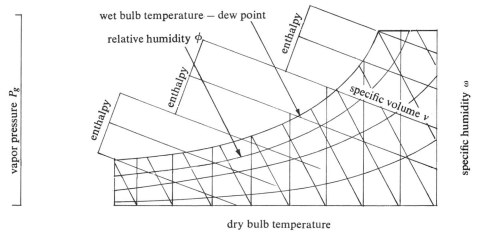

Figure 15-4　　Sketch of the basic parameters constituting a psychrometric chart.

deviates greatly from this value, the chart must be corrected. (See appendix C, references: Lee and Sears, *Thermodynamics* and Mooney, *Mechanical Engineering Thermodynamics*.)

Example 15.3

Unsaturated air is at 80°F. Find the partial pressure required of the water vapor to convert this mixture to saturated air at 80°F.

Solution

Using steam table B.1 or the psychrometric chart, we obtain $p_g = 0.5069$ psia at 80°F so the pressure of the vapor p_V must be 0.5069 psia for the conditions of saturated air at 80°F.

Example 15.4

Atmospheric air at 14.7 psia has a temperature of 85°F and a wet-bulb temperature of 70°F. Determine the vapor pressure, the relative humidity, the specific humidity, and the enthalpy of the atmospheric air.

Solution

Since the atmospheric pressure is standard, we may directly use the data of the psychrometric chart. The dry-bulb temperature can be assumed to be 85°F. Then, using the psychrometric chart we obtain the vapor pressure

$$P_V = 0.286 \text{ psia} \qquad \textit{Answer}$$

The relative humidity can also be directly read from the chart,

$$\beta = 48\% \qquad \textit{Answer}$$

and the specific humidity is

$$\omega = 86 \text{ grains/lbm}$$

Since there are 7000 grains in 1 lbm we get

$$\omega = \frac{86}{7000} \text{ lbm vapor/lbm air} = 0.0123 \qquad \textit{Answer}$$

Also, we should be able to get the above answers to correlate in equation (15–12)

$$\omega = 0.622 \, \beta \, \frac{P_g}{P_a} \, \frac{1}{100}$$

where, from equation (15–11)

$$p_g = \left(\frac{p_V}{\beta}\right) 100$$

$$= \text{saturation pressure at dry-bulb temperature}$$

From table B.1 we have $p_g = 0.5959$ psia at a temperature of 85°F. Then, using the above value of p_v from the psychrometric chart, we have

$$p_g = \frac{0.286 \text{ psia}}{48} \times 100 = 0.596 \text{ psia}$$

which agrees with our result from the saturated steam table. The enthalpy is read directly from the psychrometric chart as

$$h = 34.1 \text{ Btu/lbm air} \qquad \qquad \textit{Answer}$$

The enthalpy is based on the unit of mass of dry air and so the total enthalpy must be determined by using only the mass of dry air, not the mixture mass.

There are numerous reasons for determining the thermodynamic properties of the atmosphere, among which are *humidification* and *dehumidification processes*. The following two example problems will indicate the method of approach to these two processes.

Example 15.5   *Humidification* is a technique of increasing the specific humidity of air to provide human comfort. In figure 15–5 is shown the schematic of a typical humidifier which is also operating as a heater. Assuming that air enters the humidifier at 14.7 psia, 50°F, and 40% relative humidity and should be 72°F and 50% relative humidity upon leaving, determine the amount of heat required in the heating unit and the amount of water required per pound of dry air, assuming the entering liquid water is at 45°F.

Solution   The steady flow energy equation for the humidifier can be written as

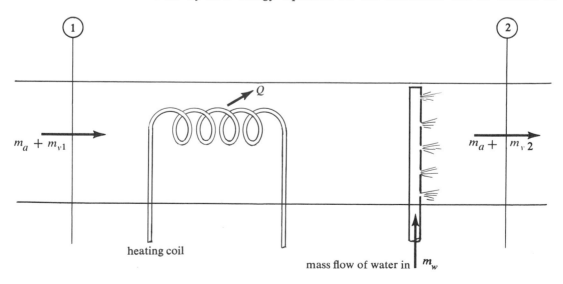

Figure 15–5   Humidifying process.

$$m_a h_2 - (m_a h_1 + m_w h_w) = Q \tag{15-13}$$

and the mass flow equation as

$$\dot{m}_a + \dot{m}_{V1} + \dot{m}_w = \dot{m}_a + \dot{m}_{V2} \tag{15-14}$$

From the psychrometric chart we read

$$h_2 = 26.6 \text{ Btu/lbm dry air}$$

$$h_1 = 15.3 \text{ Btu/lbm dry air}$$

$$\omega_2 = 59 \text{ grains/lbm dry air} \times 1 \text{ lbm vapor/700 grains}$$

$$= 0.0084 \text{ lbm vapor/lbm dry air}$$

and

$$\omega_1 = \frac{21}{7000} = 0.003 \text{ lbm vapor/lbm dry air}$$

Using equation (15–13), we find the heat needed per pound mass of dry air Q/m_a

$$\frac{Q}{m_a} = q = h_2 - h_1 - \frac{m_w}{m_a} h_w$$

$$q = 26.6 - 15.3 - \frac{m_w}{m_a} h_w$$

From equation (15–14) we note that

$$\dot{m}_w = \dot{m}_{V2} - \dot{m}_{V1} = \dot{m}_a \omega_2 - \dot{m}_a \omega_1$$

Then

$$\frac{m_w}{m_a} = \omega_2 - \omega_1 = 0.0084 - 0.003 = 0.0054 \text{ lbm water/lbm dry air}$$

and, from table B.1 we obtain the enthalpy of the 45°F water which is evaporated. We have,

$$h_w = 13.04 \text{ Btu/lbm}$$

Then

$$q = 26.6 - 15.3 - (0.0054)(13) = 11.23 \text{ Btu/lbm dry air}$$

Answer

As we have determined above, we need 0.0054 lbm of water per pound-mass of dry air.

Example 15.6   *Dehumidification* involves the reduction of the specific humidity of air, usually to make surrounding air more comfortable. The process of dehumidification is indicated in the *T-s* diagram of figure 15–6 where air

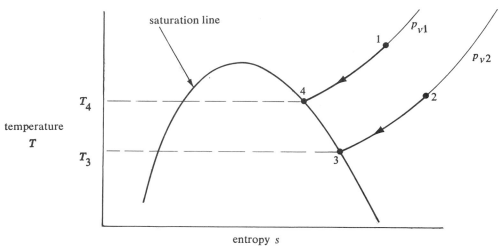

Figure 15–6   Dehumidifying process on a temperature-entropy plane.

at state (1) is reduced in its specific humidity to state (2) by first being cooled to the dew point (4), being condensed and losing moisture to state (3), and being subsequently heated to state (2). The physical operation of this is shown in figure 15–7.

Assume that air is taken in at 14.7 psia, 80°F, and 80% relative humidity. If the air is to be cooled to 72°F and reduced to 60% relative humidity, determine the heat rejected in the cooling unit, the heat added in the

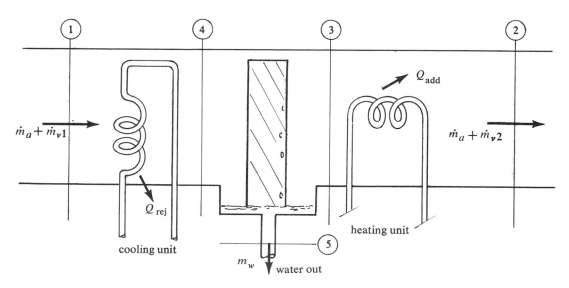

Figure 15–7   Dehumidifying process.

heating unit, and the amount of water condensed per pound-mass of dry air.

Solution

Applying the steady flow energy equation to the cooling and heating units separately, we obtain

$$m_a h_3 + m_w h_5 - m_a h_1 = Q_{rej} \tag{15-15}$$

and

$$m_a h_2 - m_a h_3 = Q_{add} \tag{15-16}$$

Per unit mass of dry air we have

$$h_3 + \frac{m_w}{m_A} h_5 - h_1 = q_{rej} \tag{15-17}$$

and

$$h_2 - h_3 = q_{add} \tag{15-18}$$

From the psychrometric chart we read

$$h_1 = 38.8 \text{ Btu/lbm dry air}$$
$$h_2 = 28.4 \text{ Btu/lbm dry air}$$

and

$$\omega_1 = \frac{125}{7000} = 0.0178 \text{ lbm/lbm dry air}$$

$$\omega_2 = \frac{70}{7000} = 0.010 \text{ lbm/lbm dry air}$$

The amount of water condensed per pound mass of dry air is the decrease in specific humidity,

$$\frac{m_w}{m_a} = \omega_1 - \omega_2 = 0.0178 - 0.010 = 0.0078 \text{ lbm condensate/lbm dry air}$$
Answer

The enthalpy at state (5), the condensed water, is given by $h_5 = h_F$ at the dew point of the incoming air mixture. The dew point is found from the psychrometric chart to be 73.4°F and then $h_F = 41.45 = h_5$ from table B.1.

The method of reading the dew point corresponding to state (1) is indicated in figure 15–8 and in the same manner we obtain the dew point of the dehumidified air at state (2); $T_3 = 57.2$°F and the enthalpy $h_3 = 24.6$. We can now calculate the heat transfer from equation (15–17);

$$q_{rej} = 24.6 + (0.0078)(41.45) - 38.8 = -13.9 \text{ Btu/lbm dry air}$$
Answer

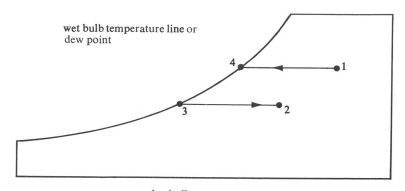

wet bulb temperature line or
dew point

dry bulb temperature

Figure 15–8   Psychrometric chart for dehumidifying process.

and, from equation (15–18)

$$q_{add} = 28.4 - 24.6 = 3.8 \text{ Btu/lbm dry air} \qquad \textit{Answer}$$

15.3 Chemical Potential

Equilibrium is a relative term which requires precise definition to become meaningful. We have earlier seen that two bodies or systems are in *thermal equilibrium* if their temperatures are equal. *Mechanical equilibrium* is achieved between two systems when their pressures are equal at the interface. (Mechanical equilibrium is static or quasi-static.) In figure 15–9 we see two systems, A and B, which we can consider as liquids, solids, or gases. They are in thermal equilibrium when $T_A = T_B$, and in static mechanical equilibrium when $p_A = p_B$. There is also one other

$$\mu_{A1} = h_{A1} - T_A s_{A1}$$

$$\mu_{A2} = h_{A2} - T_A s_{A2}$$

$$\mu_{B1} = h_{B1} - T_B s_{B1}$$

$$\mu_{B2} = h_{B2} - T_B s_{B2}$$

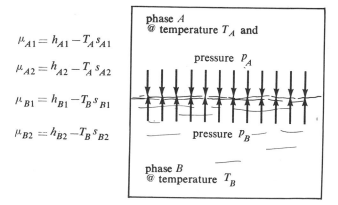

phase A
@ temperature T_A and

pressure p_A

pressure p_B

phase B
@ temperature T_B

Figure 15–9   Equilibrium between two systems.

form of equilibrium, called *chemical equilibrium*, which we must consider for a complete definition of equilibrium. These two systems, which we assume are each mixtures of the same components, 1 and 2, are said to be in chemical equilibrium when the chemical potential of component 1 in system A (μ_{A1}) is equal to the chemical potential of component 1 in system $B(\mu_{B1})$ and in a similar manner $\mu_{B2} = \mu_{B1}$. That is, two systems which interface one another attain chemical equilibrium when both have the same components and these components have the same chemical potential in both systems. It can be shown that the chemical potential of a component 1 in a state A is

$$\mu_{A1} = h_{A1} - T_A s_A$$

and for any general component i,

$$\mu_{Ai} = h_{Ai} - T_A s_{Ai} \tag{15-19}$$

Chemical equilibrium can be visualized as a balance of the mass flow of a component, say i, from a system A to another system B with the mass flow of the same component from system B to A. Chemical equilibrium between A and B is then achieved when all the components of the two systems balance the mass flow rates. It is this balance which is reflected in the criterion for chemical equilibrium

$$\mu_{Ai} = \mu_{Bi} \tag{15-20}$$

Example 15.7

Solution

Determine the chemical potential of pure saturated liquid water at 300°F.

The system here (water) is a one-component mixture and its chemical potential can be computed from equation (15–19). From table B.1 we get

$$h_{A1} = 269.\,70$$
$$s_{A1} = 0.4372$$

so that

$$\mu_{A1} = 269.70 - (760)(0.4372) = -62.57 \text{ Btu/lbm} \qquad \textit{Answer}$$

**15.4
Diffusion**

Suppose two systems are brought into contact with each other and are in thermal and mechanical equilibrium. Then, if we consider the systems A and B as shown in figure 15–10(a), the conditions follow that

$$T_A = T_B \quad \text{and} \quad p_A(\text{at } x = 0) = p_B(\text{at } x = 0)$$

Let us then assume that system A is dry air and system B is pure water so

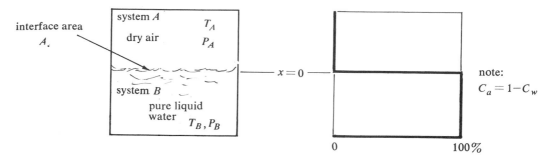

Figure 15–10(a)   Closed system of pure water and dry air, initial condition (time $\tau = 0$).

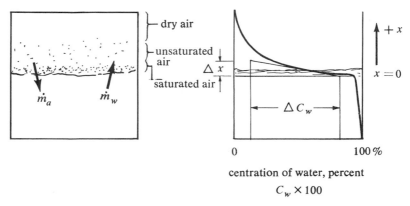

centration of water, percent

$$C_w \times 100$$

Figure 15–10(b)   Closed system of pure water and dry air after a finite period of time has elapsed.

that the mixture A can be described as 100% air and B as 100% water, as depicted in the graph of figure 15–10(a). As we have seen in section 15.3, however, for complete equilibrium to be achieved, the chemical potential of the air in A must equal the chemical potential of the air in the water $(\mu_{A(\text{air})} = \mu_{B(\text{water})})$. Likewise, $\mu_{A(\text{water})} = \mu_{B(\text{water})}$ must hold and these two statements imply that water must be present in A and air present in B. As a consequence of this natural action, water will migrate to A and air will migrate to B. The description of the mass flow rates of this migration are given by Fick's law of diffusion:

$$\text{Mass flow rate of water to } A = \dot{m}_w = -\rho_w A_i \mathfrak{D}_{wA} \frac{dc_w}{dx} \quad \textbf{(15–21)}$$

and

$$\text{Mass flow rate of air to } B = \dot{m}_a = -\rho A_i \mathfrak{D}_{aB} \frac{dc_a}{dx} \quad \textbf{(15–22)}$$

Here A_i is the cross-sectional area of the interface between the two systems; $\mathfrak{D}_{wA}$ is the diffusivity or coefficient of diffusion of water in system A; $\mathfrak{D}_{aB}$ is the diffusivity of air in system B; and c is the concentration or mass ratio such that c_w = mass of w in volume V/total mass in volume V and c_a = mass of a in V/total mass in V.

We can approximate the equation (15–21) by taking finite increments of x so that

$$\dot{m}_w \approx -\rho_w A_i \mathfrak{D}_{wA} \frac{(\Delta c_w)}{\Delta x} \qquad (15\text{–}23)$$

and

$$\dot{m}_a \approx -\rho_a A_i \mathfrak{D}_{aB} \frac{(\Delta c_a)}{\Delta x} \qquad (15\text{–}24)$$

where Δc is the change in concentration in a distance Δx. Initially we will see an infinite mass flow rate of water to the system A and air to the system B since

$$\Delta c_a = c_{aB} - c_{aA} = 0 - 1 = -1$$
$$\Delta c_{Bw} = c_{wA} - c_{wB} = 0 - 1 = -1$$

but $\Delta x = 0$ for both so that from equations (15–23) and (15–24) we get

$$\dot{m}_w \approx -\rho_w A_i \mathfrak{D}_{wA} \left(\frac{1}{0}\right) = \infty$$

and

$$\dot{m}_a \approx -\rho_a A_i \mathfrak{D}_{aB} \left(\frac{1}{0}\right) = \infty$$

This mass flow rate will only occur for an infinitesimal time as mixing of air and water will occur in both systems A and B. After some period of time, system A will be a nonuniform mixture of dry air and water. Near the interface the mixture will be saturated air, but further up, the concentration of water will progressively decrease so that the mixture could be described as almost dry air. Similarly, air will diffuse into the water, system B, but the concentration will decrease the further and further down into the system. This state is depicted by figure 15–10(b) where the concentration is plotted as a function of the distance from the interface. Notice that $\Delta c/\Delta x$ represents the slope of the curve of concentration at a particular point. Steady state conditions will be reached when the curve of concentration remains unchanged with time, but until this condition is achieved the curve can be expected to change so that the magnitude of $\Delta c/\Delta x$ decreases. That is, diffusion will continue to seek an equality of concentration between systems.

Diffusion can occur between a liquid and a gas (as we have just indicated), between a gas and a gas, a liquid and a liquid, a liquid and a solid, a gas and a solid, or a solid and a solid. Of course, the rates of mass flow vary extremely between these combinations, but the mass flow can be predicted through the coefficient of diffusion. In table 15–1 is shown the

Table 15–1

The Orders of Magnitude for Diffusivity at 70°F, 14.7 psia

Diffusing Phase (Transported Material) i	Diffused Phase, A (System Invaded by Material)	$\mathcal{D}_{ia}$ cm²/s
gas	gas	0.1
gas	liquid	1×10^{-5}
gas	solid	1×10^{-10}
liquid	gas	1×10^{-5}
liquid	liquid	1×10^{-5}
liquid	solid	1×10^{-15}
solid	gas	1×10^{-10}
solid	liquid	1×10^{-15}
solid	solid	1×10^{-20}

order of magnitudes of $\mathcal{D}$ for the various combinations of systems. Remember that the numbers are merely approximate values to describe the variation in diffusion. It can easily be seen from the table that diffusion of a solid to a solid is much less than a gas to a gas or any other combination.

In table 15–2 are given some nominal values of diffusivity $\mathcal{D}$ for common materials.

Example 15.8   A 70-cm diameter closed steel drum contains water and air at 10°C and 1 atmosphere pressure. At a given time the concentration curve is found to be that shown in figure 15–11. Determine the net mass flow rate across the liquid-air interface.

Solution   From figure 15–11 we see that the slope of the concentration curves at the interface are measured to be

$$\frac{\Delta c_w}{\Delta x} = \frac{-80}{10 \times 100} = -0.08/\text{cm}$$

$$\frac{\Delta c_a}{\Delta x} = \frac{80}{-10 \times 100} = -0.08/\text{cm}$$

Table 15-2

Some Common Materials at 20°C and 1 Atmosphere Pressure*

A	B	$\mathcal{D}_{ab}$	
Diffusing Material	Diffused Material	cm²/s	ft²/hr
Ammonia	air	0.236	0.914
Carbon Dioxide	air	0.164	0.636
Water Vapor	air	0.256	0.992
Ethyl Ether	air	0.093	0.360
Helium	SiO$_2$	3×10^{-10}	1.16×10^{-9}
Helium	pyrex	4.5×10^{-11}	1.74×10^{-10}
Bismuth	Pb	1.1×10^{-16}	4.26×10^{-16}
Mercury	Pb	2.5×10^{-15}	9.67×10^{-15}
Ethanol	water	1.13×10^{-5}	4.37×10^{-5}
Water (liquid)	n-Butanol	1.25×10^{-5}	4.80×10^{-5}

Notes: Data abstracted from M. Jakob and G. Hawkins, *Elements of Heat Transfer*, 3rd. ed. (New York, 1957), p. 276; and from R. B. Bird, W. Stewart, and E. Lightfoot, *Transport Phenomena* (New York, 1960), pp. 504–5; with permission of John Wiley & Sons, Inc.

The density of the water at the interface we assume to be 62.4 lbm/ft³ or 1 g/cm³. We shall assume the air to have a density of 0.00125 g/cm³, found from the perfect gas relation. The area of the interface A_i is obtained from

$$A_i = \pi \times \frac{D^2}{4} = \pi \times \frac{70^2}{4} \text{ cm}^2 = 3847 \text{ cm}^2$$

and we assume the diffusivity is the same for air and water, giving us

$$\mathcal{D}_{wA} = \mathcal{D}_{aB} = 0.256 \text{ cm}^2/\text{s}$$

from table 15-2. Then from equations (15–23) and (15–24) we get

$$\dot{m}_w = -(1 \text{ g/cm}^3)(3847 \text{ cm}^2)(0.256 \text{ cm}^2/\text{s})(-0.08/\text{cm})$$
$$= 78.787 \text{ g/s}$$

and

$$\dot{m}_a = -(0.00125 \text{ g/cm}^2)(3847 \text{ cm}^2)(0.256 \text{ cm}^2/\text{s})(-0.08/\text{cm})$$
$$= 0.098$$

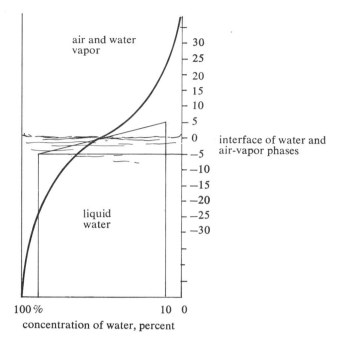

air and water
vapor

30
25
20
15
10
5
0   interface of water and
−5   air-vapor phases
−10
−15
−20
liquid   −25
water   −30

100%   10 0
concentration of water, percent

Figure 15–11   Relation of the concentration of water in a closed steel drum.

yielding, as a net mass flow rate,

$$\dot{m} = \dot{m}_w - \dot{m}_a$$
$$= 78.787 - 0.098$$
$$= 78.689 \text{ g/s} \qquad\qquad Answer$$

We see that the mass of the system is moving from the liquid to the gas. Of course, when chemical equilibrium is achieved, there will be no net mass transfer across the interface. The reader should recognize, however, that two or more systems which interact are not homogeneous, uniform mixtures, but rather are mixtures which contain gradients of concentration. Indeed, if this were not so, and if diffusion did not occur, mixtures of the kind we have considered throughout this book would not be possible.

15.5 Chemical Reactions

We have considered mixtures of inert materials; that is, the components of the mixtures were not chemically reacting. Here we will consider the mixing of components which do react chemically. Suppose that we have hydrogen gas (H_2) and oxygen gas (O_2) mixed together. The chemical reaction

$$1 H_2 + 1 O_2 \rightarrow 2 H_2O \qquad \text{(15–25)}$$

will follow, which means water will be produced. Consider the two elements carbon (C) and oxygen (O_2). The chemical reaction resulting will be

$$1 C + 1 O_2 \rightarrow 1 CO_2 \qquad \text{(15–26)}$$

Finally, if we had octane fuel (C_8H_{18}), called a *hydrocarbon* or *fossil fuel*, mixed with oxygen the reaction

$$1 C_8H_{18} + 12.5 O_2 \rightarrow 8 CO_2 + 9 H_2O \qquad \text{(15–27)}$$

follows. There are many, many more examples which we could write for chemical reactions, but let us further investigate the above three examples. In equation (15–25) we see that 1 mole of hydrogen and 1 mole of oxygen produces 2 moles of water. This is equivalent to saying that 2 lbm of hydrogen and 32 lbm of oxygen produce 34 lbm of water; that is, the molecular weight MW converts the chemical reaction equation to a mass balance equation. From equation (15–26) we get

$$(1 \text{ mole})(12 \text{ lbm C/mole}) + (1 \text{ mole})(32 \text{ lbm } O_2/\text{mole})$$
$$= (1 \text{ mole})(44 \text{ lbm } CO_2/\text{mole})$$

or

$$12 \text{ lbm} + 32 \text{ lbm} = 44 \text{ lbm}$$

Similarly, from equation (15–27) we see that a mass of $1 \; C_8H_{18} =$ $(12 \times 8) + (1 \times 18) = 114$ grams requires a mass of $(12.5)(32) = 400$ grams of oxygen. We have earlier found that 23.5% of air is normally considered to be oxygen, so that

$$400 \text{ grams} \times \frac{1}{0.235} = 1702 \text{ grams air}$$

is required to furnish sufficient oxygen to burn or react with octane fuel. The ratio of the mass of air m_A to the mass of fuel (octane in this case), m_F, is called the air-fuel ratio and is a parameter frequently used in analysis of combustion processes. We then can write

$$\text{air-fuel ratio} = \frac{m_A}{m_F} \qquad \text{(15–28)}$$

and for a complete reaction of octane fuel, or complete combustion, we have

$$\text{air-fuel ratio} = \frac{1702}{114} = 14.9 \text{ grams air/grams fuel}$$

Normally to insure combustion, an excessive amount of air is used with a fuel mixture.

A chemical reaction can be visualized as a rearrangement of atoms and molecules through a conservation of mass and energy. We have already mentioned the balance of mass in a chemical reaction so let us now consider the balance of energy. A molecule such as H_2, O_2, N_2, C_8H_{18}, H_2O, and CO_2 can be considered as having stored energy* in the amount required to form the molecule from the separate atoms. Each type of molecule has a different amount of stored energy so that during a chemical reaction there could be an imbalance in stored energy between the reactants (atoms or molecules on the left side of the chemical equation) and the products (molecules and atoms on the right side). If the reactants have more energy than the products, there will be an excess of energy after the reaction, which will be liberated. This type of reaction is called *exothermic* to describe the release of energy during the process. The combustion of octane by reacting with oxygen as described by equation (15–27) represents an exothermic reaction. The energy released, called the *heat of combustion* ΔQ_c or heating value HV, is accounted for and the energy balance fulfilled by rewriting equation (15–27) to read

$$C_8H_{18} + 12.5O_2 \rightarrow 8\,CO_2 + 9\,H_2O + \Delta Q_c$$

The energy released then represents the desired product of a combustion reaction to produce power with heat engines.

The reader should be aware that there are many exothermic reactions, some fast (called *combustion*) and some slow, and a characteristic speed can be given to each one. In designing or analyzing heat engines, we must account for these speeds of combustion if completion of reactions is to be achieved, along with a more efficient release of the energy for thermal uses.

Some chemical reactions require energy to drive them; that is, the energy of the reactants is less than the energy of the products, in which case there exists the possibility of using this reaction as a heat sink or cooling device. These types of reactions are called *endothermic* to indicate that the reaction tends to cool or lower the system's temperature.

Practice Problems

Section 15.1

15.1. Determine the partial pressures of the constituents in a mixture of 0.05 lbm oxygen, 0.10 lbm CO_2, 0.10 lbm argon, and 0.062 lbm helium at 18 psia pressure.

---

*This is called "free energy" or "free energy of formation" by many authors.

15.2. Determine the partial pressure of water vapor in a mixture of 8% (by mass) H_2O, 80% air, and 12% CO_2 if the mixture is at 14.7 psia and 80°F.

15.3. Calculate the entropy increase as 8 grams of water vapor mix with 70 grams of air at 1 atmosphere pressure and 20°C.

15.4. Determine the entropy increase due to mixing of 2 ft³ of helium at 10 psia and 60°F and 3 ft³ of argon at 20 psia and 60°F in a 3-ft³ container at 60°F.

Section 15.2

15.5. Obtain the saturation temperature of steam (water vapor) at 20 psia.

15.6. Determine the saturation pressure of water vapor at 60°F.

15.7. Air at 1 atmosphere pressure and 40°F has a wet-bulb temperature of 33°F. Determine (a) the vapor pressure, (b) ω, and (c) β.

15.8. Air has a relative humidity of 60% and a dry-bulb temperature of 90°F. Determine the enthalpy of the mixture and the specific humidity.

15.9. Atmospheric air is at a pressure of 29.4 mm of Hg and 80°F. If the wet-bulb temperature is 68°F, determine (a) ω, (b) β, and (c) dew point temperature.

15.10. A humidifier operates with air entering at 14.7 psia and 78°F. If the wet-bulb temperature is 50°F and the leaving air must be at 72°F and 50% relative humidity, determine the amount of heat transfer required and the water required per unit mass of air.

15.11. Air at 95°F, 14.7 psia, and 90% relative humidity must be brought to 75°F and 60% relative humidity. Determine the amount of heat rejected, the amount of heat added, and the amount of water condensed per pound mass of dry air for an air conditioner-dehumidifier serving this purpose.

15.12. There are 1500 ft³/min of air at 92°F, 14.7 psia, and 10% relative humidity to be conditioned to 70°F and 60% relative humidity. Determine the amount of heat transfer required and the rate of water flow necessary to humidify this air.

15.13. A dehumidifier having a maximum heater capacity of 100 Btu/min provides 10 lbm of air per minute at 68°F and 55% relative humidity. If the incoming air is at 80°F, determine the maximum relative humidity it may have and still allow the dehumidifier to furnish 10 lbm/min air at 68°F and 55% relative humidity.

Section 15.3

15.14. Determine the chemical potential of saturated water vapor at 500°F.

15.15. Determine the chemical potential of saturated ammonia vapor at 23.74 psia.

15.16. Calculate the chemical potential of saturated Freon-22 liquid at 140°F.

15.17. Determine the chemical potential of water at 70°F which is in chemical equilibrium with 70°F air and water mixture at 60% relative humidity.

Section 15.4

15.18. A 3-ft radius sphere contains helium gas at 20 psia and 95°F. Determine the amount of helium lost through diffusion in 24 hours if the sphere is made of Pyrex and if it is assumed that at $\frac{1}{8}$ inch from the inside wall, helium is totally absent.

15.19. Predict the concentration of ammonia in air 1 centimeter from an interface of liquid ammonia and air if the interface area is 15 cm² and the evaporation rate of ammonia is 2 g/min. Assume the air and ammonia system is at 20°C.

15.20. Water is evaporating at 3 lbm/hr in a 3-inch square container. Estimate the specific humidity 1 inch from the water-air surface in the air if the dry bulb temperature is 70°F.

15.21. Liquid mercury (Hg) is contained in a lead beaker as shown in figure 15–12. Calculate the amount of mercury lost through diffusion to the beaker after 48 hours, if the concentration of mercury is 2% at a distance 0.01 cm into the lead beaker wall from the inside surface. Assume the system is at a temperature of 20°C.

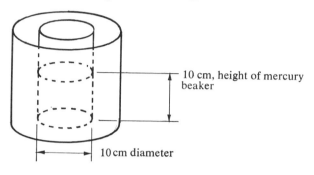

10 cm, height of mercury beaker

10 cm diameter

Figure 15–12   Problem 15.21.

Section 15.5

15.22. For the fuel C_8H_{17}, having the reaction

$$2\,C_8H_{17} + 24\tfrac{1}{2}\,O_2 \rightarrow 16\,CO_2 + 17\,H_2O$$

determine the ideal air-fuel ratio.

15.23. For propane gas C_3H_8, having the reaction

$$C_3H_8 + 5\,O_2 \rightarrow 3\,CO_2 + 4\,H_2O$$

determine the ideal air-fuel ratio.

15.24. For butane gas having the reaction

$$C_4H_{10} + 6\tfrac{1}{2}\,O_2 \rightarrow 4\,CO_2 + 5\,H_2O$$

calculate the air-fuel ratio.

16

Other Power Devices

In this chapter we are concerned with some diversified examples of systems to which the concepts of thermodynamics can be applied. Of course, any volume in space can be identified as a system and thus subject to the laws of thermodynamics, but the examples presented here are typical devices (or systems) which might be encountered by engineers and technologists and which thus should be considered from a thermodynamic approach. *Electric generators*, *motors*, and *batteries* are described, and a brief discussion is given to present the electrical analogies of work, heat, and energy. We then consider the operation of the *hydrogen-oxygen fuel cell* and some simple *thermoelectric devices*. Finally, *magnetohydrodynamic* and *electrohydrodynamic systems* are described in very brief terms.

Engineers and technologists must apply their knowledge and expertise to all phases of social needs, and to emphasize this trend, some *biological systems* are identified for thermodynamic analysis.

459

16.1 Electric Generators, Motors, and Batteries

Elementary to a consideration of electrical devices is an understanding of the concept of *current*. We visualize atoms as being comprised of a nucleus surrounded by electrons, e. Each electron, when stripped from its nuclear influence, possesses a charge, called a *coulomb;* the amount of charge possessed by each electron has been found to be 1.6×10^{-19} coulombs. If a metal or other material contains an abundance of free electrons, it is said to have a negative charge and if there is a shortage of electrons it is positive. Free electrons will migrate from an area in which they are dense to an area in which they are scarce; that is, electrons flow from negative to positive, as shown in figure 16–1, where we also see that current $\mathcal{I}$ is depicted as flowing from positive to negative. Current is defined by the equation

$$\mathcal{I} = \frac{dQ}{dt} \tag{16-1}$$

where Q is amount of charge and where the unit of current, coulomb/s, is called the *ampere*. Since electrons possess negative charges, we then say the positive current of electricity must travel from positive to negative.

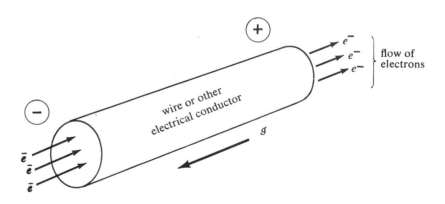

Figure 16–1   Electric current flow.

Another concept we need to consider briefly in electrical phenomena is that of the electrical potential ε. We write

$$\varepsilon = \frac{dE_e}{dQ} \tag{16-2}$$

where E_e is the electrical energy. The unit for the electrical potential can easily be seen to be Btu/coulomb or more commonly, joule/coulomb. The volt is defined as the joule/coulomb.

The essential components of a common electric generator are a *rotor* and a *stator*. The stator is a stationary housing consisting of magnets or electromagnets, as shown in figure 16–2. The rotor is situated between the stator magnets so that as it rotates, it interrupts the magnetic lines of force and thus induces an electric current through a conducting wire. Frequently the stator is composed of the conducting wires, and the rotor is then the source of magnetic fields, but here we will consider the arrangement as shown in figure 16–2. We first write the first law for the generator

$$\Delta E_e = Q_f - Wk_{\text{gen}} \tag{16–3}$$

and then, if we assume the process to be reversible, $Q_f = 0$. This gives us

$$\Delta E_e = -Wk_{\text{gen}}$$

and, from equation (16–2), we obtain

$$\Delta E_e = \int dE_e = \int \varepsilon \, dQ = -Wk_{\text{gen}} \tag{16–4}$$

Since

$$E_e = \frac{dE_e}{dt} = \varepsilon \frac{dQ}{dt}$$

we can use the definition of current, equation (16–1), to obtain

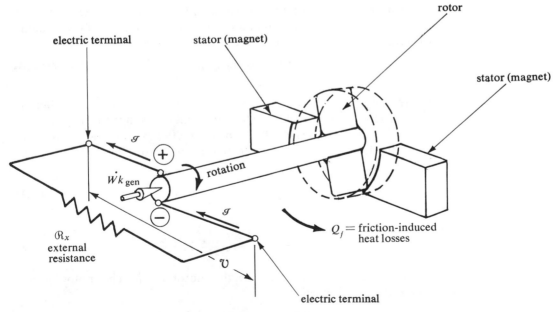

Figure 16–2   The electric generator.

$$\mathcal{E}\mathcal{I} = -\dot{W}k_{\text{gen}} \qquad (16\text{--}5)$$

If we measure the potential or voltage across the terminals, its value will be $\mathcal{E}$, provided there exists no external connection between the poles. If, on the other hand, we insert a load or resistance $\mathcal{R}_x$ between the poles, as depicted in figure 16–2, the voltage will be less than $\mathcal{E}$. Ohm's law states that

$$\text{Potential drop} = \text{resistance} \times \text{current} \qquad (16\text{--}6)$$

so that, calling the voltage $\mathcal{U}$, we have across the resistance that

$$\mathcal{U} = \mathcal{R}_x \mathcal{I} \qquad (16\text{--}7)$$

where $\mathcal{R}_x$ is the resistance measured in ohms, $\mathcal{I}$ is the current in amps, and $\mathcal{U}$ is the voltage. The current is also flowing through the generator rotor so that a potential drop, due to the conducting wire resistance $\mathcal{R}_I$, will be present. Then

$$\mathcal{U} = \mathcal{E} - \mathcal{I}\mathcal{R}_I \qquad (16\text{--}8)$$

for the closed circuit generator.

The efficiency of the generator is given by

$$\eta = \frac{\mathcal{U}\mathcal{I}}{\dot{W}k_{\text{gen}}} \times 100 \qquad (16\text{--}9)$$

and is normally in the range of 85 to 95%. Irreversibilities such as friction have been neglected here so that a more realistic correction may be made to actual generators by writing

$$\dot{W}k_{\text{gen}} = \mathcal{E}\mathcal{I} + \dot{W}k_{\text{friction}} \qquad (16\text{--}10)$$

Electric motors can be visualized as reversed generators. A voltage is applied across the terminals so that a current exists in the rotor, which in turn induces an electric field, $\mathcal{E}_{\text{ind}}$. The interaction between the electric and magnetic fields then produces a rotation and torque through the rotor. For the motor we have

$$\mathcal{E}_{\text{ind}} = \mathcal{U} - \mathcal{I}\mathcal{R}_I \qquad (16\text{--}11)$$

and the mechanical efficiency is

$$\eta_{\text{mech}} = \frac{\dot{W}k_{\text{motor}}}{\mathcal{U}\mathcal{I}} \times 100 \qquad (16\text{--}12)$$

where the electric power used by the motor is $\mathcal{U}\mathcal{I}$. The motor output power $\dot{W}k_{\text{motor}}$ is given by

$$\dot{W}k_{\text{motor}} = \mathcal{E}_{\text{ind}}\mathcal{I} - \dot{W}k_{\text{friction}} \qquad (16\text{--}13)$$

While the electric generator and the motor involve energy transfers between electrical and mechanical systems, the common electrical battery utilizes a transfer of energy between chemical and electrical systems.

The Daniel cell is a device which produces electrical energy from chemical reactions. It is not exactly like the common batteries, but its workings involve the essential concepts of the typical battery. The Daniel cell, shown in figure 16–3, is composed of zinc sulfate ($ZnSO_4$) and copper sulfate ($CuSO_4$) solutions separated by a barrier which prevents mixing of these two solutions, yet which allows ions (such as Cu^{++}, Zn^{++}, and SO_4^{--}) to pass. A solid bar of zinc (anode) is placed in the $ZnSO_4$ solution and a bar of copper (cathode) in the $CuSO_4$ solution. The chemical reaction at the anode is

$$Zn \rightarrow Zn^{++} + 2e^- \qquad (16\text{--}14)$$

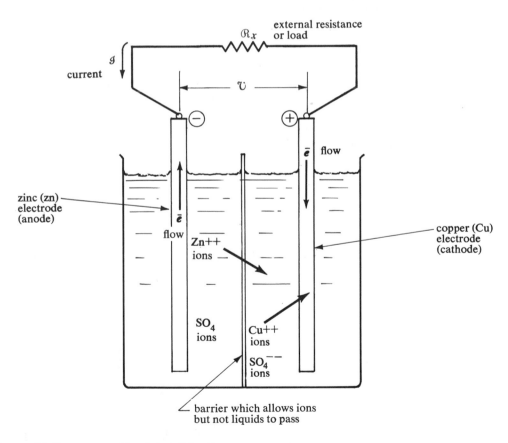

Figure 16–3   Daniel cell battery.

and at the cathode is

$$Cu^{++} + 2e^- \rightarrow Cu \qquad (16\text{--}15)$$

The total chemical reaction for the Daniel cell is found by adding equations (16–14) and (16–15)

$$\underbrace{Zn + Cu^{++}}_{\text{reactants}} \rightarrow \underbrace{Cu + Zn^{++}}_{\text{products}} \qquad (16\text{--}16)$$

with the result that the cathode becomes plated with copper, the anode loses zinc, and the electrons are allowed to flow through an external load. The maximum work or electrical energy obtainable from the chemical reaction (16–16) is given by equation (9–28)

$$Wk_{\text{use}} = -\Delta G$$

and since the useful work is also the maximum work obtainable for useful purposes, we write

$$Wk_{\text{max}} = -\Delta G$$

where the Gibbs free energy can be shown to be calculated from the summation

$$\Delta G = \underbrace{\Sigma \Delta G_f^\circ}_{\text{products}} - \underbrace{\Sigma \Delta G_f^\circ}_{\text{reactants}} \qquad (16\text{--}17)$$

The ΔG_f° term represents the amount of the Gibbs free energy required to form the particular molecule from the elements, for example, the energy required to form sodium chloride (NaCl) from sodium (Na) and chlorine (Cl) atoms. Table B.19 lists values of the Gibbs free energies of formation ΔG_f° for various materials. Of course, the value of the Gibbs free energy of formation is zero for the natural elements un-ionized.

The work done to convey the electrons from the cathode to the anode is also equal to $n\mathcal{F}\mathcal{V}^\circ$ where n is the gram-moles of electrons, $\mathcal{F}$ is the *Faraday constant* given by

$$\mathcal{F} = 96{,}500 \ \frac{\text{coulomb}}{\text{gram-mole}}$$

and $\mathcal{V}^\circ$ is the potential across the anode and cathode when the circuit is open; that is, the resistance $\mathcal{R}_x$ is disconnected. Then we have

$$n\mathcal{F}\mathcal{V}^\circ = -\Delta G \qquad (16\text{--}18)$$

and the actual output voltage $\mathcal{V}$ is

$$\mathcal{V} = \mathcal{V}^\circ - g\mathcal{R}_{\text{I}} \qquad (16\text{--}19)$$

where $\mathcal{R}_{\text{I}}$ is the Daniel cell resistance. The power obtainable from the cell

is
$$\mho\mathcal{g} = \mathcal{g}^2\mathfrak{R}_x \qquad\qquad (16\text{--}20)$$

and for the efficiency we write

$$\eta_{\text{battery}} = \frac{\mho\mathcal{g}}{\Delta\mathcal{g}} \qquad\qquad (16\text{--}21)$$

Example 16.1   Determine the maximum work obtainable from a Daniel cell, the open circuit voltage, and the operating voltage if 0.1 ampere of current is drawn and the internal resistance is 0.05 ohm.

Solution   We calculate the maximum work from the sum of the Gibbs free energy values. Using equation (16–17) and the values from table B.19, we have

$$\Delta\mathcal{g} = Cu + Zn^{++} - Zn + Cu^{++}$$

or

$$\Delta\mathcal{g} = \underbrace{0 + (-35.18)}_{\text{products}} - \underbrace{0 - 15.53}_{\text{reactants}} = -50.71$$

Then,

$$Wk_{\text{max}} = -\Delta\mathcal{g} = 50.71 \text{ kcal/g-mole} \qquad\qquad Answer$$

The open circuit voltage can be calculated from equation (16–18):

$$\mho^\circ = \frac{-\Delta\mathcal{g}}{n\mathcal{F}} = \frac{50.71}{(96,500)}$$

and $n = 2$ g-moles of electrons per g-mole of reaction so that

$$\mho^\circ = \frac{50.71 \text{ kcal/g-mole} \times 4184 \text{ joule/kcal}}{2 \text{ g-mole/g-mole} \times 96,500 \text{ coulomb/g-mole}}$$

$$= 1.098 \text{ joule/coulomb} = 1.098 \text{ volts} \qquad\qquad Answer$$

If we now use equation (16–19), we can calculate the operating voltage when 0.1 amp of current is drawn.

$$\mho = 1.098 \text{ volts} - (0.1 \text{ amp})(0.05 \text{ ohm})$$

$$= 1.093 \text{ volts} \qquad\qquad Answer$$

This example problem and the preceding discussion are items and areas which engineers and engineering technologists in particular may encounter in practice — they were presented to illustrate the application of the thermodynamic concepts to electrical devices. In no way, however, should the reader infer that this brief presentation is a sufficient treatment of electrical concepts nor should he assume that other thermodynamic ideas such as entropy cannot be applied to electrical phenomena.

**16.2
Fuel Cells**

The fuel cell has gathered increased attention due to its attractiveness as a source of electric power. The power is developed by means of a "controlled" chemical reaction during which free electrons are forced to travel an external path to complete the reaction. Many fuels have been investigated for use in fuel cells, including biological or organic fuels, but one of the most successful combinations has been hydrogen and oxygen. The device using these particular fuels is called the *hydrogen-oxygen fuel cell* and is shown in figure 16–4. As can be seen, pure hydrogen and oxygen are added to the cell, and water and electrical power are produced. Pumps recirculate the portions of the oxygen and hydrogen which do not react in the cell. The chemical reaction at the anode is

$$H_2 - 2\ OH^- \rightarrow 2\ H_2O + 2e^- \qquad\qquad (16\text{--}22)$$

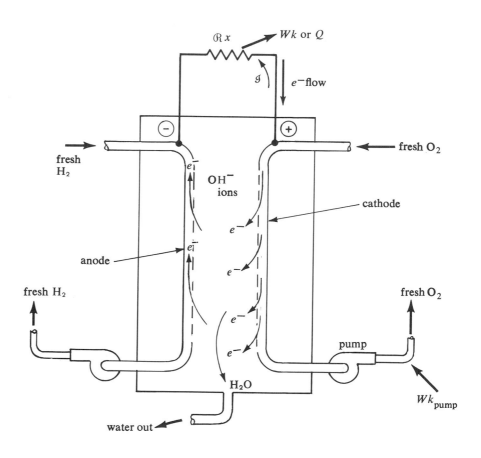

Figure 16–4   The hydrogen-oxygen fuel cell.

and at the cathode

$$H_2O + \frac{1}{2} O_2 + 2e^- \rightarrow 2 \text{ OH} \qquad \text{(16–23)}$$

The full chemical reaction of the fuel cell is then given by the addition of equations (16–22) and (16–23)

$$H_2O + H_2 + \frac{1}{2} O_2 + 2 \text{ OH}^- + 2e^- \rightarrow 2 \text{ OH}^- + 2 H_2O + 2e^-$$

or

$$H_2 + \frac{1}{2} O_2 \rightarrow \quad H_2O \qquad \text{(16–24)}$$
$$\underbrace{\qquad\qquad\qquad}_{\text{reactants}} \quad \underbrace{\qquad}_{\text{product}}$$

The maximum work is obtained from equation (9–28)

$$Wk_{\text{max}} = -\Delta\mathcal{G}$$

and the open circuit voltage from equation (16–18)

$$\mathcal{V}° = \frac{-\Delta\mathcal{G}}{n\mathcal{F}}$$

Example 16.2

Calculate the maximum work obtainable from the hydrogen-oxygen fuel cell and calculate the open circuit voltage.

Solution

From table B.19 we have

$$\Delta\mathcal{G}_f° = -56.69 \text{ for } H_2O$$
$$= 0 \text{ for } H_2 \text{ and } O_2$$

so that, using equation (16–15) we get

$$\Delta\mathcal{G} = -56.69 \text{ kcal/g-mole}$$

Since the molecular weight (MW) of water is 18 g/g-mole we have

$$\Delta\mathcal{G} = -(56.69/18) \text{ kcal/g } H_2O = -3.15 \text{ kcal/g } H_2O$$

Then

$$Wk_{\text{fuel cell}} = 3.15 \text{ kcal/g} \qquad Answer$$

The open circuit voltage is

$$\mathcal{V}° = \frac{-\Delta\mathcal{G}}{n\mathcal{F}} = \frac{56.69}{2 \times 96,500} \frac{\text{kcal}}{\text{coulomb}}$$

$$= 0.000294 \text{ kcal/coulomb} \times 4184 \text{ joule/kcal}$$

$$= 1.23 \text{ joule/coulomb}$$

$$= 1.23 \text{ volts} \qquad Answer$$

Fuel cells are attractive devices for providing power. A major tech-nological problem associated with them, however, is in the fabricating of the electrodes (anode and cathode) with reasonable effort, and until this difficulty is alleviated, the fuel cell cannot be expected to replace the battery for common uses.

16.3 Thermoelectric Devices

We have seen that resistance in a conducting wire produces a voltage drop given by Ohm's law, equation (16–7)

$$\mathcal{V} = \mathcal{I}\mathcal{R}$$

The power represented by this is given by

$$\mathcal{V}\mathcal{I} = \mathcal{I}^2\mathcal{R} \tag{16–25}$$

and this power is dissipated in a resistor $\mathcal{R}$ in the form of an increase in the temperature of the resistor. If the resistor is in equilibrium and steady state conditions with the surroundings, the power $\mathcal{I}^2\mathcal{R}$ is reflected in heat transfer to the surrounding as shown in figure 16–5. We call this effect *Joule heating* Q_J, which represents the most common means of electrical energy dissipation. Joule heating is used to advantage in any electric heater and is present in any conductor of electricity having resistance.

If two wires or conductors, A and B, composed of different materials, are joined as shown in figure 16–6 and a current is impressed through

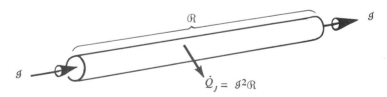

Figure 16–5   Joule heating.

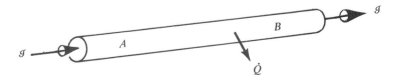

Figure 16–6   Peltier effect.

them, another heat transfer can be induced. This heat, called *Peltier heating*, is due to thermoelectric phenomena called the *Peltier* and *Seebeck* effects and is the basis for the *thermocouple*. The heat emanating from the joint $\dot{Q}$ is given by the sum of the Joule and Peltier heats

$$\dot{Q} = \mathcal{I}^2\mathcal{R} + \mathcal{I}(\alpha_A - \alpha_B) \tag{16–26}$$

where α_A and α_B are the Seebeck coefficients of wires A and B. The Seebeck coefficients are the mathematical values allowing for the Peltier heat and are generally found to be functions of the material temperature. A further discussion of the Seebeck coefficient can be found in publications devoted to direct energy conversion devices. It may be noted that the Peltier effect can be reversed by impressing heat transfer into the joint and thus inducing a current in the wires. This is the essential operation of the temperature measuring device called the *thermocouple*. In the thermocouple, the current is measured, allowing for a prediction of the joint temperature.

Another thermoelectric phenomenon found in all actual conductors is the *Thomson effect*. A wire or other electric conductor subject to a temperature change in its volume, as shown in figure 16–7, will have a current induced in it. The heat transfer due to this effect is called the *Thomson heat* and is generally written as $\mathcal{I}_{\text{ind}}\mathcal{J}\Delta T$ where $\mathcal{J}$ is the Thomson coefficient (determined by experimental means) and ΔT is $T_1 - T_2$. The heat transfer on the sides of the conductor is then

$$\dot{Q} = \mathcal{I}^2_{\text{ind}}\mathcal{R} + \mathcal{I}_{\text{ind}}\mathcal{J}\Delta T \tag{16–27}$$

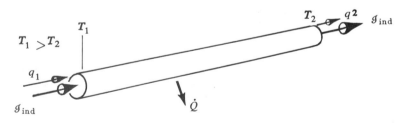

Figure 16–7   Thomson effect.

Semiconductor materials are utilized in some useful thermoelectric devices, called *thermoelectric generators*. These materials (semiconductors) exhibit a selectivity in allowing current flow, generally by prohibiting current to pass in one of the two directions and they also have wide varieties of values for Seebeck coefficients. In figure 16–8 is shown a thermoelectric device composed of two distinct types of semiconductor materials, N-type and P-type. The bar labeled A is a conductor of electricity and a

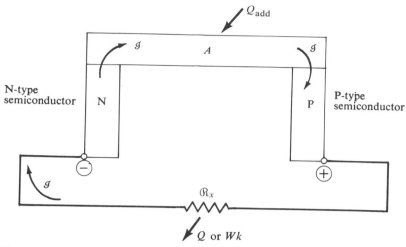

Figure 16–8   Elements of a simple thermoelectric generator.

receiver of external heat transfer. From the previously discussed concept
of Peltier heating, we see that in this case the Peltier heat is added to the
junctions between A and the N-type material and the P-type material.
At these two joints we then have, from equation (16–26),

$$\dot{Q}_{add} = \mathscr{I}_{ind}^2 \mathscr{R} + \mathscr{I}_{ind}(\alpha_n - \alpha_A) + \mathscr{I}_{ind}(\alpha_A - \alpha_p) \qquad (16\text{–}28)$$

where $\mathscr{I}_{ind}$ is the current induced in the circuit due to the *addition* of
Peltier heat. The semiconductor materials, N- and P-types, have properties
such that α_n is negative and α_p is positive. This means that the Seebeck
effect at the two junctions, instead of canceling, will add and give us,
from equation (16–28)

$$\dot{Q}_{add} = \mathscr{I}_{ind}^2 \mathscr{R} - \mathscr{I}_{ind}(|\alpha_n| + |\alpha_p|) \qquad (16\text{–}29)$$

Thermoelectric generators are attractive as sources of electric power since
they are not inherently dependent on the source of thermal energy. Heat
may be transferred from a nuclear reactor, steam, hot exhaust gas, the
sun, or numerous other sources.

16.4 Magnetohydro-dynamics (MHD)

The electric generator produces electric power by passing a conducting
wire through a magnetic field. A magnetohydrodynamic (MHD) generator, such as shown in figure 16–9, produces electric power in much the
same way except the conductor passing through the magnetic field is a
hot gas or fluid. The basic operation of this device is indicated in figure
16–10 where hot gases (ionized gases if possible) flow through a chamber

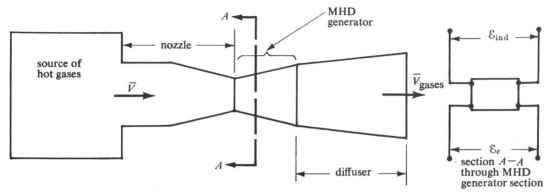

Figure 16–9   Sketch of magnetohydrodynamic power generator.

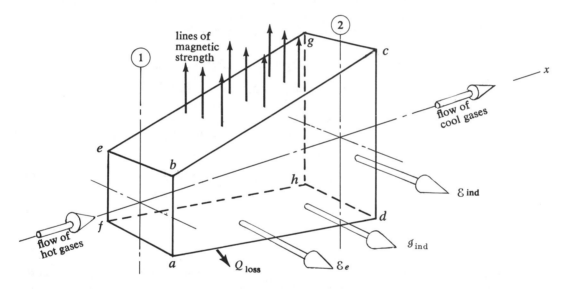

Figure 16–10   Basic operation of MHD generator.

having two sides which conduct electricity (sides *a-b-c-d* and *f-e-g-h*) and two sides which are insulated (sides *b-e-g-c* and *a-f-h-d*). An electric field $\mathcal{E}_e$ is applied across the two conducting sides and this field induces a magnetic field having lines of force perpendicular to the gas flow. The flow of hot gases through these fields induces an electric potential $\mathcal{E}_{ind}$ of significant strength. By applying the first law of thermodynamics to this system we have

$$\dot{m}\left(\frac{\bar{V}_2^2 - \bar{V}_1^2}{2g_c}\right) + \dot{m}(h_2 - h_1) = \dot{Q}_{loss} - \mathcal{g}_{ind}\mathcal{E}_{ind} + \mathcal{g}_{ind}\mathcal{E}_e$$

$$(16\text{–}30)$$

where $\mathscr{I}_{ind}$ is the induced current due to the MHD effect and $\dot{m}$ is the mass flow rate of the gases.

The MHD generator represents a device that can utilize very high temperature gases with present materials and produce significant amounts of power. In addition, the exhaust gases from the device can be used to provide a heat source for a conventional heat engine, such as a gas turbine or a steam turbine. When used tandem in this manner with other devices, the MHD generator can represent a significant improvement in thermodynamic efficiencies.

Another interesting aspect of the phenomena being considered here is the reversed procedure; that is, if magnetic and electric fields are applied across a conduit containing a fluid which can conduct electricity, a velocity or flow is induced in the fluid. This technique is responsible for the concept of pumps having no moving parts.

The primary disadvantages of the MHD device are twofold:

(1) The gas or fluid must have a sufficiently low resistance and must be a good electrical conductor. Most gases do not have these properties.
(2) The source of hot gases must be exceedingly large if significant amounts of power are to be generated.

Significant treatises exist which the reader may refer to for a more complete treatment of magnetohydrodynamic devices.

16.5 Biological Systems

Engineers and technologists have been applying engineering and scientific concepts to inorganic, inanimate objects or systems without too much hesitation. On the other hand, organic, biological systems have been avoided with few exceptions. The following two systems, the muscle and the heart, are biological devices which have been investigated from engineering viewpoints and are presented here to set examples for application of thermodynamic concepts to other biological systems, heretofore avoided.

The muscle represents an organic system which converts chemical energy to mechanical work and this is achieved at essentially constant temperature. Many have attempted to explain scientifically the details of this conversion with no great success, and indeed the muscle represents the only known device which directly converts chemical energy to work. It appears that the source of chemical energy resides in a macromolecular substance called adenosine triphosphate (ATP) and other simpler phosphates. This reaction releases energy to be used in work and heat. The muscle system is shown in figure 16–11, along with a weight W representing an external force applied to the muscle. It is viewed in a process of contracting or shortening while exerting a force; the actual process has been simplified by replacing the muscle force by a weight W. We may apply the first law to the muscle and get

$$H_2 - H_1 = Q - Wk \tag{16-31}$$

where H_2 is the enthalpy of the ATP and phosphates and H_1 is the enthalpy of the resulting chemical, adenosine diphosphate (ADP). The work gotten from a muscle is Wk and the maximum work is predicted from equation (9–28)

$$Wk_{max} = -\Delta G'$$

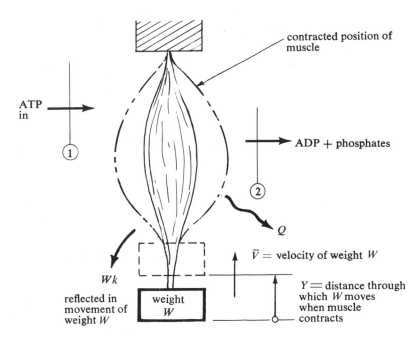

Figure 16–11   The muscle as a system.

We may consider the mechanical efficiency of the muscle system to be

$$\eta_{mech} = \frac{Wk}{-\Delta G'} \times 100 \tag{16-32}$$

and this seems to be around 30% to 40% for most healthy muscles. The actual work obtained from a muscle is, of course, dependent on the

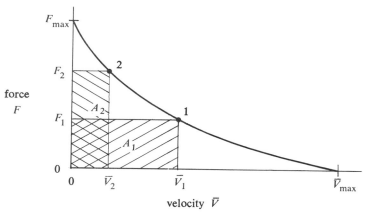

Figure 16–12

Force-velocity relationship for typical muscle contraction.

amount of contraction or distance through which force is applied. If the force applied to the muscle is more than the muscle can move, there will be no contraction of the muscle and no work done. (The muscle will use up ΔG, however.) We can also say that the weight W will have no velocity in this case. Now, if the force applied to the muscle is reduced by using a smaller weight W, the muscle will then pull the weight through a distance and with some velocity. Continuing, if the force is reduced further the weight will be pulled faster by the muscle until it happens that when no weight or force is applied to the muscle, contraction will proceed at a rapid rate. This relationship, the force applied to the muscle versus the velocity at which the weight moves, is shown in figure 16–12. The velocity $\bar{V}_{max}$ represents the velocity of muscle contraction when no force ($F = 0$) is applied and the force F_{max} represents the greatest amount of force a muscle can pull. The power gotten from a muscle pulling a constant force or weight is

$$\dot{W}k = F\bar{V} \qquad \text{(16–33)}$$

and is represented by the rectangular area under the curve in figure 16–12. As an example, the power gotten from a muscle working under a pull of F_1 and with velocity $\bar{V}_1$ is the area A_1. Similarly, for a force F_2 and velocity $\bar{V}_2$, the power is the area A_2.

The heart is an organ which is composed of various muscles acting together. While the actual configuration of the heart is quite complex and blood flowing through it is equally complicated, as indicated in figure 16–13, for our purposes we can consider it to be a form of pump. The heart functions as essentially two separate blood pumps. The solid arrows in the diagram indicate directions of blood flow. Oxygen-poor

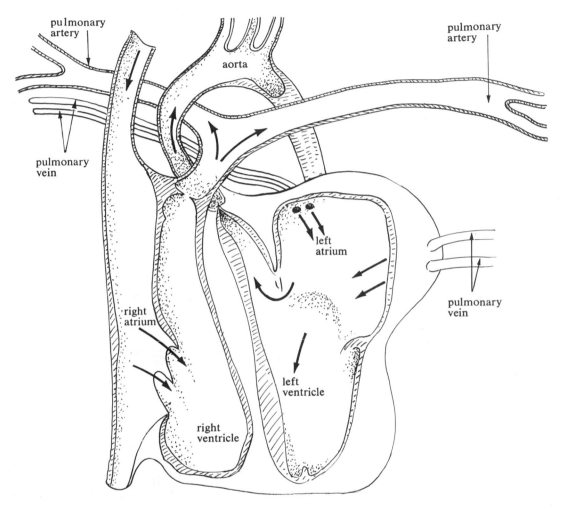

Figure 16–13   Cross section of heart. Revised from W. I. Keeton, *Biological Science* (New York, 1967), p. 246; with permission of W. W. Norton & Company, Inc.

blood comes from the body to the *right atrium*. When the heart expands this blood is drawn into the *right ventricle*. Then the blood is expelled into the pulmonary artery when the heart contracts. The pulmonary artery sends the blood to the lungs where oxygen is supplied to the blood. The oxygen-rich blood returns to the heart through the pulmonary veins and into the *left atrium*. Upon heart expansion the blood flows into the *left ventricle* and then, when the heart contracts, the *left ventricle* reduces in volume and sends the oxygen-rich blood into the body via the *aorta*.

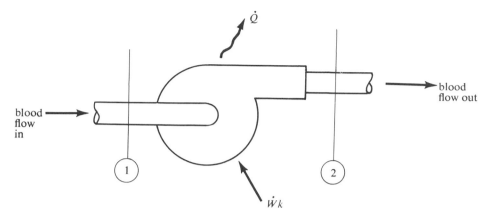

Figure 16-14   Simplistic representation of the heart by a blood pump.

We can simplify the heart to the system as sketched in figure 16–14 and the energy equation gives us, approximately

$$\dot{m}\left(\frac{\bar{V}_2^2 - \bar{V}_1^2}{2g_0}\right) + (h_1 - h_2)\dot{m} = \dot{Q} - \dot{W}k \qquad (16\text{–}34)$$

where $\dot{m}$ is the mass flow of the blood. It should be mentioned that blood flow through the heart is not steady, although over a long period of time it may be approximated as such; nor is the flow of blood as simple an analogy as water flowing through a rigid pipe. The mechanism of the heart is so complex that man will never fully understand its workings, but our simplistic approach is better than none in attempting to analyze the heart. The pressure-volume diagram of the blood passing through the heart is shown in figure 16–15. The enclosed area is frequently referred to as the *heart output*. Of course, the diagram shown is only representative and the maximum pressure shown varies from one heart to another. An average maximum blood pressure for a healthy, medium-aged human heart seems to be 120 mm of Hg. We will not here attempt to calculate the area of the diagram except to recall that the maximum work must be given by equation (9–28)

$$Wk_{\max} = -\Delta g$$

where $-\Delta g$ is the change in the Gibbs free energy of the ATP/ADP reaction previously mentioned as the source of energy for muscles. The interested reader is here directed to textbooks on biology, physiology, and bioengineering for more complete and varied discussions of biological systems.

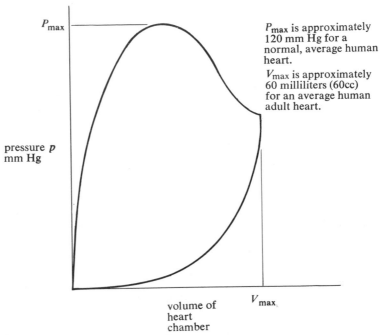

Figure 16–15   Pressure-volume diagram for heart as it operates for one complete cycle.

Practice Problems

Problems designated with an asterisk * should only be attempted by those having a background which includes the knowledge of calculus.

Section 16.1

16.1. An electric generator is found to have an efficiency of 93% when delivering 60 amps of electric current. If 15 horsepower are required to drive the generator, determine the generator's voltage output.

16.2. An electric generator has an internal resistance of 0.03 ohm and delivers 100 amps at 120 volts. Determine the induced electric potential and the generator efficiency.

16.3. An electric motor has an internal resistance of 0.06 ohm and operates at 120 volts and 60 amps. Determine the motor output power if no friction exists.

16.4. A 3-hp electric motor dissipates 2 watts of frictional resistance, and operates at 220 volts. If internal resistance is 0.09 ohm, determine the motor current and efficiency.

16.5. A Daniel cell is built and found to have an internal resistance of 0.06 ohm. If 1 amp of current is drawn, calculate the maximum work, the open circuit voltage, and the operating voltage.

16.6. An electric cell is proposed which uses a zinc anode, a silver (Ag) cathode, and zinc sulfate and silver sulfate (Ag_2SO_4) solutions. In all other respects it resembles the Daniel cell. The cathode reaction is

$$2Ag^+ + 2e^- \rightarrow Ag$$

Determine the maximum work obtainable from this cell and its open circuit voltage.

Section 16.2

16.7. If the internal resistance of a hydrogen-oxygen fuel cell is 0.05 ohm, determine the operating voltage if 2 amps are drawn.

16.8. A fuel cell is proposed which uses sodium (Na) and chlorine (Cl) and produces sodium chloride (NaCl). Determine the maximum work obtainable and the open circuit voltage.

16.9. Octane (C_8H_{18}) in the gaseous phase is utilized as a fuel in a fuel cell which also uses oxygen. The chemical reaction is $C_8H_{18} - 12\frac{1}{2} O_2 \rightarrow 8 CO_2 + 9 H_2O$. Determine the maximum work obtainable, the open circuit voltage, and the operating voltage when 16 amps are drawn with no internal resistance.

Section 16.3

16.10. A 10-ohm resistor carries 8 amps. Determine the steady state Joule heating.

16.11. A resistor radiates 50 Btu/hr when a current of 0.7 amp passes through it. Determine the resistance in ohms.

16.12. An electric heater is required to conduct 200 Btu/min. Determine the resistance and current if 120 volts are applied across the heater coil.

16.13. A P-N type thermoelectric generator receives 10 cal/s of heat from an external supply. If the resistance of the generator is 0.05 ohm and the Seebeck coefficients are $\alpha_n = -0.13$ volt and $\alpha_p = 0.20$ volt, determine the induced current and the voltage of the device.

Section 16.4

16.14. Hot gases having properties of air enter an MHD generator at 2000°F and 700 ft/s. If they leave the generator at 1050°F and 400 ft/s, estimate the work generated per lbm of gases if heat losses are neglected.

16.15. An MHD power generator receives 1200 g/s of 1600°C gases having a velocity of 35 m/s and discharges them at 10 m/s and 650°C. A potential of 300 volts is applied across the generator section where the gases have a resistance of 1.5 ohms. If heat losses are 500 cal/s and the gases have properties of air, determine the net power produced by the MHD generator.

Section 16.5

16.16. A muscle lifts 50 grams through 2 centimeters. If the efficiency is found to be 32%, what is the change in Gibbs free energy of the ATP-ADP reaction?

16.17. A heart is pumping 5000 g/min of blood which has a density of 1 g/cc. If the thermal effects are neglected (no temperature change and $Q = 0$) and if the velocity change is negligible between entering and leaving blood, estimate the power produced by the heart to pump this blood if the pressure is 110 mm Hg leaving and 0 mm Hg entering.

*16.18. A muscle is found to have the following force-velocity relationship:

$$F(\text{grams}) = 60 - 6\overline{V}$$

where $\overline{V}$ is in cm/s. Plot the F-$\overline{V}$ curve and determine the maximum power obtainable from the muscle. Then sketch the graph of power versus velocity of muscle.

16.19. A heart which has a volume of 65 cc is using 750 kg-meter/min of power to pump blood to a pressure of 170 mm of Hg. Predict the average or mean effective blood pressure in the heart if the heart is pumping at 120 beats per minute.

Appendices

A

Formulas and Equations

The following algebraic relations and definitions are useful in the manipulation of logarithmic quantities, including the algebra of powers:

$$\log X = N \qquad 10^N = X$$
$$\log Y = M \qquad 10^M = Y$$
$$\log X + \log Y = \log XY$$
$$(X)(Y) = 10^{N+M}$$
$$\log X - \log Y = \log \frac{X}{Y}$$
$$\frac{X}{Y} = 10^{N-M}$$
$$\ln X = L \qquad e^L = X$$
$$\ln Y = K \qquad e^K = Y$$
$$e = 2.303 \ldots$$
$$\ln X = (2.303 \ldots)\log X$$

Using x and y as variables, independent of the exponents n and m, we have,

$$\frac{x^n}{y^m} = x^n y^{-m}$$

$$\left(\frac{x}{y}\right)^n \left(\frac{x}{y}\right) = \left(\frac{x}{y}\right)^{n+1}$$

$$\left(\frac{x}{y}\right)^n \left(\frac{y}{x}\right) = \left(\frac{x}{y}\right)^n \left(\frac{x}{y}\right)^{-1} = \left(\frac{x}{y}\right)^{n-1}$$

A.2 Raising Numbers to Powers

Frequently in thermodynamic calculations the following problem arises: Given a value for x and n, determine the result of x^n. Using logarithms we have

$$(\log x^n) = n(\log x) = y$$
$$\text{antilog } y = x^n$$

By using a slide rule or log table, we can obtain the log of x. Then multiplying this value by n yields a new value, y. Using log tables or appropriate slide rule scales, we can obtain the antilog of y. This value corresponds to the desired quantity, x^n.

A.3 Integral Formulas

For equations of reversible work, the integrals $\int p \, dv$ or $\int v \, dp$ were evaluated. In order to complete the integration, however, the variable p must be written in terms of V or V in terms of p. To generalize to any variable, x, the following indefinite integrals (or antiderivatives) result from the given quantity:

$$\int dx = x + c_1$$

$$\int x \, dx = \frac{1}{2}x^2 + c_2$$

$$\int x^n \, dx = \frac{1}{n+1}x^{n+1} + c_n$$

$$\int \frac{dx}{x} = \ln x + c_3$$

If the integral is to be evaluated between two given values (say x_1 and x_2), the following results are obtained:

$$\int_{x_1}^{x_2} dx = x_2 - x_1$$

$$\int_{x_1}^{x_2} x \, dx = \frac{1}{2}x_2^2 - \frac{1}{2}x_1^2 = \frac{1}{2}\left(x_2^2 - x_1^2\right)$$

$$\int_{x_1}^{x_2} dx = \frac{1}{n+1}x_2^2 - \frac{1}{n+1}x_1^2 = \frac{1}{n+1}\left(x_2^2 - x_1^2\right)$$

$$\int_{x_1}^{x_2} \frac{dx}{x} = \ln x_2 - \ln x_1 = \ln \frac{x_2}{x_1}$$

B

Tables and Charts

Table B.1   Saturated Steam Table

Temp °F t	Pressure psia p	Specific Volume ft³/lbm		Enthalpy Btu/lbm			Entropy Btu/lbm-°F		
		v_f	v_g	h_f	h_{fg}	h_g	s_f	s_{fg}	s_g
32.00	0.089	0.0160	3304.7	−0.018	1075.5	1075.5	0.000	2.1873	2.1873
36.00	0.104	0.0160	2839.0	4.008	1073.2	1077.2	0.0081	2.1651	2.1732
40.00	0.122	0.0160	2445.8	8.027	1071.0	1079.0	0.0162	2.1432	2.1594
50.00	0.178	0.0160	1704.8	18.05	1065.3	1083.4	0.0361	2.0901	2.1262
60.00	0.256	0.0160	1207.6	28.06	1059.7	1087.7	0.0555	2.0391	2.0946
80.00	0.506	0.0161	633.3	48.04	1048.4	1096.4	0.0932	1.9426	2.0359
100.00	0.949	0.0161	350.4	68.0	1037.1	1105.1	0.1295	1.8530	1.9825
101.74	1.000	0.0161	333.60	69.7	1036.1	1105.8	0.1326	1.8455	1.9781
120.00	1.693	0.0162	203.26	89.0	1025.6	1113.6	0.1646	1.7693	1.9339
140.00	2.889	0.0163	123.00	108.0	1014.0	1122.0	0.1985	1.6910	1.8895
160.00	4.741	0.0164	77.29	128.0	1002.2	1130.2	0.2313	1.6174	1.8487
162.24	5.000	0.0164	73.532	130.2	1000.9	1131.1	0.2349	1.6094	1.8443
180.00	7.51	0.0165	50.225	148.0	990.2	1138.2	0.2631	1.5480	1.8111
193.21	10.000	0.0166	38.420	161.2	982.1	1143.3	0.2836	1.5043	1.7879
200.00	11.53	0.0166	33.639	168.1	977.9	1146.0	0.2940	1.4824	1.7764
212.00	14.696	0.0167	26.799	180.2	970.3	1150.5	0.3121	1.4447	1.7568
213.03	15.000	0.0167	26.290	181.2	969.7	1150.9	0.3137	1.4415	1.7552
220.00	17.19	0.0168	23.148	188.2	965.2	1153.4	0.3241	1.4201	1.7442
227.96	20.00	0.0168	20.087	196.2	960.1	1156.3	0.3358	1.3962	1.7320
240.00	24.97	0.0169	16.321	208.5	952.1	1160.6	0.3533	1.3609	1.7142
250.34	30.00	0.0170	13.744	218.9	945.2	1164.1	0.3682	1.3313	1.6995
260.00	35.43	0.0171	11.762	228.8	938.6	1167.4	0.3819	1.3043	1.6862
267.25	40.00	0.0172	10.497	236.1	933.6	1169.8	0.3921	1.2844	1.6765
280.00	49.20	0.0173	8.644	249.2	924.6	1173.8	0.4098	1.2501	1.6599
292.71	60.00	0.0174	7.174	262.2	915.4	1177.6	0.4273	1.2167	1.6440
300.00	67.01	0.0175	6.466	269.7	910.0	1179.7	0.4372	1.1979	1.6351
312.04	80.00	0.0176	5.471	282.1	901.0	1183.1	0.4534	1.1675	1.6208
327.82	100.00	0.0177	4.431	298.5	888.6	1187.2	0.4743	1.1284	1.6027
340.00	118.0	0.0179	3.788	311.3	878.8	1190.1	0.4902	1.0990	1.5892
358.43	150.0	0.0181	3.014	330.6	863.4	1194.1	0.5141	1.0554	1.5695
380.00	195.7	0.0184	2.335	353.6	844.5	1198.0	0.5416	1.0057	1.5473
400.97	250.0	0.0187	1.843	376.1	825.0	1201.1	0.5679	0.9585	1.5264
420.00	308.8	0.0189	1.500	396.9	806.2	1203.1	0.5915	0.9165	1.5080
444.60	400.0	0.0193	1.161	424.2	780.4	1204.6	0.6217	0.8630	1.4847
500.00	680.9	0.0204	0.675	487.9	714.3	1202.2	0.6890	0.7443	1.4333
518.21	800.0	0.0209	0.569	509.8	689.6	1199.4	0.7111	0.7051	1.4163
550.00	1045.4	0.0218	0.424	549.4	641.8	1191.2	0.7501	0.6356	1.3856
567.19	1200.0	0.0223	0.362	571.9	613.0	1184.8	0.7714	0.5969	1.3683
600.00	1543.2	0.0236	0.267	617.1	550.6	1167.7	0.8134	0.5196	1.3330
650.00	2208.4	0.0268	0.162	696.4	425.0	1121.4	0.8837	0.3830	1.2667
700.00	3094.3	0.0366	0.0752	822.5	172.7	995.2	0.9901	0.1490	1.1390
705.47	3208.2	0.0508	0.0508	906.0	000.0	906.0	1.0612	0.0000	1.0612

Note: Data abstracted from ASME Steam Tables, copyright 1967, and reproduced with permission of the American Society of Mechanical Engineers.

Table B.2
Steam Table
Superheated Steam

Temperature, °F, t

Pressure psia, p (Sat. Temp.)		Sat. Vapor	200	300	400	500	600	700	800	1000	1200	1400	1500
1 (101.74)	v	333.60	392.50	452.3	511.90	571.5	631.1	690.7					
	h	1105.8	1150.2	1195.7	1241.8	1288.6	1336.1	1384.5					
	s	1.9781	2.0209	2.1152	2.1722	2.2237	2.2708	2.3144					
4 (152.97)	v	90.63	97.79	112.86	127.85	142.79	157.71	172.62	187.53	217.33	247.13	276.92	291.82
	h	1127.3	1149.0	1195.0	1241.4	1288.3	1335.9	1384.3	1433.6	1534.8	1639.6	1748.0	1803.5
	s	1.8625	1.8967	1.9617	2.0190	2.0707	2.1178	2.1615	2.2022	2.2767	2.3440	2.4057	2.4347
10 (193.21)	v	38.42	38.84	44.98	51.03	57.04	63.03	69.00	74.98	86.91	98.84	110.76	116.72
	h	1143.3	1146.6	1193.7	1240.6	1287.8	1335.5	1384.0	1433.4	1534.6	1639.5	1747.9	1803.4
	s	1.7879	1.7928	1.8593	1.9173	1.9692	2.0166	2.0603	2.1011	2.1757	2.2430	2.3046	2.3337
15 (213.03)	v	26.290		29.899	33.963	37.985	41.986	45.978	49.964	57.926	65.882	73.833	77.807
	h	1150.9		1192.5	1239.9	1287.3	1335.2	1383.8	1433.2	1534.5	1639.4	1747.8	1803.4
	s	1.7552		1.8134	1.8720	1.9242	1.9717	2.0155	2.0563	2.1309	2.1982	2.2599	2.2890
20 (227.96)	v	20.087		22.356	25.428	28.457	31.466	34.465	37.458	43.435	49.405	55.370	58.352
	h	1156.3		1191.4	1239.2	1286.9	1334.9	1383.5	1432.9	1534.3	1639.3	1747.8	1803.3
	s	1.7320		1.7805	1.8397	1.8921	1.9397	1.9836	2.0244	2.0991	2.1665	2.2282	2.2572
30 (250.34)	v	13.744		14.810	16.892	18.929	20.945	22.951	24.952	28.943	32.927	36.907	38.896
	h	1164.1		1189.0	1237.8	1286.0	1334.2	1383.0	1432.5	1534.0	1639.0	1747.6	1803.2
	s	1.6995		1.7334	1.7937	1.8467	1.8946	1.9386	1.9795	2.0543	2.1217	2.1834	2.2125
50 (281.01)	v	8.515		8.769	10.062	11.306	12.529	13.741	14.947	17.350	19.746	22.137	23.332
	h	1174.1		1184.1	1234.9	1284.1	1332.9	1382.0	1431.7	1533.4	1638.6	1747.3	1802.9
	s	1.6585		1.6720	1.7349	1.7890	1.8374	1.8816	1.9227	1.9977	2.0652	2.1270	2.1561
75 (307.60)	v	5.816			6.645	7.494	8.320	9.135	9.945	11.553	13.155	14.752	15.550
	h	1181.9			1231.2	1281.7	1331.3	1380.7	1430.7	1532.7	1638.1	1746.9	1802.6
	s	1.6259			1.6868	1.7424	1.7915	1.8361	1.8774	1.9526	2.0202	2.0821	2.1113

Superheated Steam table (temperature column headers not printed on this page). For each pressure the rows are v (specific volume), h (enthalpy), and s (entropy). The first data column is the saturation value.

P, psia (Sat. temp)		Sat.									
100 (327.82)	v	4.431	4.935	5.588	6.216	6.833	7.443	8.655	9.860	11.060	11.659
	h	1187.2	1227.4	1279.3	1329.6	1379.5	1429.7	1532.0	1637.6	1746.5	1802.2
	s	1.6027	1.6516	1.7088	1.7586	1.8036	1.8451	1.9205	1.9883	2.0502	2.0794
150 (358.43)	v	3.014	3.2208	3.6799	4.1112	4.5298	4.9421	5.7568	6.5642	7.3671	7.7674
	h	1194.1	1219.1	1274.3	1326.1	1376.9	1427.6	1530.5	1636.5	1745.7	1801.7
	s	1.5695	1.5993	1.6602	1.7115	1.7573	1.7992	1.8751	1.9431	2.0052	2.0344
200 (381.80)	v	2.287	2.3598	2.7247	3.0583	3.3783	3.6915	4.3077	4.9165	5.5209	5.8219
	h	1198.3	1210.1	1269.0	1322.6	1374.3	1425.5	1529.1	1635.4	1745.0	1800.9
	s	1.5454	1.5593	1.6242	1.6773	1.7239	1.7663	1.8426	1.9109	1.9732	2.0025
250 (400.97)	v	1.843		2.1504	2.4262	2.6872	2.9410	3.4382	3.9278	4.4131	4.6546
	h	1201.1		1263.5	1319.0	1371.6	1423.4	1527.6	1634.4	1744.2	1800.2
	s	1.5264		1.5951	1.6502	1.6976	1.7405	1.8173	1.8858	1.9482	1.9776
300 (417.35)	v	1.543		1.7665	2.0044	2.2263	2.4407	2.8585	3.2688	3.6746	3.8764
	h	1202.9		1257.7	1315.2	1368.9	1421.3	1526.2	1633.3	1743.4	1799.6
	s	1.5105		1.5703	1.6274	1.6758	1.7192	1.7964	1.8652	1.9278	1.9572
350 (431.73)	v	1.326		1.4913	1.7028	1.8970	2.0832	2.4445	2.7980	3.1471	3.3205
	h	1204.0		1251.5	1311.4	1366.2	1419.2	1524.7	1632.3	1742.6	1798.9
	s	1.4968		1.5483	1.6077	1.6571	1.7009	1.7787	1.8477	1.9105	1.9400
400 (444.60)	v	1.161		1.2841	1.4763	1.6499	1.8151	2.1339	2.4450	2.7515	2.9037
	h	1204.6		1245.1	1307.4	1363.4	1417.0	1523.3	1631.2	1741.9	1798.2
	s	1.4847		1.5282	1.5901	1.6406	1.6850	1.7632	1.8325	1.8955	1.9250
500 (461.01)	v	0.928		0.9919	1.1584	1.3037	1.4397	1.6992	1.9507	2.1977	2.3200
	h	1204.7		1231.2	1299.1	1357.7	1412.7	1520.3	1629.6	1740.3	1796.9
	s	1.4639		1.4921	1.5652	1.6176	1.6578	1.7371	1.8069	1.8702	1.8998
600 (486.20)	v	0.770		0.7944	0.9456	1.0726	1.1892	1.4093	1.6211	1.8284	1.9309
	h	1203.7		1215.9	1290.3	1351.8	1408.3	1517.4	1627.0	1738.8	1795.6
	s	1.4461		1.4590	1.5329	1.5884	1.6351	1.7155	1.7859	1.8494	1.8792
700 (503.08)	v	0.656			0.7928	0.9072	1.0102	1.2023	1.3858	1.5647	1.6530
	h	1201.8			1281.0	1345.6	1403.7	1514.4	1624.8	1737.2	1794.3
	s	1.4304			1.5090	1.5673	1.6154	1.6970	1.7679	1.8318	1.8617
800 (518.21)	v	0.569			0.6774	0.7828	0.8759	1.0470	1.2093	1.3669	1.4446
	h	1199.4			1271.1	1339.3	1399.1	1511.4	1622.7	1735.7	1792.9
	s	1.4163			1.4869	1.5484	1.5980	1.6807	1.7522	1.8164	1.8464

Table B.2, Contd.

Pressure psia, p (Sat. Temp.)		Sat. Vapor	Temperature, °F, t										
			200	300	400	500	600	700	800	1000	1200	1400	1500
900 (531.95)	v	0.501					0.5869	0.6858	0.7713	0.9262	1.0720	1.2131	1.2825
	h	1196.4					1260.6	1332.7	1394.4	1508.5	1620.6	1734.1	1791.6
	s	1.4032					1.4659	1.5311	1.5822	1.6662	1.7382	1.8028	1.8329
1000 (544.58)	v	0.446					0.5137	0.6080	0.6875	0.8295	0.9622	1.0901	1.1529
	h	1192.9					1249.3	1325.9	1389.6	1505.4	1618.4	1732.5	1790.3
	s	1.3910					1.4457	1.5149	1.5677	1.6530	1.7256	1.7905	1.8207
1200 (567.19)	v	0.362					0.4016	0.4905	0.5615	0.6845	0.7974	0.9055	0.9584
	h	1184.8					1224.2	1311.5	1379.7	1499.4	1614.2	1729.4	1787.6
	s	1.3683					1.4061	1.4851	1.5415	1.6298	1.7035	1.7691	1.7996
1400 (587.07)	v	0.302					0.3176	0.4059	0.4712	0.5809	0.6798	0.7737	0.8195
	h	1175.3					1194.1	1296.1	1369.3	1493.2	1609.9	1726.3	1785.0
	s	1.3474					1.3652	1.4575	1.5182	1.6096	1.6845	1.7508	1.7815
1600 (604.87)	v	0.255						0.3415	0.4032	0.5031	0.5915	0.6748	0.7153
	h	1164.5						1279.4	1358.5	1486.9	1605.6	1723.2	1782.3
	s	1.3274						1.4312	1.4968	1.5916	1.6678	1.7347	1.7657
1800 (621.02)	v	0.219						0.2906	0.3500	0.4426	0.5229	0.5980	0.6343
	h	1152.3						1261.1	1347.2	1480.6	1601.2	1720.1	1779.7
	s	1.3079						1.4054	1.4768	1.5753	1.6528	1.7204	1.7516
2000 (635.80)	v	0.188						0.2488	0.3072	0.3942	0.4680	0.5365	0.5695
	h	1138.3						1240.9	1335.4	1474.1	1596.9	1717.0	1777.1
	s	1.2881						1.3794	1.4578	1.5603	1.6391	1.7075	1.7389

2500 (668.11)	v	0.131	0.1681	0.2293	0.3068	0.3692	0.4259	0.4529
	h	1093.3	1176.7	1303.4	1457.5	1585.9	1709.2	1770.4
	s	1.2345	1.3076	1.4129	1.5269	1.6094	1.6796	1.7116
3000 (695.33)	v	0.085	0.0982	0.1759	0.2484	0.3033	0.3522	0.3753
	h	1020.3	1060.5	1267.0	1440.2	1574.8	1701.4	1763.8
	s	1.1619	1.1966	1.3692	1.4976	1.5841	1.6561	1.6888
3200 (705.08)	v	0.0566		0.1588	0.2301	0.2827	0.3291	0.3510
	h	931.6		1250.9	1433.1	1570.3	1698.3	1761.2
	s	1.0832		1.3515	1.4866	1.5749	1.6477	1.6806

Note: Data abstracted from ASME Steam Tables, copyright 1967, and reproduced with permission of the American Society of Mechanical Engineers.

Table B.3

Steam Table
Compressed Liquid

Pressure psia p		Temperature, °F, t						
		32	100	200	300	400	500	600
200	$(v - v_f) \times 10^5$ ft³/lbm	−1.2	−1.0	−0.7	−1.0			
	$(h - h_f)$ Btu/lbm	+0.61	+0.52	+0.42	+0.26			
	$(s - s_f) \times 10^3$ Btu/lbm-°F	+0.0	−0.1	−0.2	−0.3			
400	$(v - v_f) \times 10^5$ ft³/lbm	−2.2	−1.9	−1.7	−2.0	−2.0		
	$(h - h_f)$ Btu/lbm	+1.21	+1.05	+0.88	+0.63	+0.17		
	$(s - s_f) \times 10^3$ Btu/lbm-°F	+0.0	−0.2	−0.5	−0.6	−0.4		
600	$(v - v_f) \times 10^5$ ft³/lbm	−3.2	−3.0	−3.7	−4.0	−4.0		
	$(h - h_f)$ Btu/lbm	+1.82	+0.58	+1.33	+1.00	+0.39		
	$(s - s_f) \times 10^3$ Btu/lbm-°F	+0.0	−0.3	−0.7	−1.0	−1.0		
1000	$(v - v_f) \times 10^5$ ft³/lbm	−5.2	−5.0	−5.7	−7.0	−9.0	−7.0	
	$(h - h_f)$ Btu/lbm	+3.02	+1.37	+2.24	+1.74	+0.86	−0.11	
	$(s - s_f) \times 10^3$ Btu/lbm-°F	+0.1	−0.6	−1.2	−1.7	−2.0	−1.4	
2000	$(v - v_f) \times 10^5$ ft³/lbm	−11.2	−10.0	−10.7	−14.0	−20.0	−29.0	−32.0
	$(h - h_f)$ Btu/lbm	+6.01	+5.26	+4.51	+3.62	+2.09	−0.37	−2.62
	$(s - s_f) \times 10^3$ Btu/lbm-°F	+0.2	−1.2	−2.4	−3.5	−4.6	−5.6	−4.3
3000	$(v - v_f) \times 10^5$ ft³/lbm	−16.2	−14.0	−15.7	−21.0	−31.0	−48.0	−88.0
	$(h - h_f)$ Btu/lbm	+8.97	+7.88	+6.79	+5.52	+3.37	−0.38	−7.02
	$(s - s_f) \times 10^3$ Btu/lbm-°F	+0.2	−1.8	−3.6	−5.2	−7.0	−9.4	−12.5

Note: Data abstracted from ASME Steam Tables, copyright 1967, and reproduced with permission of the American Society of Mechanical Engineers.

Table B.4 Gas Constants

Substance	Symbol	M	R ft-lb$_f$ / lb$_m$ °R	c_p Btu / lb$_m$ °R at 77°F	c_v Btu / lb$_m$ °R at 77°F	k $\frac{c_p}{c_v}$
Acetylene............	C_2H_2	26.038	59.39	0.4030	0.3267	1.234
Air.................	......	28.967	53.36	0.2404	0.1718	1.399
Ammonia............	NH_3	17.032	90.77	0.5006	0.3840	1.304
Argon..............	A	39.944	38.73	0.1244	0.0746	1.668
Benzene............	C_6H_6	78.114	19.78	0.2497	0.2243	1.113
n-Butane............	C_4H_{10}	58.124	26.61	0.4004	0.3662	1.093
Isobutane............	C_4H_{10}	58.124	26.59	0.3979	0.3637	1.094
1-Butene............	C_4H_8	56.108	27.545	0.3646	0.3282	1.111
Carbon dioxide........	CO_2	44.011	35.12	0.2015	0.1564	1.288
Carbon monoxide.......	CO	28.011	55.19	0.2485	0.1776	1.399
Carbon tetrachloride...	CCl_4	153.839				
n-Deuterium.........	D_2	4.029				
Dodecane............	$C_{12}H_{26}$	170.340	9.074	0.3931	0.3814	1.031
Ethane.............	C_2H_6	30.070	51.43	0.4183	0.3522	1.188
Ethyl ether...........	$C_4H_{10}O$	74.124				
Ethylene............	C_2H_4	28.054	55.13	0.3708	0.3000	1.236
Freon, F-12..........	CCl_2F_2	120.925	12.78	0.1369	0.1204	1.136
Helium..............	He	4.003	386.33	1.241	0.7446	1.667
n-Heptane...........	C_7H_{16}	100.205	15.42	0.3956	0.3758	1.053
n-Hexane............	C_6H_{14}	86.178	17.93	0.3966	0.3736	1.062
Hydrogen............	H_2	2.016	766.53	3.416	2.431	1.405
Hydrogen sulfide.......	H_2S	34.082				
Mercury.............	Hg	200.610				
Methane............	CH_4	16.043	96.40	0.5318	0.4079	1.304
Methyl fluoride.......	CH_3F	34.035				
Neon................	Ne	20.183	76.58	0.2460	0.1476	1.667
Nitric oxide...........	NO	30.008	51.49	0.2377	0.1715	1.386
Nitrogen.............	N_2	28.016	55.15	0.2483	0.1774	1.400
Octane..............	C_8H_{18}	114.232	13.54	0.3949	0.3775	1.046
Oxygen.............	O_2	32.000	48.29	0.2191	0.1570	1.396
n-Pentane............	C_5H_{12}	72.151	21.42	0.3980	0.3705	1.074
Isopentane............	C_5H_{12}	72.151	21.42	0.3972	0.3697	1.074
Propane.............	C_3H_8	44.097	35.07	0.3982	0.3531	1.128
Propylene............	C_3H_6	42.081	36.72	0.3627	0.3055	1.187
Sulfur dioxide.........	SO_2	64.066	24.12	0.1483	0.1173	1.264
Water vapor.........	H_2O	18.016	85.80	0.4452	0.3349	1.329
Xenon...............	Xe	131.300	11.78	0.03781	0.02269	1.667

Source: E. F. Obert, *Concepts of Thermodynamics*, copyright 1960, McGraw-Hill. Used with permission of McGraw-Hill Book Company.

Note: 1 Btu/lbm °R = 1 calorie/g°K

Note: Data selected from J.F. Masi, Trans. ASME, 76:1067 (October, 1954); National Bureau of Standards (U.S.). Circ. 500, February 1952; "Selected Values of Properties of Hydrocarbons and Related Compounds," American Petroleum Institute Research Project 44, Thermodynamics Research Center, Texas A & M University; College Station, Texas (Loose Leaf Data Sheets, extant 1972).

Table B.5
Critical Properties

Substance	Ref. date	Symbol	M	T_c °K	p_c atm	v_c cm³/g mole	v_c ft³/ mole	z_c
Acetylene..........	1928	C_2H_2	26.038	309.5	61.6	113	.....	0.274
Air................	1917	......	28.967	132.41	37.25	93.25		
Ammonia...........	1920	NH_3	17.032	405.4	111.3	72.5	1.16	0.243
Argon..............	1910	A	39.944	150.72	47.996	75	1.20	0.291
Benzene............	1948	C_6H_6	78.114	562.6	48.6	260	4.17	0.274
n-Butane...........	1939	C_4H_{10}	58.124	425.17	37.47	255	4.08	0.274
Isobutane..........	1910	C_4H_{10}	58.124	408.14	36.00	263	4.21	0.283
1-Butene...........	1950	C_4H_8	56.108	419.6	39.7	240	3.84	0.277
Carbon dioxide.....	1950	CO_2	44.011	304.20	72.90	94	1.51	0.275
Carbon monoxide.....	1936	CO	28.011	132.91	34.529	93	1.49	0.294
Carbon tetrachloride..	1931	CCl_4	153.839	556.4	45.0	276	.....	0.272
n-Deuterium.........	1951	D_2	4.029	38.43	16.421			
Dodecane...........	1953	$C_{12}H_{26}$	170.340	659	17.9	.....	11.5	0.237
Ethane.............	1939	C_2H_6	30.070	305.43	48.20	148	2.37	0.285
Ethyl ether.........	1929	$C_4H_{10}O$	74.124	467.8	35.6	282.9		
Ethylene...........	1939	C_2H_4	28.054	283.06	50.50	124	1.99	0.270
Freon, F-12........	1957	CCl_2F_2	120.925	385.16	40.6	217	3.47	0.279
Helium.............	1936	He	4.003	5.19	2.26	58	0.929	0.308
n-Heptane..........	1937	C_7H_{16}	100.205	540.17	27.00	426	6.82	0.260
n-Hexane...........	1946	C_6H_{14}	86.178	507.9	29.94	368	5.89	0.264
Hydrogen...........	1951	H_2	2.016	33.24	12.797	65	1.04	0.304
Hydrogen sulfide.....	1948	H_2S	34.082	373.7	88.8	98	1.57	0.284
Mercury............	1953	Hg	200.610					
Methane...........	1953	CH_4	16.043	190.7	45.8	99	1.59	0.290
Methyl fluoride......	1932	CH_3F	34.035	317.71	58.0			
Neon...............	1936	Ne	20.183	44.39	26.86	41.7	0.668	0.308
Nitric oxide.........	1951	NO	30.008	179.2	65.0	58	0.929	0.256
Nitrogen...........	1951	N_2	28.016	126.2	33.54	90	0.144	0.291
Octane.............	1931	C_8H_{18}	114.232	569.4	24.64	486	7.77	0.256
Oxygen.............	1948	O_2	32.000	154.78	50.14	74	1.19	0.292
n-Pentane..........	1899	C_5H_{12}	72.151	469.78	33.31	311	4.98	0.269
Isopentane.........	1910	C_5H_{12}	72.151	461.0	32.92	308	4.93	0.268
Propane............	1940	C_3H_8	44.097	370.01	42.1	200	3.20	0.277
Propylene..........	1953	C_3H_6	42.081	365.1	45.40	181	2.90	0.274
Sulfur dioxide.......	1945	SO_2	64.066	430.7	77.8	122	.....	0.269
Water..............	1934	H_2O	18.016	647.27	218.167	56	0.897	0.230
Xenon..............	1951	Xe	131.330	289.81	58.0	118.8	1.90	0.290

Source: E. F. Obert, *Concepts of Thermodynamics*, copyright 1960, McGraw-Hill. Used with permission of McGraw-Hill Book Company.

Table B.6

Air Table
Properties of Air at Low Pressure*

T °R	h Btu/lbm	p_r	u Btu/lbm	v_r	ϕ Btu/lbm-°R
200	47.67	.04320	33.96	1714.9	0.36303
250	59.64	.09415	42.50	983.6	0.41643
300	71.61	.17795	51.04	624.5	0.46007
350	83.57	.3048	59.58	425.4	0.49695
400	95.53	.4858	68.11	305.0	0.52890
450	107.50	.7329	76.65	227.45	0.55710
500	119.48	1.0590	85.20	174.90	0.58233
525	125.47	1.2560	89.48	154.84	0.59403
550	131.46	1.4779	93.76	137.85	0.60518
575	137.47	1.7269	98.05	123.34	0.61586
600	143.47	2.005	102.34	110.88	0.62607
625	149.49	2.313	106.64	100.08	0.63589
650	155.50	2.655	110.94	90.69	0.64533
675	161.54	3.032	115.26	82.47	0.65443
700	167.56	3.446	119.58	75.25	0.66321
725	173.60	3.900	123.91	68.86	0.67169
750	179.66	4.396	128.25	63.20	0.67991
800	191.81	5.526	136.97	53.63	0.69558
850	204.01	6.856	145.74	45.92	0.71037
900	216.26	8.411	154.57	39.64	0.72438
950	228.58	10.216	163.46	34.45	0.73771
1000	240.98	12.298	172.43	30.12	0.75042
1050	253.45	14.686	181.47	26.48	0.76259
1100	265.99	17.413	190.58	23.40	0.77426
1150	278.61	20.51	199.78	20.771	0.78548
1200	291.30	24.01	209.05	18.514	0.79628
1250	304.08	27.96	218.40	16.563	0.80672
1300	316.94	32.39	227.83	14.868	0.81680
1350	329.88	37.35	237.34	13.391	0.82658
1400	342.90	42.88	246.93	12.095	0.83604
1450	356.00	49.03	256.60	10.954	0.84523
1500	369.17	55.86	266.34	9.948	0.85416
1550	382.42	63.40	276.17	9.056	0.86285
1600	395.74	71.73	286.06	8.263	0.87130
1650	409.13	80.89	296.03	7.556	0.87954
1700	422.59	90.95	306.06	6.924	0.88758

Source: J. H. Keenan and J. Kaye, Gas Tables (New York, 1948), John Wiley & Sons, Inc., with permission of authors and publisher.
*Per pound-mass.

Table B.6, Contd.

T °R	h Btu/lbm	p_r	u Btu/lbm	v_r	ϕ Btu/lbm-°R
1750	436.12	101.98	316.16	6.357	0.89542
1800	449.71	114.03	326.32	5.847	0.90308
1900	477.09	141.51	346.85	4.974	0.91788
2000	504.71	174.00	367.61	4.258	0.93205
2200	560.59	256.6	409.78	3.176	0.95868
2400	617.22	367.6	452.70	2.419	0.98331
2600	674.49	513.5	496.26	1.8756	1.00623
3000	790.68	941.4	585.04	1.1803	1.04779
3500	938.40	1829.3	698.48	0.7087	1.09332
4000	1088.26	3280	814.06	0.4518	1.13334
4500	1239.86	5521	931.39	0.3019	1.16905
5000	1392.87	8837	1050.12	0.20959	1.20129
5500	1547.07	13568	1170.04	0.15016	1.23068
6000	1702.29	20120	1291.00	0.11047	1.25769
6500	1858.44	28974	1412.87	0.08310	1.28268

1960

Table B.7
Heat of Combustion ($-\Delta H°$ at 77°F)

Substance	Symbol	h (h_{fg}) of vaporization, Btu/lb$_m$	HHV H$_2$O(l) and CO$_2$(g) kcal g mole	HHV Btu lb$_m$	LHV H$_2$O(g) and CO$_2$(g) kcal g mole	LHV Btu lb$_m$
Acetylene.........	C$_2$H$_2$(g)	...	310.62	21,460	300.10	20,734
Benzene..........	C$_6$H$_6$(g)	186	789.08	18,172	757.52	17,446
n-Butane.........	C$_4$H$_{10}$(g)	156	687.65	21,283	635.05	19,655
Isobutane........	C$_4$H$_{10}$(g)	141	685.65	21,221	633.05	19,593
1-Butene.........	C$_4$H$_{10}$(g)	156	649.45	20,824	607.37	19,475
Carbon...........	C(graphite)	...	94.0518	14,086		
Carbon monoxide..	CO(g)	...	67.6361	4,343.6		
n-Decane.........	C$_{10}$H$_{22}$(g)	155	1632.34	20,638	1516.63	19,175
n-Dodecane.......	C$_{12}$H$_{26}$(g)	155	1947.23	20,564	1810.48	19,120
Ethane...........	C$_2$H$_6$(g)	...	372.82	22,304	341.26	20,416
Ethylene.........	C$_2$H$_4$(g)	...	337.23	21,625	316.20	20,276
n-Heptane........	C$_7$H$_{16}$(g)	157	1160.01	20,825	1075.85	19,314
n-Hexane.........	C$_6$H$_{14}$(g)	157	1002.57	20,928	928.93	19,391
Hydrogen.........	H$_2$(g)	...	68.3174	60,957	57.7979	51,571
Methane.........	CH$_4$(g)	...	212.80	23,861	191.76	21,502
n-Nonane........	C$_9$H$_{20}$(g)	156	1474.90	20,687	1369.70	19,211
n-Octane.........	C$_8$H$_{18}$(g)	156	1317.45	20,747	1222.77	19,256
n-Pentane........	C$_5$H$_{12}$(g)	157	845.16	21,072	782.04	19,499
Isopentane.......	C$_5$H$_{12}$(g)	147	843.24	21,025	780.12	19,451
Propane..........	C$_3$H$_8$(g)	147	530.6	21,646	488.53	19,929
Propylene........	C$_3$H$_6$(g)	...	491.99	21,032	460.43	19,683

Source: E. F. Obert, Concepts of Thermodynamics, McGraw-Hill, copyright 1960; by permission of McGraw-Hill Book Company.

Note: Data from "Selected Values of Properties of Hydrocarbons and Related Compounds," America Petroleum Institute Research Project 44, Thermodynamics Research Center, Texas A & M University, College Station, Texas (Loose Leaf Data Sheets, extant 1972).

Table B.8

Freon-12

Properties of the Saturation State

TEMP. °F	PRESSURE		VOLUME cu ft/lb		ENTHALPY Btu/lb			ENTROPY Btu/(lb)(° R)		TEMP. °F
	PSIA	PSIG	LIQUID v_f	VAPOR v_g	LIQUID h_f	LATENT h_{fg}	VAPOR h_g	LIQUID s_f	VAPOR s_g	
−100	1.4280	27.0138*	0.0099847	22.164	−12.466	78.714	66.248	−0.032005	0.18683	−100
−90	2.0509	25.7456*	0.010073	15.821	−10.409	77.764	67.355	−0.026367	0.18398	−90
−80	2.8807	24.0560*	0.010164	11.533-	− 8.3451	76.812	68.467	−0.020862	0.18143	−80
−70	3.9651	21.8482*	0.010259	8.5687	− 6.2730	75.853	69.580	−0.015481	0.17916	−70
−60	5.3575	19.0133*	0.010357	6.4774	− 4.1919	74.885	70.693	−0.010214	0.17714	−60
−50	7.1168	15.4313*	0.010459	4.9742	− 2.1011	73.906	71.805	−0.005056	0.17533	−50
−40	9.3076	10.9709*	0.010564	3.8750	0	72.913	72.913	0	0.17373	−40
−30	11.999	5.490*	0.010674	3.0585	2.1120	71.903	74.015	0.004961	0.17229	−30
−25	13.556	2.320*	0.010730	2.7295	3.1724	71.391	74.563	0.007407	0.17164	−25
−20	15.267	0.571	0.010788	2.4429	4.2357	70.874	75.110	0.009831	0.17102	−20
−15	17.141	2.445	0.010846	2.1924	5.3020	70.352	75.654	0.012234	0.17043	−15
−10	19.189	4.493	0.010906	1.9727	6.3716	69.824	76.196	0.014617	0.16989	−10
− 5	21.422	6.726	0.010968	1.7794	7.4444	69.291	76.735	0.016979	0.16937	− 5
0	23.849	9.153	0.011030	1.6089	8.5207	68.750	77.271	0.019323	0.16888	0
5	26.483	11.787	0.011094	1.4580	9.6005	68.204	77.805	0.021647	0.16842	5
10	29.335	14.639	0.011160	1.3241	10.684	67.651	78.335	0.023954	0.16798	10
15	32.415	17.719	0.011227	1.2050	11.771	67.090	78.861	0.026243	0.16758	15
20	35.736	21.040	0.011296	1.0988	12.863	66.522	79.385	0.028515	0.16719	20
25	39.310	24.614	0.011366	1.0039	13.958	65.946	79.904	0.030772	0.16683	25
30	43.148	28.452	0.011438	0.91880	15.058	65.361	80.419	0.033013	0.16648	30
40	51.667	36.971	0.011588	0.77357	17.273	64.163	81.436	0.037453	0.16586	40
50	61.394	46.698	0.011746	0.65537	19.507	62.926	82.433	0.041839	0.16530	50
60	72.433	57.737	0.011913	0.55839	21.766	61.643	83.409	0.046180	0.16479	60
70	84.888	70.192	0.012089	0.47818	24.050	60.309	84.359	0.050482	0.16434	70
80	98.870	84.174	0.012277	0.41135	26.365	58.917	85.282	0.054751	0.16392	80
90	114.49	99.79	0.012478	0.35529	28.713	57.461	86.174	0.058997	0.16353	90
100	131.86	117.16	0.012693	0.30794	31.100	55.929	87.029	0.063227	0.16315	100
120	172.35	157.65	0.013174	0.23326	36.013	52.597	88.610	0.071680	0.16241	120
140	221.32	206.62	0.013746	0.17799	41.162	48.805	89.967	0.080205	0.16159	140
160	279.82	265.12	0.014449	0.13604	46.633	44.373	91.006	0.088927	0.16053	160
180	349.00	334.30	0.015360	0.10330	52.562	38.999	91.561	0.098039	0.15900	180
200	430.09	415.39	0.016659	0.076728	59.203	32.075	91.278	0.10789	0.15651	200
233.6 (Critical)	596.9	582.2	0.02870	0.02870	78.86	0	78.86	0.1359	0.1359	233 6 (Critical)

Source: "Thermodynamic Properties of Freon-12," copyright 1956; with permission of E. I. du Pont de Nemours & Co.

V=volume in cu ft/lb;   H=enthalpy in Btu/lb;   S=entropy in Btu/(lb)(°R)   (saturation properties in parentheses)

ABSOLUTE PRESSURE, lb/sq in

1.0 — 27.8852* (−109.24 °F)

TEMP. °F	V	H	S
	(30.896)	(65.229)	(0.18977)
−110	—	—	—
−100	31.730	66.303	0.19279
−90	32.631	67.482	0.19602
−80	33.531	68.679	0.19922
−70	34.429	69.892	0.20237
−60	35.327	71.123	0.20549
−50	36.223	72.370	0.20857
−40	37.119	73.633	0.21162
−30	38.014	74.913	0.21463
−20	38.908	76.208	0.21761
−10	39.802	77.519	0.22056
0	40.695	78.845	0.22347
10	41.587	80.186	0.22636
20	42.480	81.542	0.22922
30	43.372	82.912	0.23204
40	44.263	84.297	0.23484
50	45.154	85.695	0.23761
60	46.045	87.107	0.24036
70	46.936	88.533	0.24307
80	47.826	89.971	0.24576
90	48.716	91.423	0.24843
100	49.606	92.887	0.25107
110	50.496	94.364	0.25368
120	51.386	95.853	0.25628
130	52.275	97.354	0.25884
140	53.165	98.867	0.26139
150	54.054	100.391	0.26391
160	54.943	101.926	0.26640
170	55.832	103.472	0.26888
180	56.721	105.030	0.27133

5.0 — 19.7412* (−62.35 °F)

TEMP. °F	V	H	S
	(6.9069)	(70.432)	(0.17759)
−60	6.9509	70.729	0.17834
−50	7.1378	72.003	0.18149
−40	7.3239	73.291	0.18459
−30	7.5092	74.593	0.18766
−20	7.6938	75.909	0.19069
−10	7.8777	77.239	0.19368
0	8.0611	78.582	0.19663
10	8.2441	79.939	0.19955
20	8.4265	81.309	0.20244
30	8.6086	82.693	0.20529
40	8.7903	84.090	0.20812
50	8.9717	85.500	0.21091
60	9.1528	86.922	0.21367
70	9.3336	88.358	0.21641
80	9.5142	89.806	0.21912
90	9.6945	91.266	0.22180
100	9.8747	92.738	0.22445
110	10.055	94.222	0.22708
120	10.234	95.717	0.22968
130	10.414	97.224	0.23226
140	10.594	98.743	0.23481
150	10.773	100.272	0.23734
160	10.952	101.812	0.23985
170	11.131	103.363	0.24233
180	11.311	104.925	0.24479
190	11.489	106.497	0.24723
200	11.668	108.079	0.24964
210	11.847	109.670	0.25204
220	12.026	111.272	0.25441
230	12.205	112.883	0.25677

10 — 9.561* (−37.23 °F)

TEMP. °F	V	H	S
	(3.6246)	(73.219)	(0.17331)
−40	—	—	—
−30	3.6945	74.183	0.17557
−20	3.7906	75.526	0.17866
−10	3.8861	76.880	0.18171
0	3.9809	78.246	0.18471
10	4.0753	79.624	0.18768
20	4.1691	81.014	0.19061
30	4.2626	82.415	0.19350
40	4.3556	83.828	0.19635
50	4.4484	85.252	0.19918
60	4.5408	86.689	0.20197
70	4.6329	88.136	0.20473
80	4.7248	89.596	0.20746
90	4.8165	91.067	0.21016
100	4.9079	92.548	0.21283
110	4.9992	94.042	0.21547
120	5.0903	95.546	0.21809
130	5.1812	97.061	0.22068
140	5.2720	98.586	0.22325
150	5.3627	100.123	0.22579
160	5.4533	101.669	0.22830
170	5.5437	103.226	0.23080
180	5.6341	104.793	0.23326
190	5.7243	106.370	0.23571
200	5.8145	107.957	0.23813
210	5.9046	109.553	0.24054
220	5.9946	111.159	0.24291
230	6.0846	112.774	0.24527
240	6.1745	114.398	0.24761
250	6.2643	116.031	0.24993

15 — 0.304 psig (−20.75 °F)

TEMP. °F	V	H	S
	(2.4835)	(75.028)	(0.17111)
−20	2.4885	75.131	0.17134
−10	2.5546	76.512	0.17445
0	2.6201	77.902	0.17751
10	2.6850	79.302	0.18052
20	2.7494	80.712	0.18349
30	2.8134	82.131	0.18642
40	2.8770	83.561	0.18931
50	2.9402	85.001	0.19216
60	3.0031	86.451	0.19498
70	3.0657	87.912	0.19776
80	3.1281	89.383	0.20051
90	3.1902	90.865	0.20324
100	3.2521	92.357	0.20593
110	3.3139	93.860	0.20859
120	3.3754	95.373	0.21122
130	3.4368	96.896	0.21382
140	3.4981	98.429	0.21640
150	3.5592	99.972	0.21895
160	3.6202	101.525	0.22148
170	3.6811	103.088	0.22398
180	3.7419	104.661	0.22646
190	3.8025	106.243	0.22891
200	3.8632	107.835	0.23135
210	3.9237	109.436	0.23375
220	3.9841	111.046	0.23614
230	4.0445	112.665	0.23850
240	4.1049	114.292	0.24085
250	4.1651	115.929	0.24317
260	4.2254	117.574	0.24547
270	4.2855	119.227	0.24775

20 — 5.304 psig (−8.13 °F)

TEMP. °F	V	H	S
	(1.8977)	(76.397)	(0.16969)
0	1.9390	77.550	0.17222
10	1.9893	78.973	0.17528
20	2.0391	80.403	0.17829
30	2.0884	81.842	0.18126
40	2.1373	83.289	0.18419
50	2.1858	84.745	0.18707
60	2.2340	86.210	0.18992
70	2.2819	87.684	0.19273
80	2.3295	89.168	0.19550
90	2.3769	90.661	0.19824
100	2.4241	92.164	0.20095
110	2.4711	93.676	0.20363
120	2.5179	95.198	0.20628
130	2.5645	96.729	0.20890
140	2.6110	98.270	0.21149
150	2.6573	99.820	0.21405
160	2.7036	101.380	0.21659
170	2.7497	102.949	0.21910
180	2.7957	104.528	0.22159
190	2.8416	106.115	0.22405
200	2.8874	107.712	0.22649
210	2.9332	109.317	0.22891
220	2.9789	110.932	0.23130
230	3.0245	112.555	0.23367
240	3.0700	114.186	0.23602
250	3.1155	115.826	0.23835
260	3.1609	117.475	0.24065
270	3.2063	119.131	0.24294
280	3.2517	120.796	0.24520
290	3.2970	122.469	0.24745

25 — 10.304 psig (2.23 °F)

TEMP. °F	V	H	S
	(1.5392)	(77.510)	(0.16867)
10	1.5714	78.635	0.17108
20	1.6125	80.088	0.17414
30	1.6531	81.547	0.17715
40	1.6932	83.012	0.18012
50	1.7329	84.485	0.18304
60	1.7723	85.965	0.18591
70	1.8114	87.453	0.18875
80	1.8502	88.950	0.19155
90	1.8888	90.455	0.19431
100	1.9271	91.968	0.19704
110	1.9653	93.490	0.19973
120	2.0032	95.021	0.20240
130	2.0410	96.561	0.20503
140	2.0786	98.110	0.20763
150	2.1161	99.667	0.21021
160	2.1535	101.234	0.21276
170	2.1908	102.809	0.21528
180	2.2279	104.393	0.21778
190	2.2650	105.986	0.22025
200	2.3019	107.588	0.22269
210	2.3388	109.198	0.22512
220	2.3756	110.817	0.22752
230	2.4124	112.444	0.22989
240	2.4491	114.080	0.23225
250	2.4857	115.723	0.23458
260	2.5223	117.375	0.23689
270	2.5588	119.035	0.23918
280	2.5953	120.703	0.24145
290	2.6317	122.378	0.24370
300	2.6681	124.061	0.24593

30 — 15.304 psig (11.11 °F)

TEMP. °F	V (1.2964)	H (78.452)	S (0.16789)
10	—	—	—
20	1.3278	79.765	0.17065
30	1.3625	81.245	0.17371
40	1.3969	82.730	0.17671
50	1.4308	84.220	0.17966
60	1.4644	85.716	0.18257
70	1.4976	87.219	0.18543
80	1.5306	88.729	0.18826
90	1.5633	90.246	0.19104
100	1.5957	91.770	0.19379
110	1.6280	93.302	0.19650
120	1.6600	94.843	0.19918
130	1.6919	96.391	0.20183
140	1.7237	97.948	0.20445
150	1.7553	99.513	0.20704
160	1.7868	101.086	0.20960
170	1.8181	102.668	0.21213
180	1.8494	104.258	0.21463
190	1.8805	105.857	0.21711
200	1.9116	107.464	0.21957
210	1.9426	109.079	0.22200
220	1.9735	110.702	0.22440
230	2.0043	112.333	0.22679
240	2.0351	113.973	0.22915
250	2.0658	115.620	0.23148
260	2.0965	117.275	0.23380
270	2.1271	118.938	0.23609
280	2.1576	120.609	0.23837
290	2.1882	122.287	0.24062
300	2.2186	123.973	0.24286

40 — 25.304 psig (25.93 °F)

TEMP. °F	V (0.98743)	H (80.000)	S (0.16676)
30	0.99865	80.622	0.16804
40	1.0258	82.148	0.17112
50	1.0526	83.676	0.17415
60	1.0789	85.206	0.17712
70	1.1049	86.739	0.18005
80	1.1306	88.277	0.18292
90	1.1560	89.819	0.18575
100	1.1812	91.367	0.18854
110	1.2061	92.920	0.19129
120	1.2309	94.480	0.19401
130	1.2554	96.047	0.19669
140	1.2798	97.620	0.19933
150	1.3041	99.200	0.20195
160	1.3282	100.788	0.20453
170	1.3522	102.383	0.20708
180	1.3761	103.985	0.20961
190	1.3999	105.595	0.21210
200	1.4236	107.212	0.21457
210	1.4472	108.837	0.21702
220	1.4707	110.469	0.21944
230	1.4942	112.109	0.22183
240	1.5176	113.757	0.22420
250	1.5409	115.412	0.22655
260	1.5642	117.074	0.22888
270	1.5874	118.744	0.23118
280	1.6106	120.421	0.23347
290	1.6337	122.105	0.23573
300	1.6568	123.796	0.23797
310	1.6799	125.494	0.24019
320	1.7029	127.200	0.24239

50 — 35.304 psig (38.15 °F)

TEMP. °F	V (0.79824)	H (81.249)	S (0.16597)
40	0.80248	81.540	0.16655
50	0.82502	83.109	0.16966
60	0.84713	84.676	0.17271
70	0.86886	86.243	0.17569
80	0.89025	87.811	0.17862
90	0.91134	89.380	0.18151
100	0.93216	90.953	0.18434
110	0.95275	92.529	0.18713
120	0.97313	94.110	0.18988
130	0.99332	95.695	0.19259
140	1.0133	97.286	0.19527
150	1.0332	98.882	0.19791
160	1.0529	100.485	0.20051
170	1.0725	102.093	0.20309
180	1.0920	103.708	0.20563
190	1.1114	105.330	0.20815
200	1.1307	106.958	0.21064
210	1.1499	108.593	0.21310
220	1.1690	110.235	0.21553
230	1.1880	111.883	0.21794
240	1.2070	113.539	0.22032
250	1.2259	115.202	0.22268
260	1.2447	116.871	0.22502
270	1.2636	118.547	0.22733
280	1.2823	120.231	0.22962
290	1.3010	121.921	0.23189
300	1.3197	123.618	0.23414
310	1.3383	125.321	0.23637
320	1.3569	127.032	0.23857
330	1.3754	128.749	0.24076

60 — 45.304 psig (48.64 °F)

TEMP. °F	V (0.67005)	H (82.299)	S (0.16537)
50	0.67272	82.518	0.16580
60	0.69210	84.126	0.16892
70	0.71105	85.729	0.17198
80	0.72964	87.330	0.17497
90	0.74790	88.929	0.17791
100	0.76588	90.528	0.18079
110	0.78360	92.128	0.18362
120	0.80110	93.731	0.18641
130	0.81840	95.336	0.18916
140	0.83551	96.945	0.19186
150	0.85247	98.558	0.19453
160	0.86928	100.176	0.19716
170	0.88596	101.799	0.19976
180	0.90252	103.427	0.20233
190	0.91896	105.060	0.20486
200	0.93531	106.700	0.20736
210	0.95157	108.345	0.20984
220	0.96775	109.997	0.21229
230	0.98385	111.655	0.21471
240	0.99988	113.319	0.21710
250	1.0159	114.989	0.21947
260	1.0318	116.666	0.22182
270	1.0476	118.350	0.22414
280	1.0634	120.039	0.22644
290	1.0792	121.736	0.22872
300	1.0949	123.438	0.23098
310	1.1106	125.147	0.23321
320	1.1262	126.863	0.23543
330	1.1418	128.585	0.23762
340	1.1574	130.313	0.23980

70 — 55.304 psig (57.90 °F)

TEMP. °F	V (0.57724)	H (83.206)	S (0.16489)
60	0.58088	83.552	0.16556
70	0.59793	85.196	0.16869
80	0.61458	86.832	0.17175
90	0.63087	88.463	0.17475
100	0.64685	90.091	0.17768
110	0.66256	91.717	0.18056
120	0.67803	93.343	0.18339
130	0.69329	94.969	0.18617
140	0.70836	96.597	0.18891
150	0.72325	98.228	0.19161
160	0.73800	99.862	0.19427
170	0.75260	101.500	0.19689
180	0.76708	103.141	0.19948
190	0.78145	104.788	0.20203
200	0.79571	106.439	0.20455
210	0.80988	108.095	0.20704
220	0.82397	109.756	0.20951
230	0.83798	111.424	0.21194
240	0.85191	113.096	0.21435
250	0.86578	114.775	0.21673
260	0.87959	116.459	0.21909
270	0.89335	118.150	0.22142
280	0.90705	119.846	0.22373
290	0.92070	121.549	0.22601
300	0.93431	123.257	0.22828
310	0.94788	124.972	0.23052
320	0.96142	126.693	0.23274
330	0.97491	128.419	0.23494
340	0.98838	130.152	0.23712
350	1.0018	131.891	0.23928

80 — 65.304 psig (66.21 °F)

TEMP. °F	V (0.50680)	H (84.003)	S (0.16450)
70	0.51269	84.640	0.16571
80	0.52795	86.316	0.16885
90	0.54281	87.981	0.17190
100	0.55734	89.640	0.17489
110	0.57158	91.294	0.17782
120	0.58556	92.945	0.18070
130	0.59931	94.594	0.18352
140	0.61286	96.242	0.18629
150	0.62623	97.891	0.18902
160	0.63943	99.542	0.19170
170	0.65250	101.195	0.19435
180	0.66543	102.851	0.19696
190	0.67824	104.511	0.19953
200	0.69095	106.174	0.20207
210	0.70356	107.841	0.20458
220	0.71609	109.513	0.20706
230	0.72853	111.190	0.20951
240	0.74090	112.872	0.21193
250	0.75320	114.559	0.21432
260	0.76544	116.251	0.21669
270	0.77762	117.949	0.21903
280	0.78975	119.652	0.22135
290	0.80183	121.361	0.22364
300	0.81386	123.075	0.22592
310	0.82586	124.795	0.22817
320	0.83781	126.521	0.23039
330	0.84973	128.253	0.23260
340	0.86161	129.990	0.23479
350	0.87347	131.734	0.23695
360	0.88529	133.482	0.23910

100 — 85.30 psig (80.76 °F)

TEMP. °F	V (0.40674)	H (85.351)	S (0.16389)
80	—	—	—
90	0.41876	86.964	0.16685
100	0.43138	88.694	0.16996
110	0.44365	90.410	0.17300
120	0.45562	92.116	0.17597
130	0.46733	93.814	0.17888
140	0.47881	95.507	0.18172
150	0.49009	97.197	0.18452
160	0.50118	98.884	0.18726
170	0.51212	100.571	0.18996
180	0.52291	102.257	0.19262
190	0.53358	103.944	0.19524
200	0.54413	105.633	0.19782
210	0.55457	107.324	0.20036
220	0.56492	109.018	0.20287
230	0.57519	110.714	0.20535
240	0.58538	112.415	0.20780
250	0.59549	114.119	0.21022
260	0.60554	115.828	0.21261
270	0.61553	117.540	0.21497
280	0.62546	119.258	0.21731
290	0.63534	120.980	0.21962
300	0.64518	122.707	0.22191
310	0.65497	124.439	0.22417
320	0.66472	126.176	0.22641
330	0.67444	127.917	0.22863
340	0.68411	129.665	0.23083
350	0.69376	131.417	0.23301
360	0.70338	133.174	0.23517
370	0.71296	134.937	0.23730

120 — 105.30 psig (93.29 °F)

TEMP. °F	V (0.33886)	H (86.459)	S (0.16340)
100	0.34655	87.675	0.16559
110	0.35766	89.466	0.16876
120	0.36841	91.237	0.17184
130	0.37884	92.992	0.17484
140	0.38901	94.736	0.17778
150	0.39896	96.471	0.18065
160	0.40870	98.199	0.18346
170	0.41826	99.922	0.18622
180	0.42766	101.642	0.18892
190	0.43692	103.359	0.19159
200	0.44606	105.076	0.19421
210	0.45508	106.792	0.19679
220	0.46401	108.509	0.19934
230	0.47284	110.227	0.20185
240	0.48158	111.948	0.20432
250	0.49025	113.670	0.20677
260	0.49885	115.396	0.20918
270	0.50739	117.125	0.21157
280	0.51587	118.857	0.21393
290	0.52429	120.593	0.21626
300	0.53267	122.333	0.21856
310	0.54100	124.077	0.22084
320	0.54929	125.825	0.22310
330	0.55754	127.578	0.22533
340	0.56575	129.335	0.22754
350	0.57393	131.097	0.22973
360	0.58208	132.863	0.23190
370	0.59019	134.634	0.23405
380	0.59829	136.410	0.23618
390	0.60635	138.191	0.23829

140 — 125.30 psig (104.35 °F)

TEMP. °F	V (0.28964)	H (87.389)	S (0.16299)
100	—	—	—
110	0.29548	88.448	0.16486
120	0.30549	90.297	0.16808
130	0.31513	92.120	0.17120
140	0.32445	93.923	0.17423
150	0.33350	95.709	0.17718
160	0.34232	97.483	0.18007
170	0.35095	99.247	0.18289
180	0.35939	101.003	0.18566
190	0.36769	102.754	0.18838
200	0.37584	104.501	0.19104
210	0.38387	106.245	0.19367
220	0.39179	107.987	0.19625
230	0.39961	109.728	0.19879
240	0.40734	111.470	0.20130
250	0.41499	113.212	0.20377
260	0.42257	114.956	0.20621
270	0.43008	116.701	0.20862
280	0.43753	118.449	0.21100
290	0.44492	120.199	0.21335
300	0.45226	121.953	0.21567
310	0.45955	123.709	0.21797
320	0.46680	125.470	0.22024
330	0.47400	127.233	0.22249
340	0.48117	129.001	0.22471
350	0.48831	130.773	0.22692
360	0.49541	132.548	0.22910
370	0.50248	134.328	0.23125
380	0.50953	136.112	0.23339
390	0.51654	137.901	0.23551

170 — 155.30 psig (118.94 °F)

TEMP. °F	V (0.23667)	H (88.531)	S (0.16245)
120	0.23765	88.741	0.16281
130	0.24668	90.694	0.16615
140	0.25528	92.606	0.16937
150	0.26353	94.486	0.17248
160	0.27149	96.342	0.17550
170	0.27921	98.177	0.17844
180	0.28671	99.997	0.18130
190	0.29403	101.804	0.18411
200	0.30119	103.602	0.18685
210	0.30821	105.391	0.18954
220	0.31510	107.175	0.19219
230	0.32189	108.954	0.19479
240	0.32857	110.731	0.19734
250	0.33516	112.505	0.19986
260	0.34167	114.277	0.20234
270	0.34811	116.050	0.20479
280	0.35448	117.823	0.20720
290	0.36078	119.596	0.20958
300	0.36704	121.371	0.21193
310	0.37324	123.148	0.21426
320	0.37939	124.927	0.21655
330	0.38551	126.708	0.21882
340	0.39158	128.493	0.22107
350	0.39761	130.280	0.22329
360	0.40361	132.070	0.22549
370	0.40958	133.864	0.22766
380	0.41552	135.661	0.22982
390	0.42143	137.461	0.23195
400	0.42732	139.266	0.23406
410	0.43318	141.074	0.23615

200 — 185.30 psig (131.74 °F)

TEMP. °F	V (0.19891)	H (89.439)	S (0.16195)
140	0.20579	91.137	0.16480
150	0.21370	93.141	0.16811
160	0.22121	95.100	0.17130
170	0.22842	97.024	0.17438
180	0.23535	98.921	0.17737
190	0.24207	100.795	0.18027
200	0.24860	102.652	0.18311
210	0.25496	104.494	0.18588
220	0.26117	106.325	0.18860
230	0.26726	108.147	0.19126
240	0.27323	109.962	0.19387
250	0.27911	111.771	0.19644
260	0.28489	113.576	0.19896
270	0.29060	115.378	0.20145
280	0.29623	117.178	0.20390
290	0.30180	118.977	0.20632
300	0.30730	120.775	0.20870
310	0.31275	122.574	0.21105
320	0.31815	124.373	0.21337
330	0.32350	126.173	0.21567
340	0.32881	127.974	0.21793
350	0.33408	129.778	0.22017
360	0.33932	131.583	0.22239
370	0.34452	133.392	0.22458
380	0.34969	135.202	0.22675
390	0.35483	137.016	0.22890
400	0.35994	138.832	0.23102
410	0.36503	140.652	0.23313
420	0.37010	142.475	0.23521
430	0.37514	144.302	0.23728

230 — 215.30 psig (143.20 °F)

TEMP. °F	V (0.17051)	H (90.157)	S (0.16144)
140	—	—	—
150	0.17586	91.630	0.16387
160	0.18326	93.728	0.16728
170	0.19024	95.767	0.17054
180	0.19688	97.759	0.17368
190	0.20324	99.715	0.17672
200	0.20937	101.642	0.17966
210	0.21530	103.547	0.18253
220	0.22106	105.432	0.18532
230	0.22667	107.303	0.18805
240	0.23216	109.161	0.19073
250	0.23753	111.010	0.19335
260	0.24280	112.850	0.19593
270	0.24798	114.685	0.19846
280	0.25308	116.514	0.20095
290	0.25811	118.340	0.20340
300	0.26307	120.164	0.20582
310	0.26797	121.985	0.20820
320	0.27282	123.806	0.21055
330	0.27762	125.626	0.21287
340	0.28238	127.446	0.21516
350	0.28709	129.267	0.21742
360	0.29177	131.089	0.21966
370	0.29640	132.912	0.22187
380	0.30101	134.737	0.22405
390	0.30559	136.564	0.22622
400	0.31013	138.393	0.22836
410	0.31465	140.225	0.23048
420	0.31915	142.060	0.23257
430	0.32362	143.897	0.23465

260 — 245.30 psig (153.60 °F)

TEMP. °F	V (0.14829)	H (90.716)	S (0.16091)
150	—	—	—
160	0.15311	92.175	0.16327
170	0.16015	94.370	0.16679
180	0.16672	96.488	0.17012
190	0.17292	98.547	0.17332
200	0.17882	100.560	0.17639
210	0.18449	102.539	0.17937
220	0.18995	104.488	0.18226
230	0.19524	106.415	0.18507
240	0.20038	108.323	0.18782
250	0.20539	110.216	0.19051
260	0.21028	112.097	0.19314
270	0.21508	113.967	0.19572
280	0.21979	115.829	0.19825
290	0.22442	117.685	0.20074
300	0.22898	119.535	0.20320
310	0.23347	121.382	0.20561
320	0.23791	123.225	0.20799
330	0.24229	125.067	0.21034
340	0.24662	126.907	0.21265
350	0.25091	128.746	0.21494
360	0.25516	130.585	0.21720
370	0.25937	132.425	0.21943
380	0.26354	134.265	0.22163
390	0.26769	136.106	0.22381
400	0.27180	137.949	0.22597
410	0.27589	139.793	0.22810
420	0.27995	141.639	0.23021
430	0.28399	143.487	0.23230
440	0.28800	145.338	0.23437

300 — 285.30 psig (166.18 °F)

TEMP. °F	V (0.12510)	H (91.240)	S (0.16012)
160	—	—	—
170	0.12794	92.187	0.16163
180	0.13482	94.556	0.16537
190	0.14111	96.808	0.16886
200	0.14697	98.975	0.17217
210	0.15249	101.080	0.17534
220	0.15774	103.136	0.17838
230	0.16277	105.153	0.18133
240	0.16761	107.140	0.18419
250	0.17230	109.102	0.18697
260	0.17685	111.043	0.18969
270	0.18129	112.968	0.19235
280	0.18562	114.879	0.19495
290	0.18986	116.779	0.19750
300	0.19402	118.670	0.20000
310	0.19811	120.553	0.20247
320	0.20214	122.430	0.20489
330	0.20610	124.302	0.20728
340	0.21002	126.171	0.20963
350	0.21388	128.036	0.21195
360	0.21770	129.900	0.21423
370	0.22148	131.762	0.21649
380	0.22522	133.624	0.21872
390	0.22893	135.485	0.22092
400	0.23260	137.346	0.22310
410	0.23625	139.208	0.22526
420	0.23987	141.071	0.22739
430	0.24346	142.935	0.22949
440	0.24703	144.800	0.23158
450	0.25058	146.667	0.23364

400 — 385.30 psig (192.93 °F)

TEMP. °F	V (0.085587)	H (91.513)	S (0.15755)
200	0.091005	93.718	0.16092
210	0.097497	96.500	0.16510
220	0.10316	99.046	0.16888
230	0.10828	101.443	0.17238
240	0.11300	103.735	0.17568
250	0.11743	105.950	0.17882
260	0.12163	108.105	0.18183
270	0.12564	110.214	0.18475
280	0.12949	112.286	0.18756
290	0.13320	114.327	0.19031
300	0.13680	116.343	0.19298
310	0.14030	118.339	0.19559
320	0.14372	120.318	0.19814
330	0.14705	122.282	0.20065
340	0.15032	124.235	0.20310
350	0.15353	126.177	0.20552
360	0.15668	128.112	0.20789
370	0.15979	130.039	0.21023
380	0.16285	131.961	0.21253
390	0.16587	133.878	0.21480
400	0.16885	135.792	0.21704
410	0.17179	137.703	0.21925
420	0.17471	139.612	0.22143
430	0.17759	141.519	0.22359
440	0.18045	143.424	0.22572
450	0.18328	145.330	0.22782
460	0.18609	147.235	0.22991
470	0.18888	149.140	0.23197
480	0.19165	151.046	0.23401
490	0.19440	152.952	0.23602

Table B.9

Freon-22

Properties of the Saturation State

TEMP.	PRESSURE	VOLUME cu ft/lb		ENTHALPY Btu/lb			ENTROPY Btu/(lb)(°R)	
°F	PSIA	LIQUID v_f	VAPOR v_g	LIQUID h_f	LATENT h_{fg}	VAPOR h_g	LIQUID s_f	VAPOR s_g
−100	2.3983	0.010664	18.433	−14.564	107.935	93.371	−0.03734	0.26274
− 90	3.4229	0.010771	13.235	−12.216	106.759	94.544	−0.03091	0.25787
− 80	4.7822	0.010881	9.6949	− 9.838	105.548	95.710	−0.02457	0.25342
− 70	6.5522	0.010995	7.2318	− 7.429	104.297	96.868	−0.01832	0.24932
− 60	8.818	0.011113	5.4844	− 4.987	103.001	98.014	−0.01214	0.24556
− 50	11.674	0.011235	4.2224	− 2.511	101.656	99.144	−0.00604	0.24209
− 40	15.222	0.011363	3.2957	0.000	100.257	100.257	0.00000	0.23888
− 30	19.573	0.011495	2.6049	2.547	98.801	101.348	0.00598	0.23591
− 25	22.086	0.011564	2.3260	3.834	98.051	101.885	0.00894	0.23451
− 20	24.845	0.011634	2.0926	5.131	97.285	102.415	0.01189	0.23315
− 15	27.865	0.011705	1.8695	6.436	96.502	102.939	0.01483	0.23184
− 10	31.162	0.011778	1.6825	7.751	95.704	103.455	0.01776	0.23058
− 5	34.754	0.011853	1.5177	9.075	94.889	103.964	0.02067	0.22936
0	38.657	0.011930	1.3723	10.409	94.056	104.465	0.02357	0.22817
5	42.888	01012008	1.2434	11.752	93.206	104.958	0.02645	0.22703
10	47.464	0.012088	1.1290	13.104	92.338	105.442	0.02932	0.22592
15	52.405	0.012171	1.0272	14.466	91.451	105.917	0.03218	0.22484
20	57.727	0.012255	0.93631	15.837	90.545	106.383	0.03503	0.22379
25	63.450	0.012342	0.85500	17.219	89.620	106.839	0.03787	0.22277
30	69.591	0.012431	0.78208	18.609	88.674	107.284	0.04070	0.22178
40	83.206	0.012618	0.65753	21.422	86.720	108.142	0.04632	0.21986
50	98.727	0.012815	0.55606	24.275	84.678	108.953	0.05190	0.21803
60	116.31	0.013025	0.47272	27.172	82.540	109.712	0.05745	0.21627
70	136.12	0.013251	0.40373	30.116	80.298	110.414	0.06296	0.21456
80	158.33	0.013492	0.34621	33.109	77.943	111.052	0.06846	0.21288
90	183.09	0.013754	0.20789	36.158	75.461	111.619	0.07394	0.21122
100	210.60	0.014038	0.25702	39.267	72.838	112.105	0.07942	0.20956
120	274.60	0.014694	0.19238	45.705	67.077	112.782	0.09042	0.20613
140	351.94	0.015518	0.14418	52.528	60.403	112.931	0.10163	0.20235
160	444.53	0.016627	0.10701	59.948	52.316	112.263	0.11334	0.19776
180	554.78	0.018332	0.07679	68.498	41.570	110.068	0.12635	0.19133
200	686.35	0.022436	0.047438	80.862	21.990	102.853	0.14460	0.17794
204.81	721.91	0.030525	0.30525	91.329	0.000	91.329	0.16016	0.16016

Source: "Thermodynamic properties of Freon-22", copyright 1964; with permission
of E. I. du Pont de Nemours & Co.

V = volume in cu ft/lb;   H = enthalpy in Btu/lb;   S = entropy in Btu/(lb)(°R)   (saturation properties in parentheses)

ABSOLUTE PRESSURE, lb/sq in

5.0
19.7411*
(−78.62 °F)

TEMP. °F	V	H	S
	(9.3011)	(95.871)	(0.25283)
−80	—	—	
−70	9.5237	97.018	0.25581
−60	9.7810	98.362	0.25921
−50	10.038	99.721	0.26257
−40	10.293	101.094	0.26588
−30	10.549	102.482	0.26915
−20	10.803	103.885	0.27238
−10	11.058	105.302	0.27557
0	11.311	106.735	0.27872
10	11.565	108.182	0.28183
20	11.818	109.643	0.28491
30	12.070	111.120	0.28796
40	12.323	112.611	0.29097
50	12.575	114.117	0.29396
60	12.826	115.638	0.29691
70	13.078	117.174	0.29984
80	13.329	118.724	0.30274
90	13.580	120.288	0.30561
100	13.831	121.867	0.30845
110	14.082	123.461	0.31128
120	14.333	125.069	0.31407
130	14.583	126.691	0.31685
140	14.833	128.327	0.31960
150	15.084	129.977	0.32233
160	15.334	131.642	0.32504
170	15.584	133.320	0.32772
180	15.834	135.012	0.33039
190	16.083	136.718	0.33304
200	16.333	138.437	0.33566
210	16.583	140.170	0.33827
220	16.832	141.916	0.34086
230	17.082	143.676	0.34343
240	—	—	

10.0
9.561*
(−55.59 °F)

TEMP. °F	V	H	S
	(4.8778)	(98.515)	(0.24339)
−50	4.9518	99.291	0.24590
−40	5.0838	100.690	0.24927
−30	5.2152	102.101	0.25260
−20	5.3460	103.526	0.25588
−10	5.4762	104.963	0.25911
0	5.6060	106.414	0.26230
10	5.7353	107.878	0.26545
20	5.8643	109.356	0.26856
30	5.9929	110.847	0.27164
40	6.1212	112.353	0.27468
50	6.2492	113.872	0.27769
60	6.3769	115.404	0.28067
70	6.5044	116.951	0.28362
80	6.6316	118.512	0.28654
90	6.7586	120.086	0.28943
100	6.8855	121.674	0.29229
110	7.0122	123.276	0.29513
120	7.1387	124.892	0.29794
130	7.2651	126.522	0.30073
140	7.3913	128.165	0.30349
150	7.5174	129.822	0.30623
160	7.6434	131.493	0.30895
170	7.7693	133.177	0.31165
180	7.8951	134.875	0.31432
190	8.0208	136.586	0.31697
200	8.1464	138.310	0.31961
210	8.2719	140.048	0.32222
220	8.3974	141.799	0.32482
230	8.5228	143.562	0.32739
240	8.6481	145.339	0.32995
250	8.7734	147.129	0.33249

15
0.304
(−40.57 °F)

TEMP. °F	V	H	S
	(3.3412)	(100.194)	(0.23906)
−40	3.3463	100.276	0.23925
−30	3.4365	101.712	0.24263
−20	3.5261	103.159	0.24596
−10	3.6152	104.618	0.24924
0	3.7037	106.088	0.25248
10	3.7918	107.570	0.25567
20	3.8794	109.065	0.25882
30	3.9667	110.571	0.26192
40	4.0537	112.091	0.26500
50	4.1404	113.623	0.26803
60	4.2268	115.168	0.27103
70	4.3129	116.727	0.27400
80	4.3989	118.298	0.27694
90	4.4846	119.882	0.27985
100	4.5701	121.480	0.28273
110	4.6554	123.091	0.28559
120	4.7406	124.715	0.28841
130	4.8256	126.352	0.29121
140	4.9105	128.003	0.29399
150	4.9952	129.667	0.29674
160	5.0799	131.344	0.29947
170	5.1644	133.034	0.30217
180	5.2488	134.737	0.30486
190	5.3332	136.454	0.30752
200	5.4174	138.183	0.31016
210	5.5016	139.925	0.31278
220	5.5857	141.680	0.31538
230	5.6697	143.448	0.31797
240	5.7537	145.229	0.32053
250	5.8376	147.022	0.32307
260	5.9215	148.828	0.32560
270	—	—	

20
5.304
(−29.12 °F)

TEMP. °F	V	H	S
	(2.5527)	(101.444)	(0.23566)
−30	—	—	
−20	2.6156	102.785	0.23874
−10	2.6841	104.266	0.24207
0	2.7521	105.756	0.24535
10	2.8196	107.257	0.24858
20	2.8867	108.769	0.25177
30	2.9534	110.292	0.25491
40	3.0198	111.826	0.25801
50	3.0858	113.372	0.26107
60	3.1516	114.930	0.26410
70	3.2171	116.500	0.26709
80	3.2823	118.082	0.27005
90	3.3474	119.677	0.27298
100	3.4122	121.284	0.27588
110	3.4769	122.904	0.27874
120	3.5414	124.536	0.28159
130	3.6058	126.182	0.28440
140	3.6700	127.840	0.28719
150	3.7341	129.510	0.28995
160	3.7981	131.194	0.29269
170	3.8619	132.890	0.29540
180	3.9257	134.599	0.29810
190	3.9894	136.321	0.30077
200	4.0529	138.055	0.30342
210	4.1164	139.802	0.30604
220	4.1799	141.562	0.30865
230	4.2432	143.334	0.31124
240	4.3065	145.119	0.31381
250	4.3697	146.916	0.31636
260	4.4329	148.725	0.31889
270	4.4960	150.547	0.32141
280	4.5591	152.381	0.32390

25
10.304
(−19.73 °F)

TEMP. °F	V	H	S
	(2.0704)	(102.444)	(0.23308)
−20	—	—	
−10	2.1251	103.907	0.23637
0	2.1808	105.419	0.23970
10	2.2360	106.939	0.24297
20	2.2908	108.469	0.24619
30	2.3452	110.008	0.24937
40	2.3992	111.558	0.25250
50	2.4529	113.118	0.25559
60	2.5063	114.689	0.25864
70	2.5594	116.271	0.26166
80	2.6123	117.865	0.26464
90	2.6650	119.470	0.26758
100	2.7175	121.087	0.27050
110	2.7698	122.716	0.27338
120	2.8219	124.357	0.27624
130	2.8738	126.010	0.27907
140	2.9257	127.675	0.28187
150	2.9774	129.353	0.28464
160	3.0289	131.043	0.28739
170	3.0804	132.745	0.29012
180	3.1318	134.460	0.29282
190	3.1830	136.187	0.29550
200	3.2342	137.926	0.29815
210	3.2853	139.678	0.30079
220	3.3363	141.443	0.30340
230	3.3873	143.219	0.30600
240	3.4382	145.008	0.30857
250	3.4890	146.809	0.31113
260	3.5397	148.622	0.31367
270	3.5905	150.447	0.31619
280	3.6411	152.284	0.31869
290	3.6917	154.134	0.32117

30
15.304
(−11.71 °F)

TEMP. °F	V	H	S
	(1.7439)	(103.279)	(0.23101)
−10	1.7521	103.541	0.23159
0	1.7997	105.076	0.23497
10	1.8467	106.616	0.23828
20	1.8933	108.165	0.24154
30	1.9395	109.721	0.24475
40	1.9853	111.286	0.24792
50	2.0308	112.861	0.25104
60	2.0760	114.445	0.25412
70	2.1209	116.040	0.25716
80	2.1655	117.645	0.26016
90	2.2100	119.261	0.26312
100	2.2542	120.888	0.26606
110	2.2982	122.526	0.26896
120	2.3421	124.176	0.27183
130	2.3858	125.837	0.27467
140	2.4294	127.510	0.27748
150	2.4728	129.194	0.28027
160	2.5162	130.891	0.28303
170	2.5594	132.600	0.28576
180	2.6024	134.320	0.28848
190	2.6455	136.053	0.29116
200	2.6884	137.797	0.29383
210	2.7312	139.554	0.29647
220	2.7739	141.323	0.29909
230	2.8166	143.104	0.30169
240	2.8592	144.897	0.30427
250	2.9018	146.701	0.30684
260	2.9443	148.518	0.30938
270	2.9867	150.347	0.31190
280	3.0291	152.187	0.31441
290	3.0715	154.040	0.31689

V = volume in cu ft/lb; H = enthalpy in Btu/lb; S = entropy in Btu/(lb)(°R) (saturation properties in parentheses)

ABSOLUTE PRESSURE, lb/sq in

TEMP. °F	40 25.304 (1.63 °F) V (1.3285)	H (104.627)	S (0.22780)	TEMP. °F	50 35.304 (12.61 °F) V (1.0744)	H (105.692)	S (0.22535)	TEMP. °F	60 45.304 (22.03 °F) V (0.90222)	H (106.569)	S (0.22337)
10	1.3594	105.953	0.23064	10	—	—	—	30	0.92300	107.904	0.22612
20	1.3959	107.541	0.23399	20	1.0968	106.897	0.22788	40	0.94863	109.577	0.22950
30	1.4319	109.134	0.23728	30	1.1269	108.529	0.23125	50	0.97385	111.249	0.23282
40	1.4675	110.732	0.24051	40	1.1564	110.163	0.23455	60	0.99871	112.923	0.23607
50	1.5028	112.337	0.24369	50	1.1857	111.800	0.23780	70	1.0232	114.600	0.23927
60	1.5378	113.950	0.24682	60	1.2145	113.443	0.24099	80	1.0475	116.281	0.24241
70	1.5724	115.570	0.24991	70	1.2431	115.091	0.24413	90	1.0715	117.967	0.24551
80	1.6068	117.199	0.25296	80	1.2714	116.745	0.24722	100	1.0952	119.658	0.24855
90	1.6410	118.837	0.25596	90	1.2994	118.406	0.25027	110	1.1187	121.356	0.25156
100	1.6749	120.485	0.25893	100	1.3272	120.075	0.25328	120	1.1420	123.061	0.25453
110	1.7087	122.142	0.26187	110	1.3548	121.752	0.25625	130	1.1652	124.774	0.25746
120	1.7423	123.810	0.26477	120	1.3822	123.438	0.25918	140	1.1882	126.495	0.26035
130	1.7757	125.487	0.26764	130	1.4095	125.133	0.26208	150	1.2111	128.225	0.26321
140	1.8090	127.176	0.27048	140	1.4366	126.838	0.26495	160	1.2338	129.963	0.26604
150	1.8421	128.875	0.27329	150	1.4635	128.552	0.26778	170	1.2564	131.711	0.26884
160	1.8751	130.585	0.27607	160	1.4903	130.276	0.27059	180	1.2788	133.468	0.27161
170	1.9080	132.306	0.27883	170	1.5170	132.010	0.27337	190	1.3012	135.235	0.27435
180	1.9407	134.039	0.28156	180	1.5436	133.755	0.27611	200	1.3235	137.012	0.27706
190	1.9734	135.783	0.28426	190	1.5701	135.510	0.27884	210	1.3457	138.799	0.27975
200	2.0060	137.538	0.28694	200	1.5965	137.276	0.28153	220	1.3678	140.596	0.28241
210	2.0385	139.304	0.28960	210	1.6228	139.053	0.28421	230	1.3898	142.403	0.28505
220	2.0709	141.082	0.29223	220	1.6491	140.840	0.28686	240	1.4118	144.221	0.28767
230	2.1033	142.872	0.29485	230	1.6752	142.638	0.28948	250	1.4337	146.050	0.29027
240	2.1356	144.673	0.29744	240	1.7013	144.448	0.29209	260	1.4556	147.889	0.29284
250	2.1678	146.486	0.30001	250	1.7274	146.268	0.29467	270	1.4773	149.739	0.29539
260	2.2000	148.310	0.30257	260	1.7533	148.100	0.29723	280	1.4991	151.600	0.29792
270	2.2321	150.145	0.30510	270	1.7792	149.943	0.29978	290	1.5208	153.471	0.30044
280	2.2641	151.992	0.30761	280	1.8051	151.797	0.30230	300	1.5424	155.353	0.30293
290	2.2961	153.851	0.31011	290	1.8309	153.661	0.30480	310	1.5640	157.246	0.30541
300	2.3281	155.721	0.31259	300	1.8567	155.537	0.30729	320	1.5856	159.149	0.30786
310	2.3600	157.602	0.31505	310	1.8824	157.424	0.30976	330	1.6071	161.063	0.31030
				320	1.9081	159.322	0.31221				

TEMP. °F	70 55.304 (30.32 °F) V (0.77766)	H (107.312)	S (0.22172)	TEMP. °F	80 65.304 (37.76 °F) V (0.68318)	H (107.954)	S (0.22029)	TEMP. °F	90 75.304 (44.53 °F) V (0.60897)	H (108.516)	S (0.21903)
30	—	—	—	40	0.68782	108.347	0.22107	50	0.61924	109.496	0.22096
40	0.79981	108.972	0.22507	50	0.70822	110.098	0.22454	60	0.63766	111.280	0.22443
50	0.82224	110.682	0.22846	60	0.72820	111.843	0.22793	70	0.65568	113.056	0.22781
60	0.84428	112.391	0.23178	70	0.74780	113.584	0.23125	80	0.67334	114.827	0.23112
70	0.86598	114.098	0.23503	80	0.76708	115.323	0.23450	90	0.69069	116.594	0.23437
80	0.88736	115.807	0.23823	90	0.78605	117.061	0.23770	100	0.70777	118.360	0.23755
90	0.90846	117.519	0.24137	100	0.80477	118.801	0.24083	110	0.72459	120.127	0.24068
100	0.92932	119.234	0.24446	110	0.82325	120.544	0.24392	120	0.74120	121.894	0.24376
110	0.94995	120.953	0.24751	120	0.84152	122.290	0.24696	130	0.75760	123.665	0.24678
120	0.97038	122.679	0.25051	130	0.85960	124.040	0.24995	140	0.77383	125.439	0.24977
130	0.99063	124.410	0.25347	140	0.87751	125.796	0.25290	150	0.78989	127.218	0.25271
140	1.0107	126.148	0.25639	150	0.89526	127.558	0.25582	160	0.80581	129.002	0.25561
150	1.0306	127.893	0.25928	160	0.91286	129.326	0.25869	170	0.82159	130.793	0.25848
160	1.0504	129.647	0.26213	170	0.93034	131.102	0.26154	180	0.83725	132.589	0.26131
170	1.0701	131.408	0.26495	180	0.94770	132.885	0.26435	190	0.85279	134.393	0.26411
180	1.0896	133.178	0.26774	190	0.96495	134.677	0.26712	200	0.86824	136.205	0.26687
190	1.1091	134.957	0.27050	200	0.98209	136.476	0.26987	210	0.88359	138.024	0.26961
200	1.1284	136.745	0.27323	210	0.99915	138.284	0.27259	220	0.89885	139.851	0.27232
210	1.1477	138.543	0.27594	220	1.0161	140.101	0.27529	230	0.91403	141.687	0.27500
220	1.1669	140.350	0.27862	230	1.0330	141.928	0.27795	240	0.92914	143.532	0.27766
230	1.1860	142.166	0.28127	240	1.0498	143.763	0.28060	250	0.94418	145.385	0.28029
240	1.2050	143.993	0.28390	250	1.0666	145.608	0.28322	260	0.95916	147.248	0.28289
250	1.2239	145.830	0.28650	260	1.0833	147.463	0.28581	270	0.97408	149.120	0.28548
260	1.2428	147.677	0.28909	270	1.0999	149.328	0.28838	280	0.98894	151.002	0.28804
270	1.2617	149.534	0.29165	280	1.1165	151.202	0.29094	290	1.0038	152.893	0.29058
280	1.2805	151.401	0.29419	290	1.1330	153.087	0.29347	300	1.0185	154.794	0.29309
290	1.2992	153.279	0.29672	300	1.1495	154.981	0.29598	310	1.0332	156.704	0.29559
300	1.3179	155.167	0.29922	310	1.1659	156.885	0.29847	320	1.0479	158.624	0.29807
310	1.3365	157.066	0.30170	320	1.1823	158.800	0.30094	330	1.0626	160.554	0.30053
320	1.3551	158.975	0.30416	330	1.1987	160.725	0.30339	340	1.0772	162.494	0.30297
330	1.3737	160.894	0.30661	340	1.2150	162.660	0.30583	350	1.0917	164.444	0.30540
340	1.3923	162.824	0.30904								

V = volume in cu ft/lb ;   H = enthalpy in Btu/lb ;°   S = entropy in Btu/(lb)(°R)   (saturation properties in parentheses)

ABSOLUTE PRESSURE, lb/sq in

100
85.304
(50.77 °F)

TEMP. °F	V	H	S
	(0.54908)	(109.013)	(0.21790)
60	0.56498	110.700	0.22117
70	0.58177	112.514	0.22463
80	0.59818	114.319	0.22801
90	0.61425	116.117	0.23131
100	0.63003	117.911	0.23454
110	0.64555	119.702	0.23771
120	0.66084	121.492	0.24083
130	0.67592	123.284	0.24389
140	0.69081	125.077	0.24691
150	0.70554	126.874	0.24988
160	0.72011	128.674	0.25281
170	0.73454	130.480	0.25570
180	0.74885	132.290	0.25855
190	0.76304	134.107	0.26137
200	0.77712	135.931	0.26415
210	0.79111	137.761	0.26691
220	0.80510	139.599	0.26963
230	0.81883	141.445	0.27233
240	0.83257	143.299	0.27500
250	0.84624	145.161	0.27764
260	0.85985	147.032	0.28026
270	0.87340	148.912	0.28285
280	0.88689	150.800	0.28542
290	0.90033	152.698	0.28797
300	0.91372	154.605	0.29050
310	0.92707	156.522	0.29300
320	0.94038	158.448	0.29549
330	0.95365	160.383	0.29796
340	0.96688	162.328	0.30040
350	0.98008	164.283	0.30283
360	0.99324	166.248	0.30525

110
95.304
(56.55 °F)

TEMP. °F	V	H	S
	(0.49969)	(109.456)	(0.21687)
60	0.50526	110.101	0.21812
70	0.52109	111.956	0.22165
80	0.53651	113.798	0.22510
90	0.55156	115.628	0.22846
100	0.56631	117.451	0.23174
110	0.58078	119.269	0.23496
120	0.59501	121.083	0.23812
130	0.60901	122.897	0.24122
140	0.62282	124.710	0.24427
150	0.63646	126.525	0.24727
160	0.64994	128.342	0.25023
170	0.66327	130.163	0.25314
180	0.67648	131.988	0.25602
190	0.68956	133.818	0.25886
200	0.70254	135.654	0.26166
210	0.71542	137.496	0.26443
220	0.72821	139.345	0.26717
230	0.74091	141.201	0.26989
240	0.75354	143.064	0.27257
250	0.76609	144.935	0.27522
260	0.77858	146.814	0.27785
270	0.79101	148.702	0.28046
280	0.80338	150.598	0.28304
290	0.81570	152.502	0.28560
300	0.82798	154.416	0.28813
310	0.84020	156.339	0.29065
320	0.85239	158.271	0.29314
330	0.86453	160.212	0.29561
340	0.87664	162.162	0.29807
350	0.88871	164.121	0.30050
360	0.90075	166.091	0.30292

120
105.304
(61.95 °F)

TEMP. °F	V	H	S
	(0.45822)	(109.853)	(0.21593)
70	0.47032	111.381	0.21884
80	0.48494	113.262	0.22236
90	0.49918	115.128	0.22578
100	0.51309	116.982	0.22912
110	0.52670	118.827	0.23239
120	0.54005	120.667	0.23559
130	0.55318	122.503	0.23873
140	0.56610	124.337	0.24182
150	0.57884	126.171	0.24485
160	0.59142	128.005	0.24784
170	0.60384	129.842	0.25078
180	0.61613	131.682	0.25368
190	0.62830	133.526	0.25654
200	0.64036	135.375	0.25936
210	0.65232	137.229	0.26215
220	0.66418	139.088	0.26490
230	0.67596	140.954	0.26763
240	0.68766	142.827	0.27033
250	0.69929	144.707	0.27299
260	0.71085	146.595	0.27564
270	0.72235	148.490	0.27825
280	0.73379	150.394	0.28084
290	0.74517	152.306	0.28341
300	0.75651	154.226	0.28595
310	0.76780	156.155	0.28848
320	0.77905	158.092	0.29098
330	0.79026	160.039	0.29346
340	0.80144	161.995	0.29592
350	0.81257	163.959	0.29836
360	0.82368	165.933	0.30078
370	0.83475	167.916	0.30319

140
125.304
(71.83 °F)

TEMP. °F	V	H	S
	(0.39243)	(110.535)	(0.21425)
70	—	—	—
80	0.40342	112.143	0.21725
90	0.41646	114.086	0.22082
100	0.42911	116.009	0.22428
110	0.44143	117.915	0.22766
120	0.45346	119.809	0.23096
130	0.46524	121.694	0.23418
140	0.47679	123.573	0.23734
150	0.48814	125.447	0.24044
160	0.49932	127.319	0.24348
170	0.51033	129.189	0.24648
180	0.52121	131.060	0.24943
190	0.53195	132.933	0.25233
200	0.54258	134.808	0.25520
210	0.55309	136.686	0.25802
220	0.56351	138.569	0.26081
230	0.57385	140.456	0.26357
240	0.58409	142.349	0.26629
250	0.59427	144.247	0.26899
260	0.60437	146.152	0.27165
270	0.61441	148.064	0.27429
280	0.62439	149.983	0.27690
290	0.63432	151.909	0.27949
300	0.64419	153.843	0.28205
310	0.65402	155.784	0.28459
320	0.66380	157.734	0.28711
330	0.67354	159.692	0.28960
340	0.68325	161.658	0.29208
350	0.69292	163.633	0.29453
360	0.70255	165.616	0.29696
370	0.71216	167.608	0.29938
380	0.72173	169.609	0.30178

160
145.304
(80.71 °F)

TEMP. °F	V	H	S
	(0.34249)	(111.095)	(0.21276)
80	—	—	—
90	0.35387	112.984	0.21623
100	0.36568	114.986	0.21984
110	0.37710	116.961	0.22334
120	0.38820	118.917	0.22674
130	0.39901	120.856	0.23006
140	0.40958	122.783	0.23330
150	0.41992	124.701	0.23647
160	0.43008	126.612	0.23958
170	0.44006	128.519	0.24263
180	0.44989	130.423	0.24563
190	0.45958	132.326	0.24858
200	0.46914	134.229	0.25149
210	0.47859	136.133	0.25435
220	0.48794	138.040	0.25718
230	0.49720	139.949	0.25997
240	0.50637	141.863	0.26272
250	0.51546	143.781	0.26544
260	0.52447	145.704	0.26814
270	0.53342	147.632	0.27080
280	0.54231	149.567	0.27343
290	0.55115	151.507	0.27604
300	0.55993	153.455	0.27862
310	0.56866	155.410	0.28117
320	0.57735	157.372	0.28371
330	0.58599	159.341	0.28622
340	0.59460	161.319	0.28870
350	0.60317	163.304	0.29117
360	0.61170	165.297	0.29362
370	0.62020	167.298	0.29604
380	0.62868	169.308	0.29845
390	0.63712	171.326	0.30084

180
165.304
(88.81 °F)

TEMP. °F	V	H	S
	(0.30323)	(111.555)	(0.21142)
90	0.30461	111.808	0.21188
100	0.31587	113.904	0.21566
110	0.32669	115.959	0.21930
120	0.33713	117.983	0.22282
130	0.34724	119.983	0.22624
140	0.35708	121.964	0.22957
150	0.36668	123.930	0.23282
160	0.37607	125.885	0.23600
170	0.38527	127.831	0.23912
180	0.39430	129.770	0.24217
190	0.40319	131.706	0.24518
200	0.41194	133.638	0.24813
210	0.42057	135.570	0.25103
220	0.42910	137.501	0.25390
230	0.43752	139.434	0.25672
240	0.44586	141.369	0.25951
250	0.45411	143.307	0.26226
260	0.46229	145.249	0.26497
270	0.47040	147.195	0.26766
280	0.47845	149.145	0.27031
290	0.48644	151.102	0.27294
300	0.49437	153.063	0.27554
310	0.50225	155.032	0.27811
320	0.51009	157.006	0.28066
330	0.51788	158.988	0.28319
340	0.52564	160.976	0.28569
350	0.53335	162.972	0.28817
360	0.54103	164.976	0.29063
370	0.54868	166.986	0.29307
380	0.55630	169.005	0.29549
390	0.56389	171.032	0.29789

V = volume in cu ft/lb; H = enthalpy in Btu/lb; S = entropy in Btu/(lb)(°R)   (saturation properties in parentheses)

ABSOLUTE PRESSURE, lb/sq in

200
185.304
(96.27 °F)

TEMP. °F	V	H	S
	(0.27150)	(111.934)	(0.21018)
100	0.27553	112.750	0.21165
110	0.28596	114.900	0.21545
120	0.29595	117.004	0.21911
130	0.30556	119.073	0.22265
140	0.31487	121.114	0.22608
150	0.32390	123.133	0.22942
160	0.33270	125.134	0.23268
170	0.34130	127.122	0.23586
180	0.34972	129.100	0.23898
190	0.35758	131.070	0.24203
200	0.36609	133.034	0.24503
210	0.37408	134.995	0.24798
220	0.38196	136.953	0.25089
230	0.38973	138.910	0.25374
240	0.39741	140.867	0.25656
250	0.40500	142.826	0.25934
260	0.41251	144.787	0.26209
270	0.41995	146.751	0.26480
280	0.42733	148.719	0.26747
290	0.43465	150.691	0.27012
300	0.44190	152.668	0.27274
310	0.44911	154.650	0.27533
320	0.45627	156.637	0.27790
330	0.46339	158.631	0.28044
340	0.47046	160.631	0.28296
350	0.47749	162.638	0.28545
360	0.48449	164.652	0.28792
370	0.49146	166.672	0.29037
380	0.49839	168.700	0.29280
390	0.50530	170.736	0.29521
400	0.51218	172.779	0.29760

240
225.304
(109.67 °F)

TEMP. °F	V	H	S
110	(0.22327)	(112.487)	(0.20793)
120	0.22360	112.564	0.20806
130	0.23318	114.873	0.21208
140	0.24225	117.113	0.21591
150	0.25089	119.299	0.21959
160	0.25920	121.443	0.22314
170	0.26721	123.554	0.22657
180	0.27498	125.639	0.22991
190	0.28253	127.702	0.23316
200	0.28990	129.749	0.23633
210	0.29710	131.783	0.23944
220	0.30416	133.807	0.24248
230	0.31109	135.822	0.24547
240	0.31790	137.832	0.24841
250	0.32461	139.838	0.25130
260	0.33122	141.842	0.25414
270	0.33775	143.844	0.25694
280	0.34420	145.847	0.25970
290	0.35059	147.850	0.26243
300	0.35690	149.855	0.26512
310	0.36316	151.863	0.26778
320	0.36936	153.874	0.27041
330	0.37551	155.889	0.27301
340	0.38161	157.908	0.27559
350	0.38767	159.932	0.27814
360	0.39369	161.962	0.28066
370	0.39967	163.997	0.28315
380	0.40562	166.037	0.28563
390	0.41153	168.084	0.28808
400	0.41741	170.138	0.29051
410	0.42327	172.198	0.29292
420	0.42910	174.265	0.29531
	—	—	—

280
265.304
(121.52 °F)

TEMP. °F	V	H	S
	(0.18821)	(112.814)	(0.20586)
130	0.19583	114.912	0.20944
140	0.20427	117.294	0.21345
150	0.21224	119.600	0.21726
160	0.21983	121.848	0.22092
170	0.22711	124.051	0.22445
180	0.23413	126.217	0.22786
190	0.24092	128.354	0.23118
200	0.24753	130.468	0.23441
210	0.25396	132.564	0.23756
220	0.26025	134.644	0.24064
230	0.26641	136.713	0.24366
240	0.27246	138.773	0.24663
250	0.27840	140.825	0.24954
260	0.28424	142.873	0.25241
270	0.29000	144.917	0.25523
280	0.29569	146.958	0.25801
290	0.30130	148.999	0.26075
300	0.30685	151.040	0.26345
310	0.31234	153.082	0.26612
320	0.31778	155.125	0.26876
330	0.32317	157.172	0.27137
340	0.32851	159.221	0.27395
350	0.33381	161.274	0.27650
360	0.33907	163.332	0.27902
370	0.34429	165.394	0.28152
380	0.34948	167.460	0.28400
390	0.35463	169.533	0.28645
400	0.35976	171.611	0.28888
410	0.36486	173.695	0.29129
420	0.36994	175.784	0.29368
430	0.37499	177.881	0.29605

290
275.304
(124.29 °F)

TEMP. °F	V	H	S
	(0.18088)	(112.865)	(0.20536)
130	0.18600	114.312	0.20783
140	0.19445	116.755	0.21194
150	0.20239	119.110	0.21583
160	0.20992	121.398	0.21955
170	0.21712	123.635	0.22313
180	0.22404	125.830	0.22659
190	0.23073	127.992	0.22995
200	0.23722	130.128	0.23321
210	0.24354	132.244	0.23639
220	0.24970	134.342	0.23950
230	0.25573	136.426	0.24255
240	0.26164	138.500	0.24553
250	0.26745	140.566	0.24846
260	0.27315	142.625	0.25135
270	0.27877	144.680	0.25418
280	0.28432	146.732	0.25697
290	0.28979	148.782	0.25973
300	0.29519	150.831	0.26244
310	0.30054	152.881	0.26512
320	0.30583	154.932	0.26777
330	0.31107	156.986	0.27039
340	0.31626	159.041	0.27297
350	0.32141	161.101	0.27553
360	0.32652	163.164	0.27807
370	0.33160	165.231	0.28057
380	0.33664	167.303	0.28306
390	0.34165	169.380	0.28551
400	0.34662	171.463	0.28795
410	0.35157	173.551	0.29037
420	0.35650	175.645	0.29276
430	0.36139	177.745	0.29513

300
285.304
(126.98 °F)

TEMP. °F	V	H	S
	(0.17400)	(112.904)	(0.20487)
130	0.17670	113.688	0.20620
140	0.18520	116.199	0.21042
150	0.19313	118.607	0.21441
160	0.20062	120.938	0.21820
170	0.20776	123.210	0.22184
180	0.21460	125.436	0.22554
190	0.22120	127.625	0.22874
200	0.22759	129.784	0.23204
210	0.23379	131.919	0.23525
220	0.23984	134.035	0.23839
230	0.24575	136.136	0.24145
240	0.25154	138.225	0.24454
250	0.25722	140.304	0.24741
260	0.26280	142.375	0.25031
270	0.26829	144.441	0.25316
280	0.27370	146.503	0.25597
290	0.27904	148.563	0.25873
300	0.28431	150.621	0.26146
310	0.28952	152.679	0.26415
320	0.29467	154.738	0.26681
330	0.29978	156.798	0.26944
340	0.30483	158.861	0.27203
350	0.30985	160.926	0.27460
360	0.31482	162.995	0.27714
370	0.31975	165.068	0.27965
380	0.32465	167.145	0.28214
390	0.32952	169.227	0.28460
400	0.33436	171.314	0.28705
410	0.33917	173.407	0.28947
420	0.34395	175.505	0.29186
430	0.34871	177.609	0.29424

Table B.10
Mercury Vapor Properties

Pressure, lb per sq in. abs, p	Temp deg F t	Specific vol, cu ft, per lb, v_g	Enthalpy, Btu			Entropy		
			Sat liquid, h_f	Vaporization, h_{fg}	Sat vapor, h_g	Sat liquid, s_f	Vaporization, s_{fg}	Sat vapor, s_g
0 4	402.3	114.5	13.81	128.1	141.9	0.02094	0.1486	0.1696
0 6	426.1	78.23	14.70	127.6	142.3	0.02195	0.1441	0.1660
0 8	443.8	59.71	15.36	127.2	142.6	0.02269	0.1408	0.1635
1.0	458.1	48.45	15.89	126.9	142.8	0.02328	0.1382	0.1615
1.5	485.1	33.14	16.90	126.3	143.2	0.02436	0.1337	0.1580
2	505.2	25.31	17.65	125.8	143.5	0.02514	0.1304	0.1556
3	535.4	17.34	18.78	125.2	144.0	0.02629	0.1258	0.1521
4	558.0	13.26	19.62	124.7	144.3	0.02714	0.1225	0.1497
5	576.2	10.77	20.30	124.3	144.6	0.02780	0.1200	0.1478
6	591.4	9.096	20.87	123.9	144.8	0.02834	0.1179	0.1462
7	605.0	7.882	21.37	123.6	145.0	0.02882	0.1161	0.1450
8	616.8	6.963	21.81	123.4	145.2	0.02923	1.1146	0.1439
9	627.5	6.244	22.21	123.2	145.4	0.02960	0.1133	0.1429
10	637.3	5.661	22.58	122.9	145.5	0.02993	0.1121	0.1420
15	676.5	3.892	24.04	122.1	146.1	0.03124	0.1074	0.1387
20	706.2	2.983	25.15	121.4	146.6	0.03220	0.1041	0.1363
25	730.4	2.429	26.05	120.9	146.9	0.03297	0.1016	0.1345
30	750.9	2.053	26.81	120.4	147.2	0.03360	0.09953	0.1331
35	769.0	1.781	27.49	120.0	147.5	0.03416	0.09774	0.1319
40	784.8	1.576	28.08	119.7	147.8	0.03464	0.09621	0.1308
45	799.3	1.414	28.62	119.4	148.0	0.03507	0.09486	0.1299
50	812.5	1.284	29.11	119.1	148.2	0.03546	0.09364	0.1291
60	836.1	1.086	29.99	118.6	148.6	0.03614	0.09154	0.1276
70	856.6	0.9436	30.75	118.1	148.9	0.03672	0.08976	0.1264
80	874.8	0.8349	31.43	117.7	149.1	0.03725	0.08824	0.1254
90	891.6	0.7497	32.06	117.3	149.4	0.03771	0.08687	0.1245
100	906.9	0.6811	32.63	117.0	149.6	0.03813	0.08565	0.1237
120	934.4	0.5767	33.60	116.4	150.1	0.03887	0.08353	0.1224
140	958.3	0.5012	34.55	115.9	150.4	0.03951	0.08175	0.1212
160	979.9	0.4438	35.35	115.4	150.8	0.04007	0.08019	0.1202
180	999.6	0.3990	36.09	115.0	151.1	0.04058	0.07881	0.1193

Source: *Standard Handbook for Mechanical Engineers.* 7th ed. Edited by T. Baumeister and L. S. Marks. Copyright 1967; with permission of McGraw-Hill Book Company.

Note: h_f and s_f are measured from 32°F.

Table B.11
Properties of Ammonia (NH₃) at Saturated Liquid

Temp °F	Pressure psia	Specific Volume ft³/lbm		Enthalpy Btu/lbm		Entropy Btu/lbm°R	
t	p	v_f	v_g	h_f	h_g	s_f	s_g
−100	1.24	0.022	182.90	−61.5	571.4	−0.1579	1.6025
−90	1.86	0.022	124.28	−51.4	575.9	−0.1309	1.5667
−80	2.74	0.022	86.54	−41.3	580.1	−0.1036	1.5336
−70	3.94	0.023	61.65	−31.1	584.4	−0.0771	1.5026
−60	5.55	0.023	44.73	−20.9	588.8	−0.0514	1.4747
−50	7.67	0.023	33.08	−10.5	593.2	−0.0254	1.4487
−40	10.41	0.023	24.86	0.0	597.6	0.0000	1.4242
−30	13.90	0.023	18.97	10.7	601.4	0.0250	1.4001
−20	18.30	0.024	14.68	21.4	605.0	0.0497	1.3774
−10	23.74	0.024	11.50	32.1	608.5	0.0738	1.3558
0	30.42	0.024	9.116	42.1	611.8	0.0975	1.3352
10	38.51	0.024	7.309	53.8	614.9	0.1208	1.3157
20	48.21	0.025	5.910	64.7	617.8	0.1437	1.2969
30	59.74	0.025	4.825	75.7	620.5	0.1663	1.2790
40	73.32	0.025	3.971	86.8	623.0	0.1885	1.2618
50	89.19	0.026	3.294	97.9	625.2	0.2105	1.2453
60	107.60	0.026	2.751	109.2	627.3	0.2322	1.2294
70	128.80	0.026	2.312	120.5	629.1	0.2537	1.2140
80	153.00	0.027	1.955	132.0	630.7	0.2749	1.1991
90	180.60	0.027	1.661	143.5	632.0	0.2958	1.1846
100	211.90	0.027	1.419	155.2	633.0	0.3166	1.1705
110	247.00	0.028	1.217	167.0	633.7	0.3372	1.1566
120	286.40	0.028	1.047	179.0	634.0	0.3576	1.1427
125	307.80	0.029	0.973	185.1	634.0	0.3679	1.1358

Source: "Bulletin on Thermodynamic Properties of Refrigerants," The American Society of Heating, Refrigerating, and Air Conditioning Engineers, 1969; with permission of ASHRAE.

Table B.12
Properties of Ammonia (NH₃) at Superheated Vapor State

Pressure psia		−40	−20	0	20	40	60	80	100	120	160	200	300
5.0	v	52.36	54.97	57.55	60.12	62.69	65.24	67.79	70.33	72.87	77.95	—	—
	h	600.3	610.4	620.4	630.4	640.4	650.5	660.6	670.7	680.9	701.6	—	—
	s	1.5149	1.5385	1.5608	1.5821	1.6026	1.6223	1.6413	1.6598	1.6778	1.7122	—	—
10.0	v	25.90	27.26	28.58	29.90	31.20	32.49	33.78	35.07	36.35	38.90	41.45	—
	h	597.8	608.5	618.9	629.1	639.3	649.5	659.7	670.0	680.3	701.1	722.2	—
	s	1.4293	1.4542	1.4773	1.4992	1.5200	1.5400	1.5593	1.5779	1.5960	1.6307	1.6637	—
20.0	v	—	—	14.09	14.78	15.45	16.12	16.78	17.43	18.08	19.37	20.66	—
	h	—	—	615.5	626.4	637.0	647.5	658.0	668.5	678.9	700.0	721.2	—
	s	—	—	1.3907	1.4138	1.4356	1.4562	1.4760	1.4950	1.5133	1.5485	1.5817	—
30.0	v	—	—	9.25	9.731	10.20	10.65	11.10	11.55	11.99	12.87	13.73	—
	h	—	—	611.9	623.5	634.6	645.5	656.2	666.9	677.5	698.8	720.3	—
	s	—	—	1.3371	1.3618	1.3845	1.4059	1.4261	1.4456	1.4642	1.4998	1.5334	—
40.0	v	—	—	—	7.203	7.568	7.922	8.268	8.609	8.945	9.609	10.27	11.88
	h	—	—	—	620.4	632.1	643.4	654.4	665.3	676.1	697.7	719.4	774.6
	s	—	—	—	1.3231	1.3470	1.3692	1.3900	1.4098	1.4288	1.4648	1.4987	1.5766
50.0	v	—	—	—	—	5.988	6.280	6.564	6.843	7.117	7.655	8.185	9.489
	h	—	—	—	—	629.5	641.2	652.6	663.7	674.7	696.6	718.5	774.0
	s	—	—	—	—	1.3169	1.3399	1.3613	1.3816	1.4009	1.4374	1.4716	1.5500
60.0	v	—	—	—	—	4.933	5.184	5.428	5.665	5.897	6.352	6.787	7.892
	h	—	—	—	—	626.8	639.0	650.7	662.1	673.3	695.5	717.5	773.7
	s	—	—	—	—	1.2913	1.3152	1.3373	1.3581	1.3778	1.4148	1.4493	1.5281
70.0	v	—	—	—	—	4.177	4.401	4.615	4.822	5.025	5.420	5.807	6.750
	h	—	—	—	—	623.9	636.6	648.7	660.6	671.8	694.3	716.6	772.7
	s	—	—	—	—	1.2688	1.2937	1.3166	1.3378	1.3579	1.3954	1.4302	1.5095
80.0	v	—	—	—	—	—	3.812	4.005	4.190	4.371	4.722	5.063	5.894
	h	—	—	—	—	—	634.3	646.7	658.7	670.4	693.2	715.6	772.1
	s	—	—	—	—	—	1.2745	1.2981	1.3199	1.3404	1.3784	1.4136	1.4933

Temperature °F

Table B.12, Contd.

Pressure psia		−40	−20	0	20	40	60	80	100	120	160	200	300
90.0	v	—	—	—	—	—	3.353	3.529	3.698	3.862	4.178	4.484	5.228
	h	—	—	—	—	—	631.8	644.7	657.0	668.9	692.0	714.7	771.5
	s	—	—	—	—	—	1.2571	1.2814	1.3038	1.3247	1.3633	1.3988	1.4789
100.0	v	—	—	—	—	—	2.985	3.149	3.304	3.454	3.743	4.021	4.695
	h	—	—	—	—	—	629.3	642.6	655.2	667.3	690.8	713.7	770.8
	s	—	—	—	—	—	1.2409	1.2661	1.2891	1.3104	1.3495	1.3854	1.4660
120.0	v	—	—	—	—	—	—	2.576	2.712	2.842	3.089	3.326	3.895
	h	—	—	—	—	—	—	638.3	651.6	664.2	688.4	711.8	769.6
	s	—	—	—	—	—	—	1.2386	1.2628	1.2850	1.3254	1.3620	1.4435
160.0	v	—	—	—	—	—	—	—	1.969	2.075	2.272	2.457	2.895
	h	—	—	—	—	—	—	—	643.9	657.8	683.5	707.9	767.1
	s	—	—	—	—	—	—	—	1.2186	1.2429	1.2859	1.3240	1.4076
200.0	v	—	—	—	—	—	—	—	1.520	1.612	1.780	1.935	2.295
	h	—	—	—	—	—	—	—	635.6	650.9	678.4	703.9	764.5
	s	—	—	—	—	—	—	—	1.1809	1.2077	1.2537	1.2935	1.3791
250.0	v	—	—	—	—	—	—	—	—	1.240	1.386	1.518	1.815
	h	—	—	—	—	—	—	—	—	641.5	671.8	698.8	761.3
	s	—	—	—	—	—	—	—	—	1.1690	1.2195	1.2617	1.3501
300.0	v	—	—	—	—	—	—	—	—	—	1.123	1.239	1.496
	h	—	—	—	—	—	—	—	—	—	664.7	693.5	758.1
	s	—	—	—	—	—	—	—	—	—	1.1894	1.2344	1.3257

Temperature °F

Source: "Bulletin on Thermodynamic Properties of Refrigerants," The American Society of Heating, Refrigerating, and Air Conditioning Engineers, 1969, with permission of ASHRAE.

Table B.13 Internal Energy of Gases (Btu/lb-mole)(datum, 520°R)

°R	O_2	N_2	Air	CO_2	H_2O	H_2	CO	C_8H_{18}	$C_{12}H_{26}$	$pv/778.16$
520	0	0	0	0	0	0	0	0	0	1,033
536.7	83	81	81	115	101	80	81	640	980	1,066
540	100	97	97	139	122	96	97	756	1,181	1,072
560	200	196	196	280	244	193	196	1,536	2,491	1,112
580	301	295	295	424	357	291	295	2,340	3,931	1,152
600	402	395	395	570	490	390	396	3,167	5,481	1,192
700	920	896	897	1,320	1,110	887	896	7,668	13,223	1,390
800	1,449	1,399	1,403	2,120	1,734	1,386	1,402	12,768	22,044	1,589
900	1,989	1,905	1,915	2,965	2,366	1,886	1,913	18,471	31,771	1,787
1000	2,539	2,416	2,431	3,852	3,009	2,387	2,430	24,773	42,277	1,986
1100	3,101	2,934	2,957	4,778	3,666	2,889	2,954	31,677	53,468	2,185
1200	3,675	3,461	3,492	5,736	4,339	3,393	3,485	39,182	65,290	2,383
1300	4,262	3,996	4,036	6,721	5,030	3,899	4,026	47,288	77,706	2,582
1400	4,861	4,539	4,587	7,731	5,740	4,406	4,580	55,995	90,688	2,780
1500	5,472	5,091	5,149	8,764	6,468	4,916	5,145	65,303	104,209	2,979
1600	6,092	5,652	5,720	9,819	7,212	5,429	5,720	74,825	118,240	3,178
1700	6,718	6,224	6,301	10,896	7,970	5,945	6,305	84,901	132,757	3,376
1800	7,349	6,805	6,889	11,993	8,741	6,464	6,899	95,503	147,735	3,575
1900	7,985	7,393	7,485	13,105	9,526	6,988	7,501	......	....	3,773
2000	8,629	7,989	8,087	14,230	10,327	7,517	8,109	......	....	3,972
2100	9,279	8,592	8,698	15,368	11,146	8,053	8,722	......	....	4,171
2200	9,934	9,203	9,314	16,518	11,983	8,597	9,339	......	....	4,369
2300	10,592	9,817	9,934	17,680	12,835	9,147	9,961	......	....	4,568
2400	11,252	10,435	10,558	18,852	13,700	9,703	10,588	......	....	4,766
2500	11,916	11,056	11,185	20,033	14,578	10,263	11,220	......	....	4,965
2600	12,584	11,682	11,817	21,222	15,469	10,827	11,857	......	....	5,164
2700	13,257	12,313	12,453	22,419	16,372	11,396	12,499	......	....	5,362
2800	13,937	12,949	13,095	23,624	17,288	11,970	13,144	......	....	5,561
2900	14,622	13,590	13,742	24,836	18,217	12,549	13,792	......	....	5,759
3000	15,309	14,236	14,394	26,055	19,160	13,133	14,443	......	....	5,958
3100	16,001	14,888	15,051	27,281	20,117	13,723	15,097	......	....	6,157
3200	16,693	15,543	15,710	28,513	21,086	14,319	15,754	......	....	6,355
3300	17,386	16,199	16,369	29,750	22,066	14,921	16,414	......	....	6,554
3400	18,080	16,855	17,030	30,991	23,057	15,529	17,078	......	....	6,752
3500	18,776	17,512	17,692	32,237	24,057	16,143	17,744	......	....	6,951
3600	19,475	18,171	18,356	33,487	25,067	16,762	18,412	......	....	7,150
3700	20,179	18,833	19,022	34,741	26,085	17,385	19,082	......	....	7,348
3800	20,887	19,496	19,691	35,998	27,110	18,011	19,755	......	....	7,547
3900	21,598	20,162	20,363	37,258	28,141	18,641	20,430	......	....	7,745
4000	22,314	20,830	21,037	38,522	29,178	19,274	21,107	......	....	7,944
4100	23,034	21,500	21,714	39,791	30,221	19,911	21,784	......	....	8,143
4200	23,757	22,172	22,393	41,064	31,270	20,552	22,462	......	....	8,341
4300	24,482	22,845	23,073	42,341	32,326	21,197	23,143	......	....	8,540
4400	25,209	23,519	23,755	43,622	33,389	21,845	23,823	......	....	8,738
4500	25,938	24,194	24,437	44,906	34,459	22,497	24,503	......	....	8,937
4600	26,668	24,869	25,120	46,193	35,535	23,154	25,186	......	....	9,136
4700	27,401	25,546	25,805	47,483	36,616	23,816	25,868	......	....	9,334
4800	28,136	26,224	26,491	48,775	37,701	24,480	26,533	......	....	9,533
4900	28,874	26,905	27,180	50,069	38,791	25,418	27,219	......	....	9,731
5000	29,616	27,589	27,872	51,365	39,885	25,819	27,907	......	....	9,930
5100	30,361	28,275	28,566	52,663	40,983	26,492	28,597	......	....	10,129
5200	31,108	28,961	29,262	53,963	42,084	27,166	29,288	......	....	10,327
5300	31,857	29,648	29,958	55,265	43,187	27,842	29,980	......	....	10,526
5400	32,607	30,337	30,655	56,569	44,293	28,519	30,674	......	....	10,724
5500	33,386	31,026	31,353	57,875	45,402	29,298	31,369	......	....	10,923
5600	34,161	31,726	32,051	59,183	46,513	29,978	32,065	......	....	11,121
5700	34,900	32,428	32,750	60,491	47,627	30,659	32,762	......	....	11,320
5800	35,673	33,130	33,449	61,891	48,744	31,342	33,461	......	....	11,519
5900	36,412	33,833	34,150	63,293	49,863	32,026	34,161	......	....	11,717
6000	37,149	34,537	34,852	64,297	50,985	32,712	34,863	......	....	11,916
6500	......	......	38,364	......	......	......	......	......	....	12,908
7000	......	......	41,893	......	......	......	...	......	....	13,901

Source: L. C. Lichty, "Internal Combustion Engines," McGraw-Hill Book Company, Inc., New York, 1939, and based upon data of A. Hershey, J. Eberhardt, and H. Hottel, *Trans. SAE*, 39: 409 (October, 1936). Used with permission of McGraw-Hill Book Company.

Specific Heat Equations for Some Common Gases

Gas or vapor	Equation c_p, Btu/(mole)(°R)	Range, °R	Maximum % error
O_2	$c_p = 11.515 - \dfrac{172}{\sqrt{T}} + \dfrac{1{,}530}{T}$	540–5000	1.1
N_2	$c_p = 9.47 - \dfrac{3.47(10^3)}{T} + \dfrac{1.16(10^6)}{T^2}$	540–9000	1.7
CO	$c_p = 9.46 - \dfrac{3.29(10^3)}{T} + \dfrac{1.07(10^6)}{T^2}$	540–9000	1.1
H_2O	$c_p = 19.86 - \dfrac{597}{\sqrt{T}} + \dfrac{7{,}500}{T}$	540–5400	1.8
CO_2	$c_p = 16.2 - \dfrac{6.53(10^3)}{T} + \dfrac{1.41(10^6)}{T^2}$	540–6300	0.8

$$c_p = a + b(10^{-3})T + c(10^{-6})T^2 + d(10^{-9})T^3 \qquad (T, \,°\text{K})$$
$$= \text{cal/(g mole)}(°\text{K}) \text{ or closely Btu/(lb}_m \text{ mole)}(°\text{R})$$

Gas or vapor	a	b	c	d	Range, °K	Maximum % error
Air	6.713	0.4697	1.147	−0.4696	273–1800	0.72
	6.557	1.477	−0.2148	0	273–3800	1.64
CO	6.726	0.4001	1.283	−0.5307	273–1800	0.89
	6.480	1.566	−0.2387	0	273–3800	1.86
CO_2	5.316	14.285	−8.362	1.784	273–1800	0.67
	$c_p = 18.036 - 0.00004474T - 158.08(T)^{-\frac{1}{2}}$				273–3800	2.65
H_2	6.952	−0.4576	0.9563	−0.2079	273–1800	1.01
	6.424	1.039	−0.07804	0	273–3800	2.14
H_2O	7.700	0.4594	2.521	−0.8587	273–1800	0.53
	6.970	3.464	−0.4833	0	273–3800	2.03
O_2	6.085	3.631	−1.709	0.3133	273–1800	1.19
	6.732	1.505	−0.1791	0	273–3800	3.24
N_2	6.903	−0.3753	1.930	−6.861	273–1800	0.59
	6.529	1.488	−0.2271	0	273–3800	2.05
NH_3	6.5846	6.1251	2.3663	−1.5981	273–1500	0.91
CH_4	4.750	12.00	3.030	−2.630	273–1500	1.33
C_3H_8	−0.966	72.79	−37.55	7.580	273–1500	0.40
C_4H_{10}	0.945	88.73	−43.80	8.360	273–1500	0.54
C_6H_6	−8.650	115.78	−75.40	18.54	273–1500	0.34
C_2H_2	5.21	22.008	−15.59	4.349	273–1500	1.46
CH_3OH	4.55	21.86	−2.91	−1.92	273–1000	0.18

Source: E.F. Obert, *Concepts of Thermodynamics*, copyright 1960, Mcgraw-Hill. With permission of McGraw-Hill Book Company.

Table B.15
Universal Constants and Conversion Factors

Universal Constants

g_c Gravitational constant $= 32.17$ lbm ft/lbf-s^2 $= 980.7$ gm-cm/dyne-s^2

R_u Universal gas constant $= 1544$ ft-lbf/lb-mole°R
$$= 8.31 \text{ joule/gm-mole °K}$$

σ Stefan-Boltzmann constant $= 0.174 \times 10^{-8}$ Btu/hr-ft^2°R^4
$$= 5.67 \times 10^{-8} \text{ watt/m}^2\text{°K}^4$$

N Avogadro's number $= 2.72 \times 10^{24}$ atoms/lb-mole
$$= 6.02 \times 10^{23} \text{ atoms/gm-mole}$$

Conversion Factors

Energy

3.414 Btu $= 1$ watt-hour
3414 Btu $= 1$ kW-hour
778 ft-lbf $= 1$ Btu
3.76×10^{-7} kW-hour $= 1$ ft-lbf
3600 joules $= 1$ watt-hour
1054 joules $= 1$ Btu
1 calorie $= 4.184$ joules
1 joule $= 1$ watt-second
0.737 ft-lbf $= 1$ newton-meter
1×10^7 dyne-cm $= 1$ newton-meter
1×10^7 ergs $= 1$ joule
1 dyne-cm $= 1$ erg
252 calories $= 1$ Btu

Power

1.34 horsepower (hp) $= 1$ kW
0.746 kW $= 1$ hp
1.41 hp $= 1$ Btu/second
1054 watts $= 1$ Btu/second
550 ft-lbf/second $= 1$ hp
$33,000$ ft-lbf/minute $= 1$ hp
2545 Btu/hr $= 1$ hp
4.19 joules $= 1$ calorie

Volume

231 in^3 $= 1$ gal (U. S. liquid)
0.1337 ft^3 $= 1$ gal (U. S. liquid)
3.785 liters $= 1$ gal (U. S. liquid)
3785.4 cc $= 1$ gal (U. S. liquid)
0.001 m^3 $= 1$ liter
1000 cc $= 1$ liter
61.02 in^3 $= 1$ liter
0.0353 ft^3 $= 1$ liter

Table B.15, Contd.

Pressure

0.1934 psi = 1 cm Hg (0°C) (on earth's surface)
0.014 psi = 1 cm H_2O (4°C) (on earth's surface)
0.036 psi = 1 in H_2O (4°C) (on earth's surface)
27.68 in H_2O (4°C) = 1 psi (on earth's surface)
33,864 dyne/cm² = 1 in Hg (4°C) (on earth's surface)
10^6 dyne/cm² = 1 bar
1 bar = 0.9869 atm
14.696 psi = 1 atm
0.49 psi = 1 in Hg (4°C) (on earth's surface)

Force

2000 lbf = 1 ton
2.248×10^{-6} lbf = 1 dyne
8.89644×10^8 dynes = 1 ton
0.2248 lbf = 1 newton
105 dynes = 1 newton

Mass

454 grams = 1 lbm
2.2 lbm = 1 kg = 1000 grams
1 slug = 32.174 lbm

Length

3.28 ft = 1 meter
39.37 in = 1 meter
100 cm = 1 meter
91.44 cm = 1 yd
5280 ft = 1 mile
3 ft = 1 yd
12 in = 1 ft

100 cm = 3.28 ft
10000 cm² = 10.7584 ft²
1000000 cm³ = 35.2875 ft³

Table B.16
U.S. Standard Atmospheric Conditions*

Altitude feet	Temperature °Rankine	Pressure inches mercury	Density lbm/ft³
—1000	522.236	31.0185	0.078737
0	518.670	29.9213	0.076474
500	516.887	29.3846	0.075362
1000	515.104	28.8557	0.074261
2000	511.538	27.8212	0.072098
4000	504.408	25.8426	0.067917
6000	497.279	23.9798	0.063925
8000	490.152	22.2276	0.060116
10,000	483.025	20.5808	0.056483
20,000	447.415	13.7612	0.040773
40,000	389.970	5.5584	0.018895
60,000	389.970	2.1354	0.007259
80,000	397.693	0.8273	0.002758
100,000	408.572	0.3290	0.001068
120,000	433.578	0.1358	0.000415
140,000	463.923	0.0595	0.000170
200,000	456.999	0.0058	0.000017

Source: Abstracted from *U.S. Standard Atmosphere*, prepared under sponsorship of NASA, USAF, and USWB. Printed by U.S. Government Printing Office, 1962.

*U.S. Standard Atmosphere: Ideal air devoid of moisture or dust rotating with the earth. The data represent approximately 45° latitude annual average conditions in the U. S.

Table B.17

Gravitational Acceleration near the Earth*

Altitude above Sea Level Feet	Location on Earth's Surface Degrees Latitude					
	Equator 0°	20°	40°	60°	80°	90°
0	32.088	32.108	32.158	32.215	32.253	32.258
1000	32.085	32.105	32.155	32.212	32.250	32.255
2000	32.082	32.102	32.152	32.209	32.247	32.252
4000	32.076	32.096	32.146	32.203	32.241	32.246
6000	32.070	32.090	32.140	32.203	32.241	32.246
8000	32.064	32.084	32.134	32.191	32.229	32.234
10,000	32.058	32.078	32.128	32.185	32.226	32.228
100,000	31.780	31.808	31.858	31.915	31.953	31.958

Source: U.S. Coast and Geodetic Survey, 1912.

*Feet per second per second.

Table B.18
Table of the Elements

IA	IIA	IIIB	IVB	VB	VIB	VIIB	VIII(B)	VIII(B)	VIII(B)	IB	IIB	IIIA	IVA	VA	VIA	VIIA	O
1 H 1.008																	2 He 4.003
3 Li 6.94	4 Be 9.01											5 B 10.8	6 C 12.0	7 N 14.0	8 O 16.0	9 F 19.0	10 Ne 20.2
11 Na 23.0	12 Mg 24.3											13 Al 27.0	14 Si 28.1	15 P 31.0	16 S 32.1	17 Cl 35.45	18 Ar 40.0
19 K 39.1	20 Ca 40.1	21 Sc 45.0	22 Ti 47.9	23 V 50.9	24 Cr 52.0	25 Mn 54.9	26 Fe 55.8	27 Co 58.9	28 Ni 58.7	29 Cu 63.5	30 Zn 65.4	31 Ga 69.7	32 Ge 72.6	33 As 74.9	34 Se 79.0	35 Br 79.9	36 Kr 83.8
37 Rb 85.5	38 Sr 87.6	39 Y 88.9	40 Zr 91.2	41 Nb 92.9	42 Mo 95.9	43 Tc (97)	44 Ru 101.1	45 Rh 102.9	46 Pd 106.4	47 Ag 107.9	48 Cd 112.4	49 In 114.8	50 Sn 118.7	51 Sb 121.8	52 Te 127.6	53 I 126.9	54 Xe 131.3
55 Cs 132.9	56 Ba 137.3	(57–71)	72 Hf 178.5	73 Ta 180.9	74 W 183.9	75 Re 186.2	76 Os 190.2	77 Ir 192.2	78 Pt 195.1	79 Au 197.0	80 Hg 200.6	81 Tl 204.4	82 Pb 207.2	83 Bi 209.0	84 Po (210)	85 At (210)	86 Rn (222)
87 Fr (223)	88 Ra (226)	(89–103)															

57 La 139	58 Ce 140	59 Pr 141	60 Nd 144	61 Pm (145)	62 Sm 150	63 Eu 152	64 Gd 157	65 Tb 159	66 Dy 163	67 Ho 165	68 Er 167	69 Tm 169	70 Yb 173	71 Lu 175
89 Ac (227)	90 Th 232	91 Pa (231)	92 U 238	93 Np (237)	94 Pu (244)	95 Am (243)	96 Cm (247)	97 Bk (247)	98 Cf (251)	99 Es (254)	100 Fm (253)	101 Md (256)	102 No (254)	103 Lw (257)

Table B.19
Gibbs Free Energies of Formation*

Compound	ΔG_f° (kcal/gram-mole)	Ion	ΔG_f° (kcal/gram-mole)
AgCl (s)	−26.22	Ag$^+$ (aq)	18.43
AgCl (g)	16.79	Br$^-$ (aq)	−24.57
CO (g)	−32.81	Cl$^-$ (aq)	−31.35
CO$_2$ (g)	−94.26	CO$_3^{--}$ (aq)	−126.22
CO$_2$ (aq)	−92.31	Cu^{++} (aq)	15.53
CH$_4$ (g)	−12.14	H$^+$	0
C$_8$H$_{18}$ (g)	4.14	Na^{++} (aq)	−62.59
C$_8$H$_{18}$ (l)	1.77	OH$^-$ (aq)	−37.60
H$_2$SO$_4$ (aq)	−177.34	SO$_4^{--}$ (aq)	−177.34
H$_2$O (l)	−59.69	Zn^{++} (aq)	−35.18
H$_2$O (g)	−54.64		
NaCl (aq)	−93.94		

Source: Rossini, F. D.; Wagman, D. D.; Evan, W. H.; Levine, S.; and Jaffe, I., "Selected Values of Chemical Thermodynamic Properties," Circular of the National Bureau of Standards 500, Washington, D. C., 1952.

Note: s = solid phase; l = liquid; g = gaseous; aq = aqueous.
*At standard state of 298°K and 1 atm.

Chart B.1
Mollier diagram

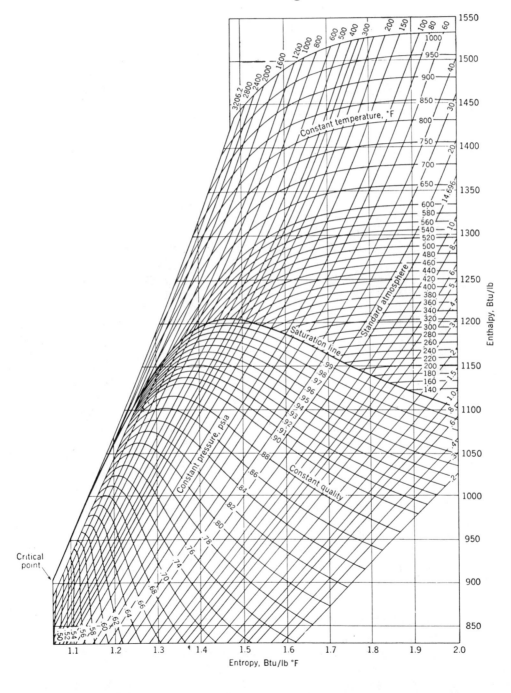

Entropy, Btu/lb °F

Enthalpy, Btu/lb

Chart B.2 Psychrometric chart

Barometric pressure, 14.696 lb/sq in.

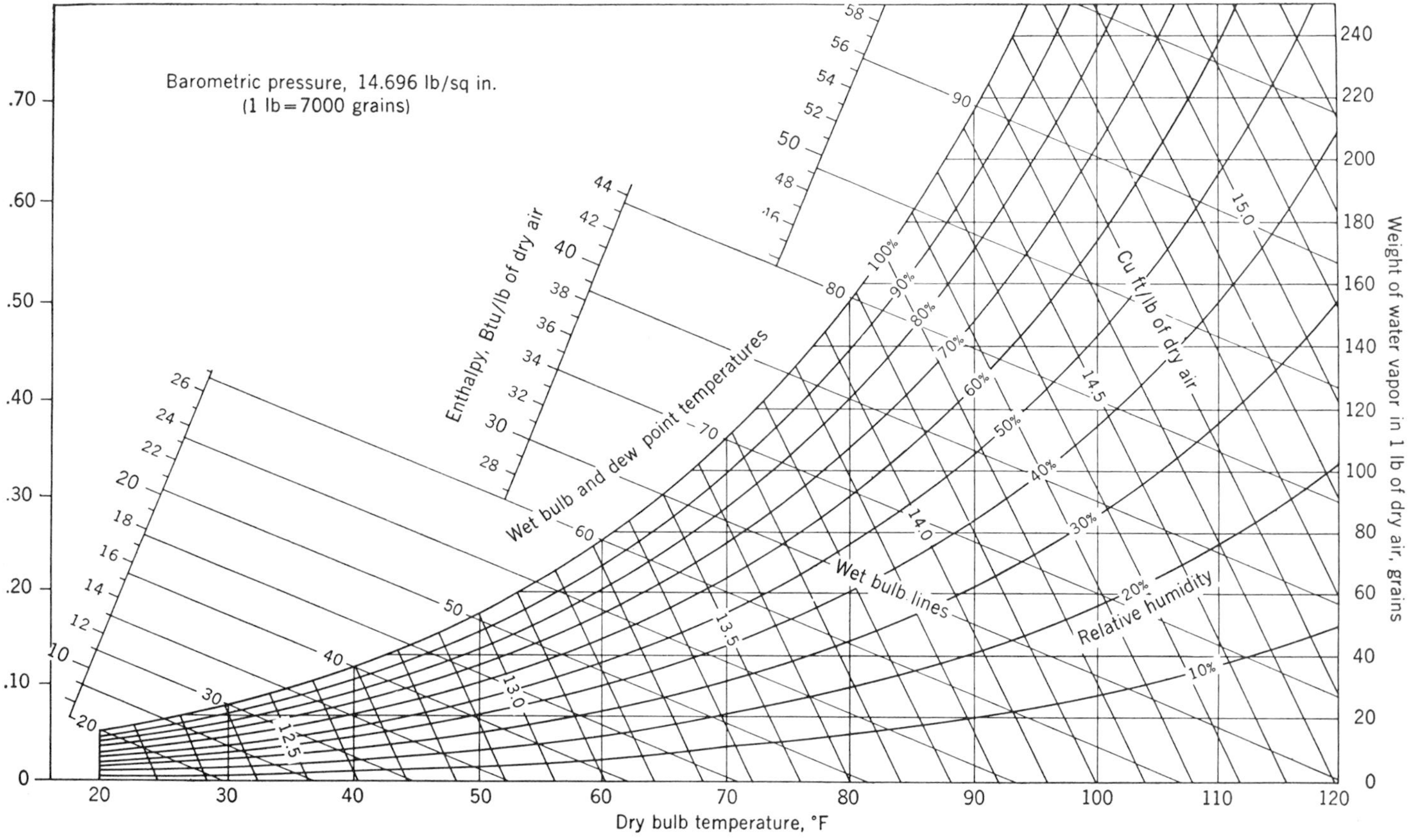

Barometric pressure, 14.696 lb/sq in.
(1 lb=7000 grains)
Enthalpy, Btu/lb of dry air
Wet bulb and dew point temperatures
Wet bulb lines
Cu ft/lb of dry air
Relative humidity
100%
90%
80%
70%
60%
50%
40%
30%
20%
10%
12.5
13.0
13.5
14.0
14.5
15.0
Pressure of water vapor, lb/sq in.
Dry bulb temperature, °F
Weight of water vapor in 1 lb of dry air, grains

C

Selected References

The following publications were used directly in the preparation of this text. This list is not intended as a complete bibliography of the literature of thermodynamics but can be used as a source for alternate readings for the interested student.

C.1 General Thermo- dynamics

Bent, H. 1965. *Second law: an introduction to classical & statistical thermodynamics*. New York: Oxford U Pr.

Crawford, F. H., and Van Vorst, W. D. 1968. *Thermodynamics for engineers*. New York: HarBrace J.

Faires, V. M. 1970. *Thermodynamics*. New York: Macmillan.

Hatsopoulos, G. N., and Keenan, J. H. *Principles of general thermodynamics*. New York: Wiley.

Hawkins, G. A. 1951. *Thermodynamics*. New York: Wiley.

Keenan, J. H. 1951. *Thermodynamics*. New York: Wiley.

Keenan, J. H., and Kaye, J. 1948. *Gas tables*. New York: Wiley.

Keenan, J. H., and Keyes, F. G. 1937. *Thermodynamic properties of steam*. New York: Wiley.

Lee, J. F., and Sears, F. W. 1963. *Thermodynamics*. Reading: A-W.

Mooney, D. A. 1953. *Mechanical engineering thermodynamics*. Englewood Cliffs: P-H.

Obert, E. F. 1960. *Concepts of thermodynamics*. New York: McGraw.

Skrotzki, B. G. A. 1963. *Basic thermodynamics: elements of energy systems*. New York: McGraw.

Solberg, H. L.; Cromer, O. C.; and Spalding; A. R. 1956. *Elementary heat power*. 2nd ed. New York: Wiley.

Van Wylen, G. J. 1959. *Thermodynamics*. New York: Wiley.

Zemanski, M. W. 1957. *Heat and thermodynamics: an intermediate textbook*. New York: Wiley.

C.2 Fluid Mechanics and Heat Transfer

Binder, R. C. 1962. *Fluid mechanics*. 4th ed. Englewood Cliffs: P-H.

Hirschfelder, J. O.; Curtis, C. F.; and Bird, R. B. 1966. *Molecular theory of gases and liquids*. New York: Wiley.

Jacob, M., and Hawkins, G. A. 1957. *Elements of heat transfer*. 3rd ed. New York: Wiley.

MacAdams, W. H. 1954. *Heat transmission*. 3rd ed. New York: McGraw.

Owczarek, J. A. 1964. *Fundamentals of gas dynamics*. Scranton: Intl Textbook.

Pao, R. H. F. 1961. *Fluid mechanics*. New York: Wiley.

C.3 History and General Discussions of Matter and Energy

Simon and Schuster. 1963. *The way things work: an illustrated encyclopedia of modern technology*. New York.

Toulmin, S., and Goodfield, J., 1962. *The architecture of matter: the physics, chemistry & physiology of matter, both animate & inanimate, as it has evolved since the beginnings of science*. New York: Har-Row.

Ubbelohde, A. R. 1963. *Man and energy*. Baltimore: Penguin.

C.4 Internal Combustion Engines

Lichty, L. C. 1951. *Internal combustion engines*. 6th ed. New York: McGraw.

Obert, E. F. 1959. *Internal combustion engines*. Scranton: Intl Textbook.

Taylor, C. F. 1960. *The internal-combustion engine in the theory and practice: thermodynamics, fluid flow, performance*. Cambridge: MIT Pr.

Taylor, C. F., and Taylor, E. S. 1948. *The internal combustion engine*. Scranton: Intl Textbook.

C.5
Gas Turbines,
Jet Propulsion

Bonney, E. A.; Zucrow, M. J.; and Besserer, C. W. 1956. *Aerodynamics, propulsion, and design practice.* New York: Van N-Rein.
Sawyer, R. T. 1945. *The modern gas turbine.* Englewood Cliffs: P-H.
Zucrow, M. J. 1948. *Principles of jet propulsion.* New York: Wiley.

C.6
Steam
Turbines
and Power
Generation

Babcock and Wilcox Co. 1955. *Steam, its generation and use.* 7th ed.
Combustion Engineering. 1957. *Combustion engineering.* New York.
Morse, F. T. 1953. *Power plant engineering and design.* New York: Van N-Rein.

C.7
Steam
Engines

Bruce, A. W. 1952. *The steam locomotive in America.* New York: Norton.
Morrison, L. H. 1930. Valve setting. *Power.* New York: McGraw.
Peabody, C. H. 1909. *Thermodynamics of the steam engine.* New York: Wiley.
Sinclair, A. 1970. *Development of the locomotive engine.* Annotated edition prepared by John H. White. Cambridge: MIT Pr.

C.8
Refrigeration

Jordan, R. C., and Priester, G. B. 1948. *Refrigeration.* Englewood Cliffs: P-H.
Sparks, N. R. 1938. *Theory of mechanical refrigeration.* New York: McGraw.

C.9
Direct Energy
Conversion
and Electrical
Theory

Angrist, S. W. 1965. *Direct energy conversion.* Boston: Allyn.
Bridgeman, P. W. 1961. *The thermodynamics of electrical phenomena in metals.* New York: Dover.
Kraus, J. D. 1953. *Electromagnetics.* New York: McGraw.
Shercliff, J. A. 1965. *A textbook of magnetohydrodynamics.* Elmsford: Pergamon.
Sutton, G. W., and Sherman, A. 1965. *Engineering magnetohydrodynamics.* New York: McGraw.

C.10
Biology,
Natural
Science

Keeton, W. T. 1967. *Biological science*. New York: Norton.

King, B. G., and Showers, M. J. 1963. *Human anatomy and physiology*. 5th ed. Philadelphia: Saunders.

Langley, L. L., and Cheraskin, E. 1958. *The physiology of man*, 2nd ed. New York: McGraw.

Tuttle, W. W., and Schottelius, B. A. 1969. *Textbook of physiology*. 16th ed. St. Louis: Mosby.

D

Answers to Selected Problems

Chapter 2

2.2 96.3 lbf

2.4 5.44 lbf

2.6 (a) 28.26 ft³ (b) 56.6 lbf/ft³
(c) 56.57 lbm/ft³, (d) 0.906

2.8 (a) 0.0993 lbm/ft³,
(b) 0.0991 lbf/ft³, (c) 0.00159

2.10 0.339 ft

2.12 (a) 2 psig, (b) 5.25 psig,
(c) 16.8 psia, 20.05 psia

2.18 $T_L = \log \frac{9}{5} T°K =$

$\log \frac{9}{5} + \log T°K$

2.22 (a) zero, (b) 12,175 ft-lbf

2.28 163.7 ft-lbf/lbm

2.34 150 ft-lbf/lbm, 100 ft-lbf/lbm

Chapter 3

3.4 210 in-lbf

3.8 86,400 ft-lbf

3.14 1.75 Btu/ft²-hr

3.18 $\dot{Q}/L = 2 (T_o - T_i)/\ln(r_o/r_i)$

3.20 640,500 Btu/hr-ft²

3.22 101.1 Btu/hr-ft²

Chapter 4

4.2 49 ft/min

4.6 (a) 2.45, (b) 1,873 ft/s, (c) 360 ft/s

4.10 10.1 ft/s

4.16 4.17 s

4.24 −3,000 Btu/lbm

4.26 0 (zero)

4.30 27.7 Btu = 21,600 ft-lbf

4.36 +114 ft-lbf

4.40 (a) 2,014 Btu/s

4.48 7,930 hp

4.50 1,232 Btu/lbm

Chapter 5

5.4 $v = 3750$ cm³/g

5.8 32 psig

5.12 0.2 liter

5.18 147.6 cal/g

5.24 2,224.5 Btu

5.30 278.61 Btu/lbm, 119.48 Btu/lbm

5.34 58.588 Btu/lbm

Chapter 6

6.2 280°R, 0.0173 ft³, −44.84 ft-lbf

6.6 30.8 psia

6.8 1.25×10^4 erg/s

6.14 54.1 Btu

6.20 88.8 psia

6.28 52,447 ft-lbf

6.32 456 Btu/lbm, 1,368 Btu/s,
−187 Btu/s

Chapter 7

7.2 +0.0056 cal/g°K

7.6 28.5 psia

7.10 722°R

7.14 0.0285 cal/g°K, 42.75 cal/°K

7.18 (a) −0.00901 Btu/lbm°R

7.22 1.37

Chapter 8
8.2   19,000 Btu, 5,600 Btu
8.8   71.9%, 0.281
8.14 (a) 72.2%, (c) 297 Btu, 82.8 Btu
8.16 (b) 74.8 Btu/min, (c) 69.3%

Chapter 9
9.2   (a) 9.79 Btu
9.6   (c) 1,909 Btu
9.10 2,794 Btu/s
9.14 (b) −534.4 Btu, −649.6 Btu

Chapter 10
10.4   2.85 in
10.8   (a) 2,573 hp, (b) 50.6%,
          (c) 106,000 Btu/min
10.14 53.5%
10.18 (b) 343.33 Btu/lbm
          (d) 7,140 Btu/min
10.22 (a) 0.4 lbm/hp-hr, (b) 185.4 psi,
          (c) 37.2%   57.9 psi
10.26 (a) 70.5%, (c) 46.3 psi
10.28 3.11
10.30 (a) 13.4, (b) 1.41,
          (c) 55.9 Btu/cycle

Chapter 11
11.2   56.2%
11.8   350,600 Btu/min
11.12 159.6 hp
11.18 36×10^6 Btu/hr
11.24 0.221 lbm/s
11.28 727 ft/s, 357 lbm/s
11.34 (a) 320 Btu/lbm,
          (c) 225.4 Btu/lbm
11.38 (a) 294.8 Btu/lbm,
          (c) 221.5 Btu/lbm
11.46 459.6 ft-lbf-s/lbm

Chapter 12
12.4   1,236.1 Btu/lbm
12.6   422.9 lbm/hr
12.10 262 Btu/lbm, 72.0%

12.16 1.4 Btu/lbm
12.24 1,087 Btu/lbm
12.28 (c) 357.5 Btu/lbm, (f) 23.1%
          (g) 9.56 lbm/kW-hr
12.32 (a) 94,200 kW,
          (c) 6.34 lbm/kW-hr
12.34 (c) $53,134 \times 10^3$ Btu/min,
          (e) $73,400 \times 10^3$ Btu/min

Chapter 13
13.2   1,176.1 Btu/lbm steam
13.8   197 hp, 11.28 psia, 20.3 in
13.14 (a) 435 lbm/s, (c) 1,460 psi

Chapter 14
14.2   6.22, 5.22
14.4   (a) 795°R, (b) 1.56
14.14 (a) 681°R, (b) 357°R,
          (c) 1291 Btu/min, (d) 2.71
14.16 466.9 Btu/lbm
14.18 (a) 91 Btu/lbm,
          (b) 58.16 Btu/lbm,
          (c) 67.26 Btu/lbm
14.24 (a) 50.3 lbm/hr, (b) 2,635 Btu/hr

Chapter 15
15.2   1.176 psia
15.4   0.000288 Btu/°R
15.8   42 Btu/lbm, 129 grains/lbm
15.10 6.08 Btu/lbm,
          0.007 lbm water/lbm air
15.16 −8.45 Btu/lbm
15.20 0.9355 g/g
15.24 15.23 lbm air/lbm fuel

Chapter 16
16.2   123 volts, 97.6%
16.4   10.2 amps, 99.5%
16.8   2.03 volts
16.12 4.93 ohms, 29.3 amps
16.16 −306,000 ergs
16.19 9.43×10^5 dynes/cm²

Index

Index